PURINE AND PYRIMIDINE METABOLISM IN MAN VII

Part B: Structural Biochemistry, Pathogenesis, and Metabolism

ADVANCES IN EXPERIMENTAL MEDICINE AND BIOLOGY

Recent Volumes in this Series

Volume 303
IMMUNOBIOLOGY OF PROTEINS AND PEPTIDES VI:
Human Immunodeficiency Virus, Antibody Immunoconjugates, Bacterial Vaccines, and Immunomodulators
Edited by M. Zouhair Atassi

Volume 304
REGULATION OF SMOOTH MUSCLE CONTRACTION
Edited by Robert S. Moreland

Volume 305
CHEMOTACTIC CYTOKINES: Biology of the Inflammatory Peptide Supergene Family
Edited by J. Westwick, I. J. D. Lindley, and S. L. Kunkel

Volume 306
STRUCTURE AND FUNCTION OF THE ASPARTIC PROTEINASES: Genetics, Structures, and Mechanisms
Edited by Ben M. Dunn

Volume 307
RED BLOOD CELL AGING
Edited by Mauro Magnani and Antonio De Flora

Volume 308
CELLULAR AND MOLECULAR MECHANISMS IN HYPERTENSION
Edited by Robert H. Cox

Volume 309A
PURINE AND PYRIMIDINE METABOLISM IN MAN VII, Part A:
Chemotherapy, ATP Depletion, and Gout
Edited by R. Angus Harkness, Gertrude B. Elion, and Nepomuk Zöllner

Volume 309B
PURINE AND PYRIMIDINE METABOLISM IN MAN VII, Part B:
Structural Biochemistry, Pathogenesis, and Metabolism
Edited by R. Angus Harkness, Gertrude B. Elion, and Nepomuk Zöllner

---

PURINE AND PYRIMIDINE METABOLISM IN MAN VII

Part B: Structural Biochemistry, Pathogenesis, and Metabolism

Edited by

R. Angus Harkness

Institute of Child Health
London, United Kingdom

Gertrude B. Elion

Burroughs Wellcome Co.
Research Triangle Park, North Carolina

and

Nepomuk Zöllner

Universität München
München, Germany

PLENUM PRESS • NEW YORK AND LONDON

Library of Congress Cataloging in Publication Data

International Symposium on Purine and Pyrimidine Metabolism in Man (7th: 1991: Bournemouth, England)

Purine and pyrimidine metabolism in man VII / edited by R. Angus Harkness, Gertrude B. Elion, and Nepomuk Zöllner.

p. cm.—(Advances in experimental medicine and biology; v. 390A–B)

Proceedings of a joint meeting of the Seventh International Symposium on Purine and Pyrimidine Metabolism in Man and the Third European Symposium on Purine and Pyrimidine Metabolism in Man, held June 30–July 5, 1991, in Bournemouth, England.

Includes bibliographical references and index.

Contents: Pt. A. Chemotherapy, ATP depletion, and gout—Pt. B. Structural biochemistry, pathogenesis, and metabolism.

ISBN 0-306-44092-X (hardbound): pt. A.—ISBN 0-306-44093-8 (hardbound): pt. B

1. Purines—Metabolism—Disorders—Congresses. 2. Pyrimidines—Metabolism—Disorders—Congresses. 3. Purines—Therapeutic use—Congresses. 4. Pyrimidines—Therapeutic use—Congresses. I. Harkness, R. A. (Robert Angus) II. Elion, Gertrude B. III. Zöllner, Nepomuk. IV. European Symposium on Purine and Pyrimidine Metabolism in Man (3rd: 1991: Bournemouth, England) V. Title. VI. Series.

[DNLM: 1. Purines—metabolism—congresses. 2. Pyrimidines—metabolism—congresses. W1 AD559 v. 309 / QU 58 I609p 1991]

RC632.P87I59 1991

616.3′9—dc20

DNLM/DLC 91-39326

for Library of Congress CIP

Proceedings of the Seventh International/Third European Joint Symposium on Purine and Pyrimidine Metabolism in Man, held June 30–July 5, 1991, in Bournemouth, United Kingdom

ISBN 0-306-44093-8

Printed in the United States of America

JOSEPH FRANKLIN HENDERSON

These proceedings are dedicated to Dr. Joseph Franklin Henderson whose achievements in the biochemistry and pharmacology of purines and pyrimidines over the years have been outstanding and have markedly influenced progress in this area of science. How many of us faced with an unusual finding have turned, for example, to the book entitled 'Nucleotide Metabolism: an Introduction' and found with gratitude the explanation we were seeking there.

Frank, as he is known affectionately to his many friends throughout the world, has contributed immeasurably to our understanding of the basic pathways of purine and pyrimidine metabolism and their interaction. His most important contributions to the field of purine biochemistry have included quantitative studies of both normal and alternative pathways of purine metabolism in intact cells and the regulation of purine metabolism in a variety of tissues. Moreover, he has alerted us all to the fact that there can be pitfalls in the extrapolation of results in cell extracts to the behaviour of enzymes in intact cells and tissues. The work of his group has focussed on anticancer chemotherapy as well as inherited diseases and, together, they have published many studies on the biochemical basis of purine-related anticancer drug action and resistance. This has also resulted in the development of a large number of new methods.

Frank was educated in the University of Arizona where he gained his Bachelor's and Master's degrees in Science and subsequently the University of Wisconsin where he was awarded a Ph.D. He has been associated for more than thirty years with the Cancer Research Unit and Department of Biochemistry of the University of Alberta, Canada. He has published over two hundred papers and, in addition, has written many excellent review articles and books. He has communicated his enthusiasm and critical approach to research to a large number of young investigators now working throughout the world. During his long and distinguished career he has been editor of the Canadian Journal of Biochemistry, invited participant at many international symposia and member of scientific committees in both Canada and the USA. He has, of course, been a long-standing member of the International Scientific Committee in the 'Purine and Pyrimidine in Man' series and a consistent contributor to successive Symposia. He has accumulated many honours for his achievements, which include the Ayerst Award of the Canadian Biochemical Society and the Geigy Prize in Rheumatology.

Frank is now retiring from active research. We hope that this will give him time to update his Nucleotide Metabolism book, as well as pursue the many other interests which he shares with his wife, Ruth. Frank is a truly remarkable man. Besides his achievements in purine biochemistry, his intellectual curiosity extends to diverse areas of science and philosophy. His enthusiasm for research in the field of purine and pyrimidine metabolism has enriched basic research in the area immeasurably. His repeated exhortations that we should temper that enthusiasm with caution, and recognise the limitations of our experimental models and working assumptions, should be heeded by all young investigators.

PREFACE

These two volumes record the scientific and clinical work presented at the VIIth International and 3rd European joint symposium on purine and pyrimidine metabolism in man held at the Bournemouth International Conference Centre, Bournemouth, UK, from 30th June to 5th July 1991. The series of international meetings at three yearly intervals have previously been held initially in 1973 in Israel, then Austria, Spain, the Netherlands, USA and Japan. The European Society for the Study of Purine and Pyrimidine Metabolism in Man (ESSPPM) which has its own executive and some finance first met in Switzerland in 1987, then in Germany in 1989.

The steady evolution of the science in this series of meetings is intellectually satisfying; the subsequent clinical progress is emotionally and economically reassuring. As befits the position of purines and pyrimidines at the centre of biochemistry, there has been steady scientific development into molecular genetics and now onto developmental controls and biochemical pharmacology. The complexities of the immune system are being unravelled but an understanding of the human brain largely eludes us. Laboratory based scientists now predominate over those who work as clinical specialists in rheumatology, immunology, oncology and paediatrics.

However, there continue to be major clinical objectives since large sections are concerned with major causes of death like ATP depletion, cancer and now AIDS; the laboratory work is providing clinical solutions.

Because basic science is often linked to a clinical problem these volumes do not separate into clinical and laboratory sections. It is clear that some articles in these two books are preliminary short communications on major work which allows priority to be established and should not prevent further full publication of major new results. However, other articles are worthwhile additional records which are adequate in themselves as contributions to our accummulated experience and do not require additional recording.

Recognition has been given to this area of science. The other participants at the meeting had great pleasure in being able to express their pleasure at the award of the 1988 Nobel Prize for Medicine to Gertrude Elion and George Hitchings for their work in this area of science; the award was announced just after the previous meeting in the international series, in Japan. The European editors are honoured by being accompanied by Dr. Elion as coeditor of these two volumes.

The international scientific community is especially endebted to the Mayor and Council of the Borough of Bournemouth, representing its citizens for the use of the superb Bournemouth International Conference Centre and the services of its efficient and friendly staff. The list of generous sponsors of the meeting must be led by the Arthritis and Rheumatism Council of the UK and include Amersham International plc, Burroughs Wellcome Co (USA) Division, Cancer Research Fund, Edward Arnold, Glaxo Group Research Ltd, Hodder & Stoughton, Millipore (UK) Ltd, Napp Laboratories, National Institute of Health (USA), Parke-Davis Pharmaceutical Research, Pfizer Ltd, The Royal Society, Sandoz Pharmaceuticals, Sigma Chemical Co Ltd, Warner Lambert and The Wellcome Foundation Ltd. Our plenary lecturers deserve special thanks for providing their own support in order to reduce costs for other participants.

The success of the meeting can in part be judged from the fact that 27 countries contributed. The Europeans especially enjoyed the freer communications amongst the peoples of our small planet.

Although the meeting was the product of many teams' work, including that of the International Committee, the untiring efforts of Anne Simmonds from 1987 onwards tower over other contributions. The rest of the local committee were Tom Scott (treasurer), Dave Perrett (printing), Simon Jarvis & George Nuki (workshops), Richard Watts & Neopomuk Zollner (abstracts) and Angus Harkness (editor) with Francoise Roch-Ramel (ESSPPM finance) who was co-opted for budgeting. At the meeting itself the volunteer administrators, Catherine Potter, Faith Scott, Helen Lees, Maureen Morris and Eveline Harkness were crucial to the success of PP 91.

We look forward to the 8th International meeting to be held in the University of Indiana, USA, in 1994 and to the 4th European meeting in Nijmegen, the Netherlands, in 1993.

G B Elion
Nobel Laureate (1988)
Burroughs Wellcome, USA

*R A Harkness
Institute of Child Health
University of London, UK

N Zöllner
University of Munich, Germany

---

*Contact Editor, 15 St Thomas's Drive, Hatch End, Pinner, Middlesex HA5 4SX, UK.

CONTENTS

PART B

STRUCTURAL BIOCHEMISTRY, PATHOGENESIS AND METABOLISM

D. MEASUREMENT OF PURINE AND PYRIMIDINES IN TISSUES AND BIOLOGICAL FLUIDS

E. STRUCTURAL BIOCHEMISTRY

1. Molecular Genetics

1.1 Gene therapy

1.2.a APRT mutations

1.2.d PNP Mutations

1.2.e Muscle Enzyme Mutations

2. Enzymology

2.1 5′-Nucleotidase and Immune Function

2.2 Characterization of 5′-Nucleotidase

2.7 Pyrimidine Metabolising Enzymes

F. PATHOGENESIS

1. Importance of GTP

2. Animal and Cellular Models

G. METABOLIC BIOCHEMISTRY

1.1 Interconnections

1.2 IMP Dehydrogenase

1.3 Pyrimidine Metabolism

1.4 Pyridine Nucleotide Metabolism

1.5 Nucleotide Metabolism in the Human Erythrocyte

1.6 Purine Metabolism and its alteration

CAPILLARY ELECTROPHORESIS FOR THE ANALYSIS OF CELLULAR NUCLEOTIDES

David Perrett and Gordon Ross

Dept of Medicine, St Bartholomew's Hospital Medical College
West Smithfield, London EC1A 7BE UK

INTRODUCTION

The electrophoretic separation of charged compounds, such as nucleotides, analysis has over the last 20 years been neglected and HPLC, although not ideal for charged species, dominates their quantitative analysis. Prior to HPLC many workers used high voltage electrophoresis with and without chromatography in the second dimension for the resolution of nucleotides, nucleosides and bases.

Capillary electrophoresis (CE) was introduced in 1981 when Jorgenson and Lukacs (1) described spectacular separations of peptides using zone electrophoresis in glass capillaries of 75μm i.d., with electrokinetic injection and fluorescence detection. One of the earliest follow-up papers on the technique reported the separation of nucleotides (2) but this went largely unnoticed.

CE is characterised by its ability to resolve with analytical precision the components of complex aqueous samples with very high resolution (N > 100,000) using less than 10nl of sample. Additionally it can separate cation, anions and uncharged molecules simultaneously. The technique has now attracted over 600 publications and has been reviewed many times (see 3) but the best introduction is Wallingford and Ewing (4) but for a more definitive review see Kuhr (5).

With the recent development of commercial instruments, CE has become more widely available. However detailed analytical applications are still relatively few and those in the field of biochemical analysis in biological extract even fewer. Here we report our preliminary observations with regard to the application of CE to the separation of nucleotides in cell extracts.

Purine and Pyrimidine Metabolism in Man VII, Part B
Edited by R.A. Harkness *et al.*, Plenum Press, New York, 1991

METHODS

Our early work used a CE apparatus assembled in the laboratory, mainly from components. It consisted of a 30kV HV source (Bertan 230R, Hicksville, NY, USA) mains powered via a safety interlock which was connected to 3 microswitches on a perspex isolation box. Closure of the box's lid allowed high voltage to be applied to two platinum electrodes dipping into two electrolyte chambers. The chambers were connected by an electrolyte-filled polyimide-coated fused silica capillary (typically 50μm i.d. x 70cm, SGE Ltd, UK). 50cm from the loading end a detection window was created by removing the coating. This window was positioned in the cell aperture of an CV^4 variable wavelength detector (Isco, Nebraska USA) usually set at 260nm. The output was to a chart recorder or a Hewlett-Packard 3396 integrator.

Recently we have employed an automated SpectraPHORESIS 1000 (Spectro-Physics Hemel Hempstead UK) CE system with rapid scanning detection and controlled by a IBM PS2 computer.

EXPERIMENTAL and DISCUSSION

Nucleotides are anions whilst nucleosides and bases can carry a positive charge but are also hydrophobic. All three groups have been separated by electrophoretic methods and CE approaches are therefore equally valid and have been demonstrated. However only studies on nucleotides will be reported here.

The following approaches for the CE separation of nucleotides have been investigated:

1. Free solution CE with normal polarity and enhanced electroosmotic flow
2. Free solution CE with reversed polarity
3. Micellar CE with reversed polarity (MECC)

Optimisation of Electrophoretic Conditions

Our initial studies up-dated those of Tsuda (2) i.e. high electroosmotic flow to overcome the natural migration of the negative nucleotides. Fig 1 shows the separation of a nucleotide mixture under conditions of high electroosmotic flow i.e 50mM borate buffer pH 9 in a fused silica capillary (75μm i.d. x 70cm) at +20kV and using electrokinetic injection. Good resolution of nucleotide standards was readily achieved. However when cell extracts were injected we observed up to 30% variations in the mobilities of peaks. This we presumed was due to the high salt concentrations in the extracts disturbing their separation compared to standards. Although Tsuda has reported the separation of such extracts with no apparent problems (6).

If the polarity is reversed so that the applied voltage is -20kV then the nucleotides migrate down the potential gradient against the electroosmotic flow however under such circumstances rapid separations were achieved but with only poor resolution.

Use of MECC

Charged micelles can enhance the resolution of both uncharged molecules and those with substantial hydrophobicity. Liu et al. (7) showed that dodecyltrimethylammonium bromide (DTAB) at above its critical micellar concentration gave improved separations of nucleotides. We have optimised the resolution of a mixture of 15 naturally occurring nucleotides with respect to pH, DTAB concentration, applied potential, presence/absence of metal chelators and running temperature. Other separation variables not yet optimised include electrolyte molarity, the type of electrolyte salt, role of organic modifiers and capillary dimensions. In all studies capillary dimensions were fused silica 50μm i.d. x 70cm (63cm to detector) and the base electrolyte was 50mM disodium phosphate pH 7.

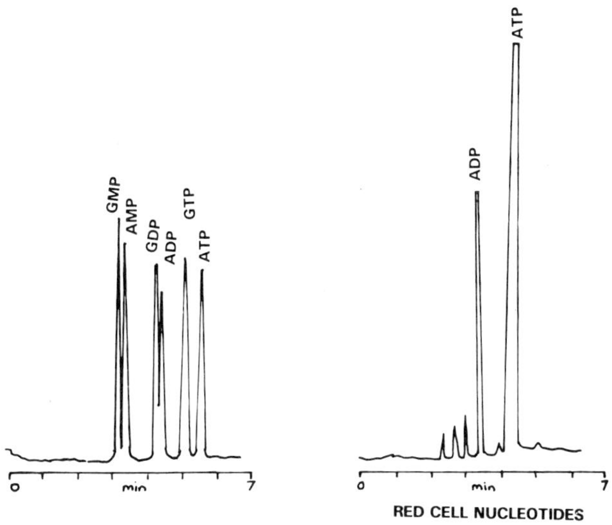

Fig 1 Separation of nucleotides using high electroosmotic flow a) Standard mixture b) TCA extract of red cells.

Increasing the DTAB concentration increased the resolution of the nucleotides by reducing their mobility. The most marked changes were observed for cXMP (our usual internal standard for HPLC) as would be expected from a strong interaction with the hydrophobic core of the micelles. 100mM DTAB was an optimal concentration. Using these conditions the effect of pH from 1.88-6.88 was studied. At the lowest values there was electrical shorting and only the higher values gave good results. pH 7 gave the best separation todate.

Increases in field strength reduced the analysis times. Increasing temperature reduced analysis times but with variable changes in resolution. Subsequent runs were performed at 35°C. In these studies it was noted that there was a tendency for peaks tailing particularly the NTP's possibly due to their interaction with metal ions. The effect of adding 0-5mM EDTA to the electrolyte was studied. In the presence of EDTA peak efficiency was increased. Subsequently, 1mM EDTA was used

Fig 2 shows the optimised separation of 15 nucleotides in <9 minutes with electrokinetic loading from a solution containing 28μM of each nucleotide. Calculated efficiencies for the nucleotide peaks exceed 100,000 theoretical plates compared to <5000 on a typical anion exchange HPLC separation. The overall order of migration was related to the charge on the nucleotides but within groups i.e. the triphosphates the separation was similar to that observed in reversed-phase HPLC i.e. based on the hydrophobicity of the ring structure. Exact conditions are given in the caption.

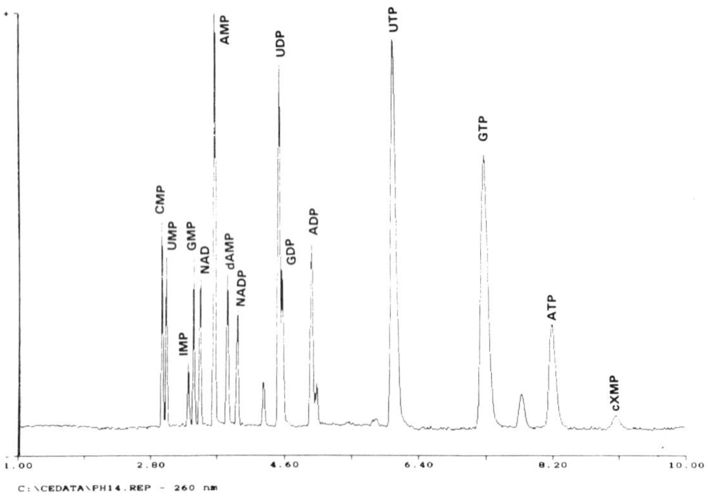

Fig 2 Optimised separation of 15 nucleotides using micellar electrokinetic capillary electrophoresis.
Fused silica capillary: 70 (63)cm x 50μm Buffer: 50mM Na_2HPO_4 1mM NaEDTA pH 7 100mM DTAB Load: electrokinetic 5sec 5kV Run: -25kV Detection 260nm

Sensitivity and precision

Although mass sensitivity is exceptionally high in CE, concentration sensitivity maybe lower than for HPLC. Using the present optimum conditions the response to increasing concentration of most nucleotides was linear from 0-200μM but remember that electrokinetic loading gives differential loading of ions. Area reproducibility varied from 3.3-6.1% (n=6) whilst Rt reproducibility varied from 2.2% at the front to 5.5% for ATP.

With low isocratic flowrates high sensitivity levels were possible. The concentration used in Fig 2 is that normally used in our standard HPLC system for cellular nucleotides when 25μl is injected whilst the amount electrokinetically loaded in Fig 2 was approximately 1nl. Absolute concentration sensitivity was 180nM for ATP using 260nm.

Separation of biological extracts

We have used the CE system to determine the nucleotides quantitatively and qualitatively in cell extracts. Unlike the free solution method there is little shift in electrophoresis times with sample status. Fig 3 shows a the nucleotide pattern in a typical electropherogram of TCA extracts from human red blood cells. Compared to anion exchange HPLC, CE separations of nucleotides general show the same number of major components but due to CE's higher resolving power a number of as yet unidentified minor components are revealed.

CONCLUSIONS

CE offers a new approach to the quantitative and qualitative determination of nucleotides as well as nucleosides and bases in biological extracts. Technically it is simpler than gradient elution HPLC, it is gives better resolution of nucleotides, it is equally fast if not faster, it requires much less sample and buffer and can match HPLC in terms of

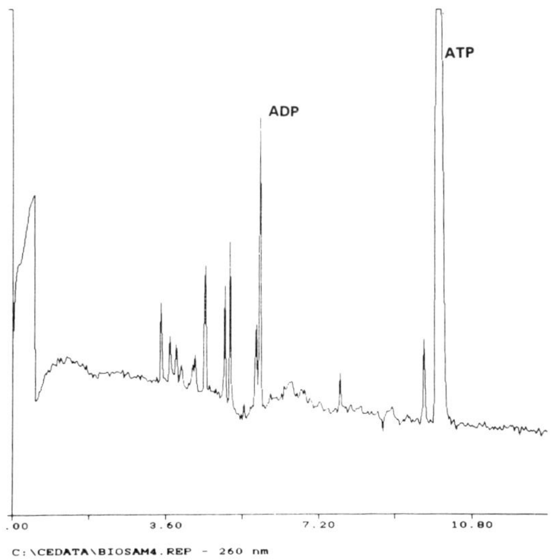

Fig 3 Separation of nucleotides in a neutralised TCA extract of human red cells. Conditions as Fig 2.

sensitivity. It is though a new technique and much work is necessary to develop robust analytical methods. However there is the potential for exciting developments such as isotachophoretic loading (8) and even two-dimensional analysis coupling to CE to HPLC.

ACKNOWLEDGEMENTS

This research was supported by grants from the Joint Research of St Bartholomew's Hospital and NE Thames Regional Health Authority.

REFERENCES

1. J W Jorgenson, & K D Lukacs, (1981) Zone electrophoresis in open tubular glass capillaries. Anal Chem 53, 1298-1302
2. T. Tsuda, G. Nakagawa, M. Sato, & K. Yag. (1983) Separation of nucleotides by high-voltage capillary electrophoresis. J. Appl Biochem, 5, 330-336
3. D. Perrett, G Ross & Goodall D (1991) Capillary Electrophoresis, A Bibliography The Chromatographic Society Nottingham UK
4. R.A. Wallingford & A.G. Ewing. (1989) Capillary Electrophoresis. Advances in Chromatography 29: 1-77.
5. W.G. Kuhr, (1990) Capillary Electrophoresis Anal Chem 62, 403R-414R
6. T. Tsuda, K. Takagi, T. Watanabe, & T. Satake (1988) Separation of nucleotides in organs of guinea pig by capillary zone electrophoresis. HRC & CC J High Resolut Chromatogr Chromatogr Comm, 11, 721-723
7. J. Liu, F. Banks, & M. Novotny (1989) High-Speed Micellar Electrokinetic Capillary Chromatography of Common Phosphorylated Nucleosides. J Microcol Sep, 1, 136-141
8. F. Foret, V. Sustacek, & P. Bocek (1990) On-Line Isotachophoretic Sample Preconcentration for Enhancement of Zone Detectability in Capillary Zone Electrophoresis. J Microcol Sep, 2, 229-233

AN IMPROVED SCREENING METHOD FOR INHERITED DISORDERS OF PURINE AND PYRIMIDINE METABOLISM BY HPLC

P.M. Davies, M.B. McBride, H.A. Simmonds

Purine Research Laboratory, Clinical Science Laboratories
UMDS Guy's Hospital, London, UK

INTRODUCTION

HPLC has been used extensively in the identification of genetic disorders of purine and pyrimidine metabolism. Some of the problems which may be encountered in assessing patients, for example on high caffeine intakes, or antibiotics such as septrin, have been reported (1). The spectrum of clinical presentation is broad, ranging from cerebral palsy, immunodeficiency, anaemia, acute renal failure to kidney stones. This means that many patients can be on multiple-drug regimes, which may include purine or pyrimidine analogues, or have recently undergone some clinical manipulation which could severely restrict the chance of making a diagnosis by direct analysis of body fluids. This paper reports an improved method of screening involving better fractionation of the urine and demonstrates how some of the above problems have been overcome.

PATIENTS AND METHODS

Patients, should preferably not be on any form of therapy (or on minimal essential therapy),and kept on a low-purine caffeine-free diet for three days prior to investigation.Therafter a 24 hour urine (toluene preservative) is required with a blood sample taken at the end of the 24 hour period. Data supplied with the sample must include: a) an adequate case history, b) previous and current drug therapy, including blood transfusion if any and date, c) a full family history.
Sample preparation: Venous blood is collected into heparin at room temperature and centrifuged at 1500g and plasma separated immediately. The urine is warmed at 60'C and processed as described below.

Methods: All reference standards (purines, pyrimidines and associated compounds) and other chemicals were of the highest analytical grade available. Solvents were all HPLC grade; water for HPLC was produced by double glass distillation.
Urine fractionation: The amount of urine loaded onto each mini-column depends on the concentration as determined by direct dilution (1/31) with acetate buffer and reverse-phase HPLC (RPLC), and on osmolarity (500 mOsms load 1-2mls urine; 500 mOsms load 3-4mls).
Anion exchange resin (AG 1-X8 100-200 mesh in the acetate form) was added to a depth of 6cm to mini glass columns plugged with cotton wool. Urine (e.g.1ml) (pH adjusted to 10 with 1M NaOH) was loaded to the column and eluted as follows and the corresponding 5ml fractions collected. Fractions:

After the urine, 1ml 0.01M ammonia solution, then distilled water to 5mls (frac 1); then 5ml distilled water (frac 2); then 0.04M HCL (frac 3); then 0.06M HCL (frac 4); then eluent containing 0.02M HCL + 29,2g NaCl/litre (frac 5). Subsequent analysis of the fractions was achieved by RPLC, and conventional chromatography when necessary. Fractions 2-4 may also be freeze-dried then taken up in 100-200ul ammonia solution (fracs 1 and 5 are not suitable due to the high salt content), and these concentrated fractions were then subjected to 2-dimensional electrophoresis and chromatography as described elsewhere.(2)

RPLC: The HPLC used was a dual wave-length (254/280nm) Millipore Waters trimodule system (1) with on-line Diode-array (Waters 990). A 250 x 4.5mm Hypersil 5u ODS-2 column (Hichrom) was used together with the 40mM acetate buffer system and a linear gradient to 20% organic solvent as described previously (1).

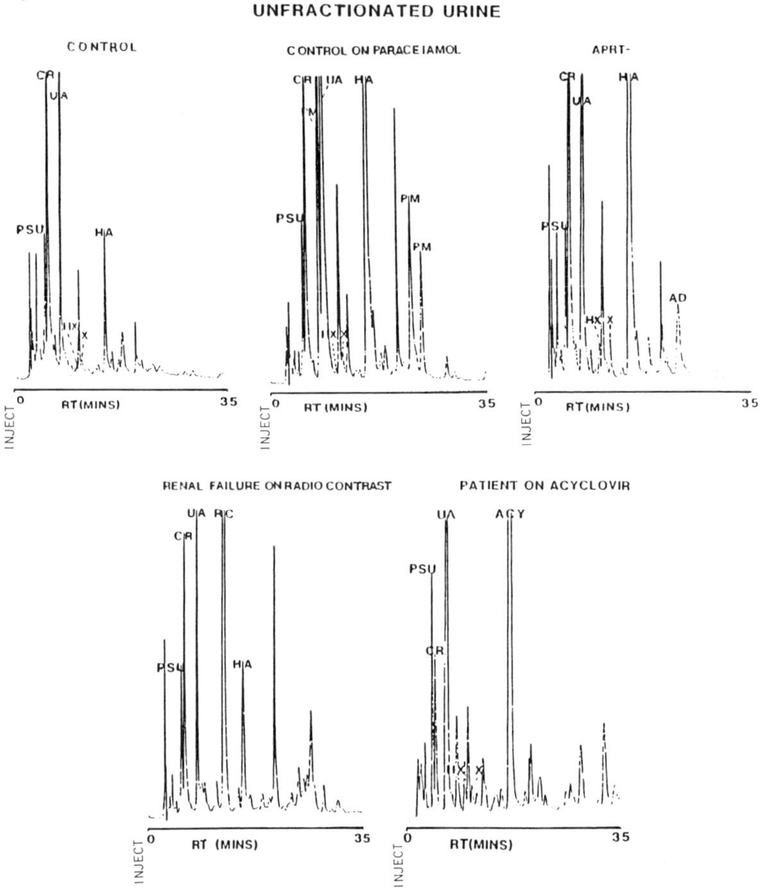

Fig. 1

(Abbreviations: Psu pseudouridine; CR creatinine; URA urocanic acid; UA uric acid; HX hypoxanthine; X xanthine; HA hippuric acid; 2PY,4PY pyridone carboxamides; 7MG, 8,7MG 7methyl & 8hydroxy-7methyl guanine; MN methylated nucleosides; AD adenine; PM paracetamol metabolites; RC radiocontrast agent)

RESULTS

Fig. 1 shows chromatograms (254nm absorbance) of 5 diluted (unfractionated) urines run on RPLC, illustrating the presence of extra peaks due to paracetamol metabolites, radiocontrast agents used in a renal failure patient, acyclovir, and in a patient with adenine phosphoribosyl transferase (APRT) deficiency.

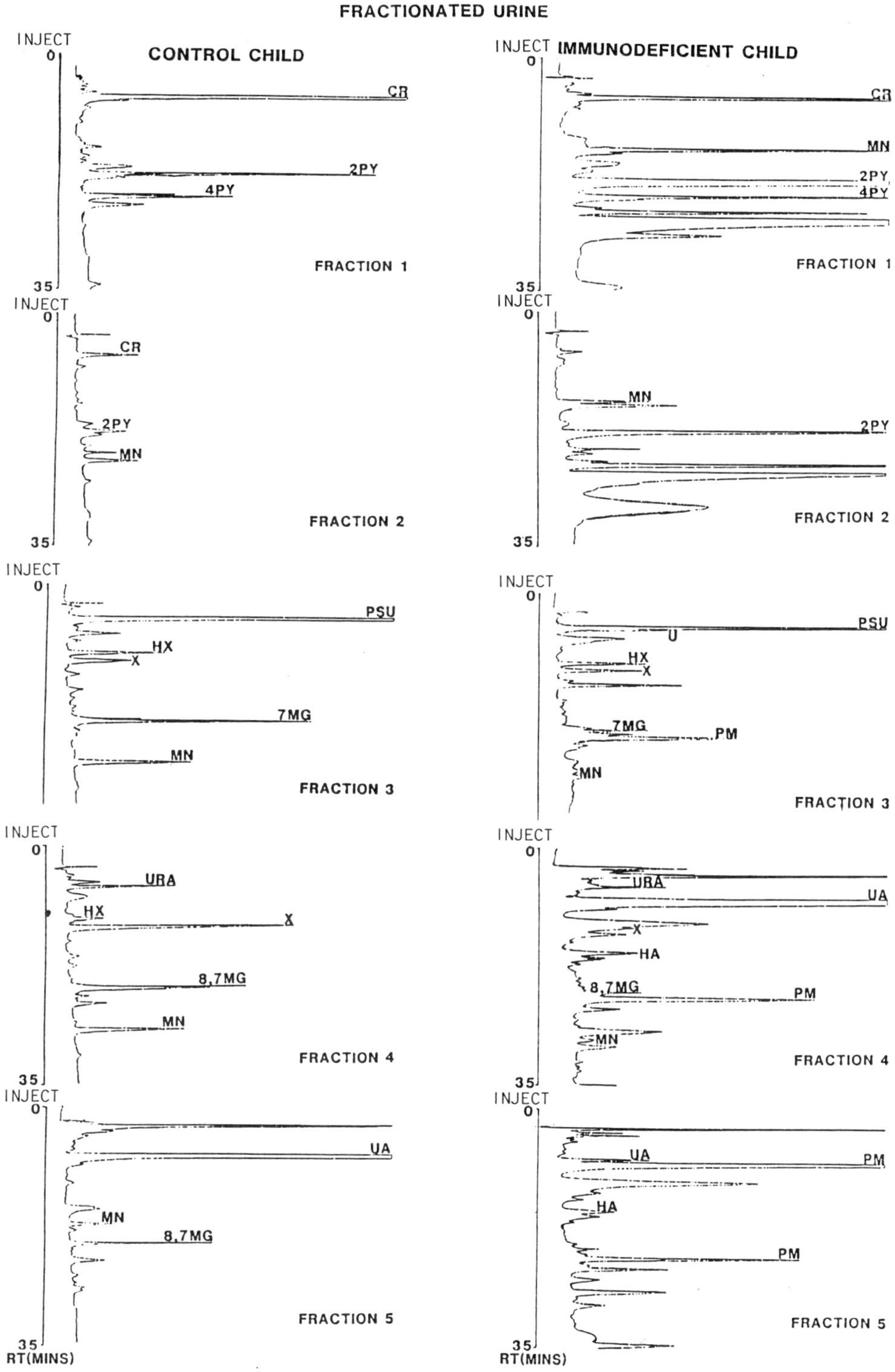

FRACTIONATED URINE
CONTROL CHILD
IMMUNODEFICIENT CHILD
INJECT
0
35
CR
2PY
4PY
MN
FRACTION 1
CR
2PY
MN
FRACTION 2
PSU
U
HX
X
7MG
PM
MN
FRACTION 3
URA
UA
HX
X
HA
8,7MG
PM
MN
FRACTION 4
UA
PM
HA
MN
8,7MG
FRACTION 5
RT(MINS)

Fig. 2

Fig. 2 demonstrates the chromatograms for the 5 different urine fractions obtained for a normal child on a caffeine-free diet. Composition of each fraction is listed in Table 1. Note the excellent separation of minor purine bases such as 7-methylguanine (frac 3) and 8-hydroxy 7-methylguanine (frac 4). A component with a similar retention time and 280/254 absorbance ratio to hypoxanthine (Hx) occurred in the direct urine dilution but eluted in frac 5. Fig.2 (right) also compares the traces of the 5 fractions from an immunodeficient child on paracetamol and indicates the peaks which could be mistaken for adenine (peak 'PM', fraction 3) and uric acid ('PM' peak, fraction 5). Abbreviations as for Fig.1

DISCUSSION

Biological fluids from patients referred for suspected purine or pyrimidine disorders may contain many u.v. absorbing components with similar HPLC characteristics, including retention time, 280/254 ratio and even u.v.spectrum. Resolution of these problems has required development of an improved method of urine fractionation with greater sensitivity, which has proved particularly useful in the following situations: 1) separating methyl-xanthines derived from the diet, or therapeutically from theophylline; 2) recognising the difference between the metabolites of Acyclovir, an anti-viral guanine-derivative, which could be misinterpreted as indicating purine nucleoside phosphorylase deficiency; 3) the possibility of mistaking Paracetamol metabolites for adenine, inferring APRT deficiency, or 4) co-elution of Furosemide or Paracetamol metabolites with uric acid, either of which could erroneously imply purine overproduction; 5) in recognising that increased levels of a component with the characteristics of hypoxanthine in patients with renal disease was really derived from radio contrast agents used in the clinical work-up and that such components are not readily cleared from the circulation and may be present for many days.

Peak-shift studies with specific enzymes has provided a useful backup in assessing difficult HPLC profiles, but only in a limited number of instances. By far the most useful recent advance has been the development of in-line diode-array analysis. This facility has greatly enhanced the speed as well as accuracy of peak identification and also enabled assessment of peak purity. Diode-array anaylsis, coupled with the improved urine fraction method described has enabled the identification and quantification of many more endogenous but minor purine and pyrimidine components, than was possible hitherto by HPLC and equals the sensitivity of a quantitative liquid chromatographic method which required 1-2 days for a single sample(2).

The difficulty in distinguishing components derived from the diet, or from treatment, from endogenous purines or pyrimidines, has underlined the need to have an accurate indication of previous and current drug therapy at referral. A sound knowledge of drug metabolism, is equally essential. This, coupled with the improved methods of analysis described here, now means that a diagnosis can be made in a much shorter time and with much greater confidence.

REFERENCES

1. Morris, G.S., Simmonds, H.A., and Davies, P.M., 1986. Use of Biological Fluids for the Rapid Diagnosis of Potentially Lethal Inherited Disorders of Human Purine and Pyrimidine Metabolism. Biomed.Chrom. 13:109.
2. Simmonds, H.A., 1969. Clin.Chim.Acta. 23:319.

A SINGLE HPLC SYSTEM FOR THE EVALUATION OF PURINE AND PYRIMIDINE METABOLITES IN BODY FLUIDS

I. A. Rivera, I. Tavares de Almeida and C. Silveira

Centro de Metabolismos e Genètica (I.N.I.C.)
Faculdade de Farmàcia, 1600 Lisboa
Portugal

INTRODUCTION

During the last few years several studies[1] in the field of inborn errors of metabolism have revealed disorders due to primary deficiency in purine and pyrimidine metabolism. Moreover, abnormal excretory patterns of intermediary metabolites of those compounds have been described in neoplastic and hereditary immunodeficiency diseases as well as in pathophysiological states which lead to primary or secondary hyperammonemia. The determination of pyrimidine metabolites also showed to give clues in the monitoring of the effectiveness of dietary therapy being used to treat patients with hyperammonemia.

Recently, several methods[2,3,4] have been described for the analysis of bases and nucleosides by high-performance liquid chromatography (HPLC). However, some of them deal with a limited number of compounds in one analytical run or are unsuitable for routine use because of their exceedingly long time of analysis

Development of methods for the study of the excretion patterns in inborn errors of metabolism has been a major thrust in our laboratory during the past few years, in order to set up a reliable screening program.

We describe a simple, selective and reliable method for qualitative and quantitative analysis of purine and pyrimidine metabolites by a reversed-phase HPLC procedure using a single column, gradient elution and UV detection, applicable to the analysis of the neutral or the acidic metabolites previously isolated by anion-exchange chromatography. Data concerning the validation of the method as well as its applicability to the study of urinary excretory patterns is presented.

METHODS

The urinary bases and nucleosides were isolated by anion-exchange chromatography as described by van Gennip[5]. The isolated fractions were then screened by two-dimensional thin layer chromatography (TLC)[5]. The information obtained was then used to complement the HPLC procedure which was used to confirm the identity of the compounds and for their quantitative analysis. Plasma after deproteinization with 70% perchloric acid and neutralization of the supernatant solution with 5N potassium carbonate, was injected without further purification. The fully developed protocol used for the HPLC analysis of the 20 purines, pyrimidines and related compounds studied is listed in table 1. HPLC analysis was performed on a system equiped with the following elements: two Model LC-6A pumps, a Model SCL-6A system programmer, a Model CR4A data system with integrator and two channel printer plotter (Shimadzu Corp., Kyoto, Japan), a Rheodyne Model 7125 injector and a Model 440 dual wavelength absorbance detector (Waters Assoc.,Mildford,MA,U.S.A.).

Table 1. Operating details of the HPLC procedure for the analysis of bases, nucleosides and related compounds

Column:	Lichrospher 60 RP-Select B, 5 uM, 250 x 4 mm (E. Merck)
Pre-column:	Lichrospher 60 RP-Select B, 5 uM
Eluents:	A: 0.01M KH_2PO_4; pH 5.00 ± 0.01 B: 0.02M KH_2PO_4: Methanol (50:50; V/V);pH 5.87 ± 0.01, adjusted with o-phosphoric acid after methanol has been added.
Gradients:	Eluents: 0% to 60% eluent "B" over 28 min by step gradient (Fig.1); 60% to 0% eluent "B" over 2 min linearly. Flow Rate: step gradient from 0.7ml to 1.4ml and back to 0.7ml (Fig.1).
Equilibration Period:	10 mins eluent "A" between runs. Overnight at 0.2ml / min of eluent "A", between days of sucessive work.
Colunm Temperature:	Room temperature (ca. 25° C)
Injected Volume:	5 ul for urinary fractions; 30 ul for plasma.
Detector:	Dual 280/254 nm U.V.detection

RESULTS

Chromatography System. Fig.1 shows a profile of a mixture of twenty authentic standards obtained under the chromatographic conditions set up. 3-Methylguanine was used as an internal standard. All the compounds tested can be separated in 28 min with good resolution, except for adenine. The tailing peak shape of adenine may be related to the interaction of its amino group with the residual free sylanol groups of the stationary phase.

Relative Retention Times and 280/254 nm Absorbance Ratios. The identity and purity of the peaks was established through the use of the relative retention times (Tr') and 280/254 nm absorbance ratios and when necessary by the addition of authentic standards to the sample. Those chromatographic parameters were determined by six independent analyses of a standard mixture. Good precision was achieved with respect to the reproducibility of the Tr' (R.S.D. (%)$\leq$ 0.62) and of the 280/254 nm absorbance ratios (R.S.D. (%)$\leq$ 5.24), except for adenine (8.06 %) and allopurinol (11.0 %). Daily confirmation was made during routine analytical work. Those parameters remained essentially constant over the life time of the column.

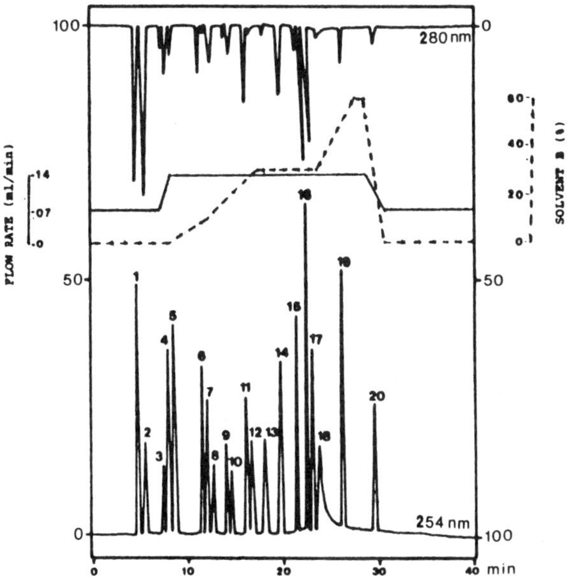

1 – Orotidine
2 – Orotic acid
3 – Pseudouridine
4 – Hydroxymethyluracil
5 – Uracil
6 – Uridine
7 – Hypoxanthine
8 – Xanthine
9 – Oxypurinol
10– Thymine
11– Methylguanine (I.S.)
12– Allopurinol
13– Inosine
14– Guanosine
15– Deoxyinosine
16– Deoxyguanosine
17– Thymidine
18– Adenine
19– Adenosine
20– Deoxyadenosine

Fig. 1. Chromatogram of a mixture of authentic standards, 1.25 nmoles each except for pseudouridine (0.05 nmoles) obtained under the conditions described in table 1.

<u>Precision of the HPLC analysis</u>. The reversed-phase HPLC internal standard method gives excellent precision for standards. Repeated injections (N=6) of 5 ul of each mixture solution (1.5 nmoles each) of the twenty compounds tested gave an average relative standard deviation (R.S.D.) of 0.6 – 3.8 %, except for orotidine (6.6 %) and orotic acid (6.1 %).This could be due to stronger interaction of these compounds with the stationary phase due to their acidic characteristics.

<u>Recovery, Linearity and Detection Limit</u>. Due to the complex nature of urine samples, a preliminary class separation ("neutral" and "acidic" compounds) was performed prior to HPLC analysis. A pooled control urine has been spiked with all compounds (250 nmoles each) tested before it was subjected to anion-exchange chromatography. The mean overall recovery of the analytical procedure was 99.8 ± 8.6 % with a range from 89.9 % for deoxyadenosine to 11.8 % for xanthine. Response (relative peak height "versus" concentration) for all compounds was found to be linear over the ranges (5–50 uM; 50–500 uM) studied (correlation coefficient ≥ 0.994). The detection limit, estimated on the basis of a signal to noise ratio of five, ranged from 3–5 ng for uracil, adenosine, deoxyguanosine, hypoxanthine and 5-hydroxy-methyluracil to 11–13 ng for inosine and adenine.

<u>Applications</u>. Fig. 2 a,b shows typical chromatograms of the "neutral" and "acidic" fractions of a pooled control urine, previously separated by anion-exchange chromatography. Fig. 2c is an HPLC profile of normal plasma. Fig. 2d represents the profile of the urinary fraction ("neutral" compounds) of a gouty subject under allopurinol therapy. Fig. 2e , f illustrates the profiles of the urinary fractions from a case of argininemia.

The HPLC method reported has been in continuous use for over one year and has proved to be reliable and reproducible throughout the life of the columns (average 500 injections) if a pre-column is used and the urinary fractions are filtered by a 0.45 um filter before injection. It proved to be suitable for routine screening purposes. However, some improvement will be necessary for the determination of normal orotic acid values. A significant

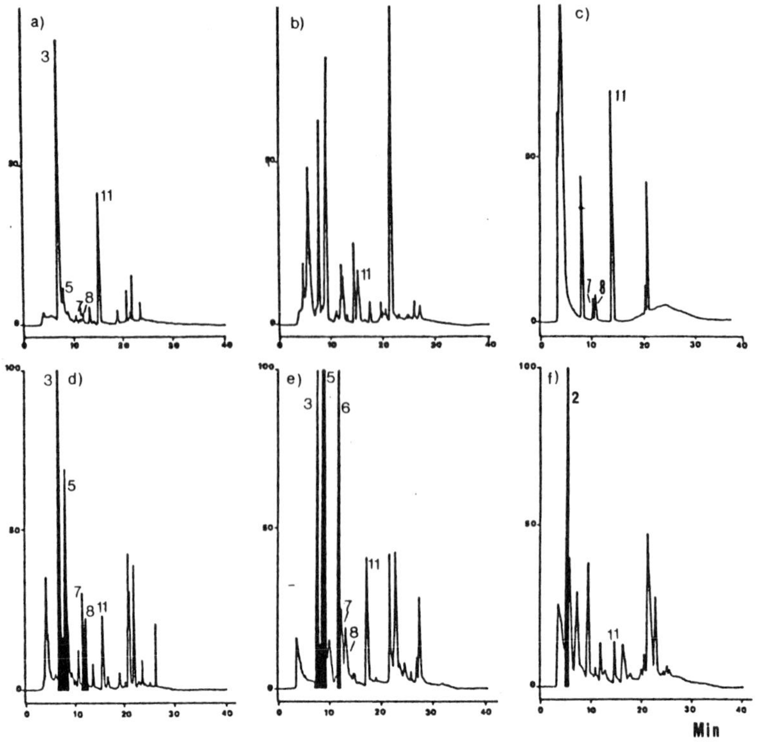

Fig. 2. Chromatograms of pooled control urine, fractions III and IV (a,b) normal plasma (c); urine fraction III (d) from a gouty subject and urine fractions III and IV (e,f) from a case of argininemia: Peak identity: 2: orotic acid; 3: pseudouridine; 5: uracil; 6: uridine; 7: hypoxanthine; 8: xanthine; 11: 3-methylguanine.

difference of the 280/254 nm absorbance ratio for urinary orotic acid from that of authentic standard was assumed to be due to the presence of an interfering compound. This problem is overcome in the presence of high excretory urinary levels of orotic acid, through the dilution of the sample before HPLC analysis.The method proved to be suitable for the purposes for which it has been intended for, namely, the screening of inborn errors and therapy monitoring.

ACKNOWLEDGEMENTS

We would like to thank Dr. A. H. van Gennip from the Academisch Medisch Centrum, Amsterdam and Dr. L. M. J. Spaapen from the Rijksuniversiteit, Limburg, Maastricht, who introduced us in the field of purine and pyrimidine metabolism, for their scientific support and technical advice.

REFERENCES

1 H.A. Simmonds, Purine and pyrimidine disorders in "The inherited metabolic diseases", J.B. Holton, ed., Churchill Livingstone, Inc., New York (1987).
2. A.H. van Gennip, D.Y. van Nordenbung-Huistra, P.K. de Bree and S.K. Wadman, Clin. Chim. Acta 86: 7 (1978)
3. G.S. Morris and H.A. Simmonds, J. Chromatogr. 344: 101 (1985).
4. R.A. de Abreu, J.M. van Baal, C.H.M.M. de Bruyn, J.A.J.M. Bakkeren and E. D. A. M. Schretlen, J. Chromatogr. 229: 67 (1982).
5. R.J. Simmonds and R.A. Harkness, J. Chromatogr. 226: 369 (1981).

SIMPLE METHOD FOR THE QUANTITATIVE ANALYSIS OF DIHYDRO-PYRIMIDINES AND N-CARBAMYL-ß-AMINO ACIDS IN URINE

Albert H. van Gennip, Sandra Busch, Eline G. Scholten, Lida E. Stroomer and Nico G. Abeling

Depts. of Pediatrics and Clinical Chemistry
University Hospital 'AMC', Meibergdreef 9
1105 AZ AMSTERDAM, The Netherlands

INTRODUCTION

In man pyrimidines are degraded in four steps catalysed by dihydropyrimidine dehydrogenase (DHPD, EC 1. 3. 1. 2.), dihydropyrimidinase (DHP, EC 3. 5. 2. 2.), ureidopropionase (UP, EC 3. 5. 1. 6.) and a transaminase (Fig. 1). As yet only deficiencies of the first and the last enzymes have been described. Deficiency of R-ß-aminoisobutyrate aminotransferase (EC 2. 6. 1. 22.) is easily detected by classical amino acid analysis[1,2],DHPD-deficiency by TLC or HPLC of pyrimidines in urine[3]. A method is presented for the detection of the two other defects.

MATERIALS

Chemicals

Dihydrouracil, dihydrothymine, N-carbamyl-ß-alanine and N-carbamylnorvaline were purchased from Sigma Chemical Co, St Louis, MO, USA. Sulphosalicylic acid, hydrochloric acid and formic acid were from Merck, Darmstadt, Germany. Norvaline was obtained from Serva, Heidelberg, Germany, ß-alanine from BDH, Poole, England and ß-aminoisobutyric acid from Fluka, Buchs, Switzerland. The ion-exchange resins Dowex 50W x 8, 50-100 mesh and Dowex 1 x 8, 100-200 mesh were purchased from Fluka and Sigma respectively. All chemicals were of the highest quality commercially available.

Apparatus

Amino acid analyses were performed by automated high performance cation-exchange column chromatography using the Chromakon-500 system (Tegimenta AG, Rotkreuz, Switzerland).

Purine and Pyrimidine Metabolism in Man VII, Part B
Edited by R.A. Harkness *et al.*, Plenum Press, New York, 1991

EXPERIMENTAL PROCEDURES

In DHP deficiency urinary dihydropyrimidines in UP deficiency urinary N-carbamylamino acids will be increased. Therefore a procedure had to be developed by which both groups of compounds could be analysed. Preferably, the method had to be applicable in the clinical chemistry laboratory equipped with routine chromatographic instrumentation. The possibility to convert the compounds of interest into amino acids led us to make the method suitable for an amino acid analyser.

Our approach was to first isolate the dihydropyrimidines, N-carbamylamino acids and amino acids from urine in three different fractions. Subsequently, the first two fractions had to be hydrolysed to obtain the corresponding

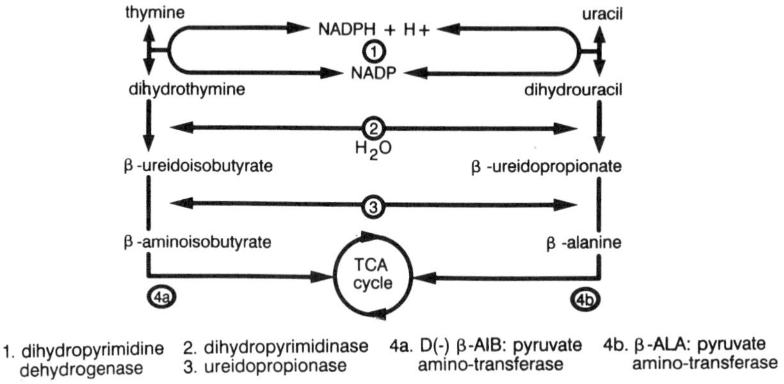

Fig. 1. Pyrimidine degradation pathways. Enzymes: dihydropyrimidine dehydrogenase(1); dihydro pyrimidinase(2); ureidopropionase(3); D(-)ß-AIB aminotransferase(4a); ß-ALA aminotransferase(4b).

amino acids. Finally, the amino acids in the fractions could be quantified using an amino acid analyser.

For isolation, we used a dual-column system consisting of a cation and an anion exchanger as shown in Fig. 2. Urine 5 ml, if necessary adjusted to pH 6 à 7, was applied onto the top column. The column system was subsequently eluted with 40 ml doubly distilled water: the effluent (45 ml) contained the dihydropyrimidines. Elution of the cation-exchange column (I) with 2 M ammonia 40 ml yielded the amino acids. The anion-exchange column (II) was eluted with 2 M formic acid 40 ml to obtain the N-carbamylamino acids. The effluents containing the dihydropyrimidines or

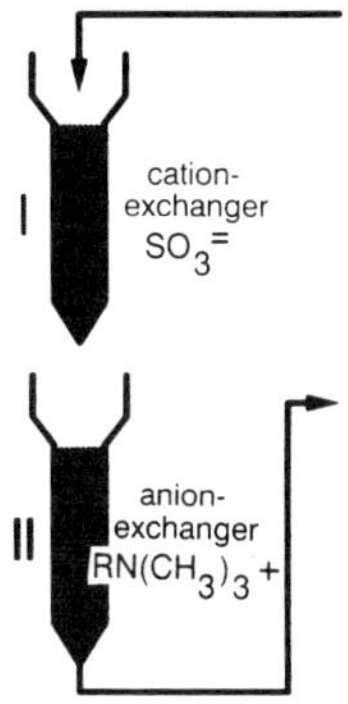

Fig. 2. Dual-column system for the differential isolation of urinary dihydropyrimidines, N-carbamylamino acids and amino acids.

N-carbamylamino acids were evaporated to dryness under reduced pressure at 37°C after the addition of internal standard (N-carbamylnorvaline). The residues were redissolved in 1.5% sulphosalicylic acid solution, 1 ml. To 0.5 ml of these extracts 6 M HCl, 2.5 ml, was added and the solutions were hydrolysed at 150°C for 18 hrs. After cooling to room temperature the hydrolysates were evaporated to dryness under reduced pressure at 37°C. The residues were redissolved in 1.5% sulphosalicylic acid, 0.5 ml, for amino acid analysis.

RECOVERIES AND REPRODUCIBILITIES

Recoveries were determined with the standard addition method using urine samples enriched with synthetic compounds to various concentrations. The same samples were used to establish the within-run reproducibilities. The results are shown in table 1. As can be seen recoveries were good and acceptable reproducibilities were obtained. Carry-over of amino acids, dihydropyrimidines and N-carbamylamino acids between their respective fractions was negligible (<3%).

APPLICATION OF THE METHOD

The potentialities of the method developed for the differential analyses of dihydropyrimidines and N-carbamylamino acids can best be demonstrated by its application on normal urine samples and samples from a patient suspected of DHP deficiency and his brother. The results are summarised in table 2. As can be seen urinary dihydropyrimidines and N-carbamylamino acids were strongly increased in the first urine of the patient which was contaminated by bacteria, but in a second, uncontaminated urine the N-carbamylamino acids were normal. The last pattern is to be expected in DHP deficiency. The excretion profile of the patient's brother is shown to be completely normal.

Table 1. Recoveries and within-run reproducibilities

compound	conc. calc. µM	conc. meas. µM	rec. %	compound	conc. calc. µM	conc. meas. µM	rec. %
DHU	1000	937	93.7	DHT	1000	1035	103.5
	500	464	92.9		500	493	98.6
	250	227	90.9		250	254	101.6
	50	41	82.7		50	56	111.9
Mean rec.:			90	Mean rec.:			104
Reprod.	:CV 10% (475 µM, n=10)			Reprod.	:CV 8% (368 µM, n=10)		
NC-ß-ALA	1000	923	92.3	NC-ßAIB	:not available		
	500	480	96.1	DHU	:dihydrouracil		
	250	251	100.5	DHT	:dihydrothymine		
	50	51	102.0	NC-ßALA	:N-carbamyl-ß-alanine		
Mean rec.:			98				
Reprod.	:CV 11% (223 µM, n=10)						

Table 2. Normal urine samples and a sample of a patient suspected of DHP deficiency.

	Excretion values in µmol/g creat.			
	DHU	DHT	NC-ß-ALA	NC-ß-AIB
E.K. (I)	7966	5643	2333	1799
E.K. (II)	5505	3592	37	0
B.K. (br.)	183	13	66	23
cont. (n=6)	138-972	14-249	52-627	14-56

E.K.=index patient; B.K.=brother; cont.=controls.

DISCUSSION

The method presented appeared to be reproducible and specific. Moreover the recovery for the various compounds is shown to be satisfactory. A great advantage of this method is the applicability in a clinical chemistry laboratory equipped with an amino acid analyser. The clinical usefullness is illustrated by the analysis of the urine of a patient with presumably DHP deficiency[4]. Moreover, analysis of the bacterially contaminated urine of the patient showed that also elevated excretion of N-carbamylamino acids can easily be detected. The last finding is relevant to the diagnosis of patients with ureidopropionase deficiency.

REFERENCES

1. Y. Kakimoto, A. Kanazawa, K. Taniguchi and T. Kappe, ß-aminoisobutyrate-α-ketoglutarate transaminase in relation to ß-aminoisobutyric aciduria, Biochim. Biophys. Acta 156: 374 (1968).
2. A.H. van Gennip, J.P. Kamerling, P.K. de Bree and S.K. Wadman, Linear relationship between the R- and S-enantiomers of ß-aminoisobutyric acid in human urine, Clin. Chim. Acta 116: 261 (1981).
3. A.H. van Gennip, E.J. van Bree-Blom, S.K. Wadman, M. Duran and F.A. Beemer, Liquid chromatography of urinary pyrimidines for the evaluation of primary and secondary abnormalities of pyrimidine metabolism, in: Biological/

Biomedical applications of Liquid Chromatography III, G.L. Hawk, P.B. Champlin, R.F. Hutton and C. Mol, eds., Marcel Dekker Inc., New York and Basel (1981).
4. M. Duran, P. Rovers, P.K. de Bree, C.H. Schreuder, H. Beukenhorst, L. Dorland and R. Berger, Dihydropyrimidinuria, Lancet 336: 817 (1990).

DIAGNOSTIC POTENTIAL OF HPLC: EXPERIMENTAL AND CLINICAL TRIALS

R.T.Toguzov and Yu.V.Tikhonov

Department of Biochemistry, Central Research Laboratory, N.I.Pirogov 2nd Moscow Medical Institute, Moscow, U.S.S.R.

INTRODUCTION

Diagnosis means complete knowledge. At present, however, the modern arsenal of diagnostic procedures lacks a universal and effective method that would provide a clinician with an overall picture of biochemical alterations responsible for a certain pathology. Of all the methods, which can furnish this valuable information, HPLC allows to perform qualitative and quantitative analyses of metabolic profiles of organs and tissues in desease. From the clinical point of view, that means a possibility to reveal specific biochemical disorders underlying the onset and development of pathology, to trace the dynamics of the pathological process in each individual case, decide upon an appropriate drug therapy, and, finally, prognosticate furthe course of disease.

However, chromatography cannot be turned into a perfect diagnostic tool without a sound theoretical basis for its application in this field. First of all, one has to reveal the class of compounds of which the profile should be analysed. Naturally, each pathology is accompanied with specific metabolic alterations. So, it's possible to find a group of compounds which reflect the pathogenesis of a disease and analyze them using various chromatographic procedures. This way, though worthy of researchers' attention, may turn into procrustean bed for real diagnosis if it is confined to investigation of biochemical disorders only in the affected organ or tissues. As is known, any disease involves a transformation of biochemical interactions at the level of the whole organism, and the final outcome of the pathological process is determined by these interactions. So, the problem is to find such compounds whose metabolic profile reflects not only the evolvement of a pathological process at the site of lesion but systemic, i.e. interorganic and intertissue, interactions, too. Futhermore, this indication must be general enough, i.e. it has to undergo certain transformations in most diseases. In our view, there is a class of compounds which can meet these requrements: namely, free components of nucleic acids and their derivatives. Purine and pyrimidine compounds are involved in almost every area of metabolic pathways and alterations in nucleotide concentrations are associated with numerous pathological conditions. Moreover, many drugs have been developed to affect one or another aspect of nucleotides metabolism.

Multiple cell populations, e.g., blood cells, bone marrow, small intestine etc. are with little or no ability to synthesize purine (or pyrimidine) de novo which makes them metabolize nucleosides and bases predominantly via the salvage pathway (1,2). On the other hand, it was demonstrated that erythrocytes act as carrier of these compounds (3). Consequently, a coupling of metabolic pools of free nucleosides and bases must take place between various tissues of the organism which includes their transport with blood and through the cell membranes, metabolic pathways, regulation, etc. Then, any pathological process apparently leads to specific alterations in the formation of pools of free purines and pyrimidines not only in the lesioned tissue but other systems, too.

In other words, according to our working hypothesis, in the whole body there must be a dynamic system of purine/pyrimidine substrates cooperating between tissues and, accordingly, disorders of this cooperations generated by various pathological processes.

Space does not permit a discussion of this, but we chose the rapidly growing transplanted Hepatoma 22 for testing this concept. This model system makes it possible to elucidate the link between the growth rate (malignancy) and the distribution and fluctuation of the free purine and pyrimidine compounds in different organs and tissues.

METHODS

Full details of all sample handling and analitical procedures are given and discussed elsewhere (4, 5, 6, 7), as a complete description of the development of the HPLC methodology (8,9).

<u>In vitro studies: Radioisotope detection in HPLC</u>

Erythrocytes were incubated with [^{14}C]Hyp in saturated conditions (50 μCi/mL, 6.4 * 10^8 cells/mL, 25°C)in 199 medium during 90 min. Erythrocytes were removed from incubation medium, washed twice in Henks buffer, transferred into the 199 medium in concentration 1.6 * 10^9 cells/mL, and mixed with equal volume of thymocytes (3.2 * 10^8 cells/mL) in RBC/Thymocytes ration 5:1. After 20 and 60 minutes of incubation at 37°C under continious shaking conditions (to avoid gravitate sedimentation) cells were separated using ficoll gradient tecnique, and acid-soluble fractions (ASF) of both kind of cells and incubation medium were analyzed by HPLC method with UV- and radioactive detections.

Reversed phase HPLC separation was carried out on ODS-XL cartridge column (Beckman, USA) with mobile phase consisted of 40 mM KH_2PO_4, 1 mM TBA, 2% CH_3CN, pH 5.1, flow rate 0.5 mL/min. IBM PC AT with GOLD Software and Chromatographics was used for device control, collection and calculation of row chromatographic data.

RESULTS AND DISCUSSION

Experimental trials

If there exists an intertissue coupling system which coordinates the transport and utilization of purine and pyrimidine derivatives, then specific transformation of their metabolism must take place in the cells and tissues of the host during tumour growth.

We previously described, that liver and RBC purine derivitives compositions were affected by the implanted hepatoma (4,5). Moreover, the obtained experimental evidens led us to the conclusion that the products of purine catabolism are released into the blood-stream in the log phase of hepatoma growth. The erithrocytes then act as a buffer, taking up the precursors, which in turn are delivered to peripheral tissues and especially to growing hepatoma 22. This conclusion is confirmed by the exhaustion of purine metabolities in the host erythrocytes within 5-7 days during the maximal rate of hepatoma growth.

A relation between dissorders of purine metabolism and immune disfunction is well known. It is quite clear that an intact purine metabolic pathway is necessary for normal immune function (10).

During our own investigations on the effect of hepatoma growth on the purine metabolism in thymocytes and splenocytes, we used another approach (6).

With this aim in view we developed the equation for quantative analysis of real relationship between erythrocytes and lymphocytes purine transport and metabolism during hepatoma growth. This equation was used for calculation of the real uptake of metabolites by host thymocytes from erythrocytes during tumor growth.

It was found that in the log phase of hepatoma growth hypoxanthine (Hyp) incorporation into acid-soluble fraction of thymocytes increased 30 and 140 fold within 3-5th day. By contrast, the concentration of Hyp in thymocytes increased 50-fold only on the 2nd day and then declined on the 5th day of hepatoma growth.

The correlation between inosine (Ino) transport and concentrations in the thymocytes was different for the Hyp. For example, the increased incorporation of Ino in thymocytes was registered within the first - third days and on the 11th day; Ino concentrations also increased during 1-3 day and then ASF thymocytes exhausted.

In other words, we have found that the relation between the rate of Hyp incorporation and Hyp concentration in thymocytes ASF is inversely proportional during the log phase of hepatoma growth, while these parameters for Ino are directly proportional.

As has been shown, adenosine can be toxic to lymphoid cells. Our data indicate that no Ado was detected in thymocytes within 1-3d days, but sharply increased within 5-9th days during the maximal rate of hepatoma growth. The concentrations of adenine and guanine decreased 2-fold on the 5th day.

Data on the activitives of purine salvage pathway in thymocytes and the data presented indicate that the drastic alterations in the metabolism of purine derivitives is closely associated with the impairments of thymocytes differentiation and maturation that result in reduced immune responsiveness during hepatoma growth.

Consequently, disturbances of purine metabolism in the liver, erythrocytes and thymus of the host mice, induced by the growing hepatoma, can be responsible for the several physiological and biochemical functions.

The next series of experiments was focused on using of HPLC radioisotope detection for the investigation of erythrocyte-dependent purine transport to mouse thymocytes during hepatoma growth. We tried to follow pathways and particularities of Hyp transport and metabolism in thymocytes *in vitro* (see "Methods"). Cells were taken at 3rd and 7th day of hepatoma 22 development as exponential period and maximum tumor cells growth rate period subsequently.

The most quantitives of labeled metabolites were found in the peak of intracellular AMP for both types of cells incubated. 60 min of joint incubation of erythrocytes/thymocytes mixture taken at 3rd day of hepatoma growth resulted in the complete distortion of [14-C] metabolites in mouse erythrocytes. In contrast, at the 7th day of hepatoma growth the both [14-C]AMP and [14-C]Ado appeared in erythrocytes after 60 min of joint incubation. The [14-C]AMP content in thymocytes did not change from 3rd to 7th day of hepatoma growth, while the content of [14-C]Ado decreased significantly at the same time. The changes in whole pool of thymocyte adenylates were similar to those described above. It is nessesary to note, there was not found [14-C]Ade in thymocytes ever, where as the concentrations of Ade in thymocytes were compatible with those of AMP. The concentrations of Hyp/Gua metabolites in thymocytes decreased more then twice from 3rd to 7th day.

The in vitro pattern of labeled purines between both kinds of incubated cells demonstrates the transport processes depend on erythrocytes being nearly. Moreover, the relatively decreased content of labeled purine nucleobases instead whole pool of those (for nucleosides and nucleoside monophosphates the inverse picture observed) may be indicates the slow rate of bases turnover inside the thymocytes.

As a consequence one can suppose that *in vitro* Hyp transfer from erythrocytes was assotiated with its rapid phosphorilation inside thymocytes and father metabolic processes.

More detailed picture of purine transport and metabolism from erythrocytes to thymocytes *in vitro* suggest the using a extended set of labeled substrates, and additional information about intracellular di- and triphosphate in real time kinetic of joint incubation process.

Clinical trials

To estimate the potential and prospects for the application of this approach to diagnosis and prognosis of disseases in man we chose several ailments with a completely unlike etiology and pathogenesis, and performed an HPLC analysis of patients'tissue.

For example, we analysed alterations in the levels of oxypurine compounds in erythrocytes from umbilical cord blood of newborn infants with perinatal hypoxia (11). We observed a reciprocal relation between hypoxanthine and IMP in the erythrocytes of infants from two groups (favourable and complicated periods of early adaptation), that can be due to effects of hypoxia upon oxypurine cycle. In both groups, the xanthine and uric acid levels were practically no different from control. This study has demonstrated the usefulness and efficiency of the HPLC separation of purine derivatives for the evaluation of perinatal hypoxia.

We also studied the purine metabolites in human intestinal mucosa (7). The small intestinal mucosa is one of the most actively regenerating mammalian tissues. In gluten-sensitive coeliac disease the intestinal mucosa shows marked structural abnormalities with flattening and loss of villi. Simultaneously, the increase in depth of the intestinal crypts is associated with the number of cells in mitosis in the crypts and a widening of the mitotic zone (12). However, little is known about biochemical abnormalities in coeliac mucosa. The differences in the profiles of nucleotides, nucleosides and bases in normal and coeliac mucosa could provide important information on the nucleic acid metabolism in mucosa of patients with coeliac desease. It was found that the levels of ADP and particulary of ATP markedly increased in coeliac mucosa with no apparent changes in the pool of AMP. A similar pattern was observed with the guanine nucleotides.
The concentration of IMP in coeliac mucosa was 50% of that of normal mucosa. The concentrations of adenosine and adenine changed relatively little. By contrast, in subtotal villous atrophy, the sharp increase in xantine level and slight increase in the uric acid concentration is attributed to the high activity of enzyme xanthine oxidase present in the mucosa. The increase in inosine concentration with the decrease in IMP indicates the functioning of IMP nucleotidase in coeliac mucosa.

Since intestinal mucosa has a limited capacity for de novo synthesis of purine nucleotides (2) this tissue is dependent on salvage pathway activity for the provision of purine nucleotides. Apparently, the accumulations of ATP and GTP may be ascribed to the very high activity and amount of adenylate and guanylate kinases in coeliac mucosa and is directly related to the marked increase in the rate of cellular proliferation. On the other hand, high level of the end products of purine catabolism in coeliac mucosa may be correlated with the changes in the cellular turnover.

Thus, the obtained experimental and clinical evidence allows to regard the metabolic profiles and streams of free nucleic acid constituents as universal characteristics of pathological processes occurring in the organism which can be used to diagnose and prognosticate the development of various ailments.

ACKNOWLEDGEMENT

We would like to thank Mr. E.Grohman of Beckman Instruments for providing research facilities.

REFERENCES

1. B.A.Lowy, B.Ramot and I.M.London, J.Biol.Chem. 235:2920 (1960).

2. A.M.Mackinon and D.J.Deller, Biochim.Biophis.Acta 319:1 (1972).

3. Y.Konishi and A.Ichihara, J.Biochem. 85:295 (1979).

4. Yu.V.Tikhonov, I.S.Meisner, A.M.Pimenov and R.T.Toguzov, Vopr.Med.Chim. 35:125 (1989).

5. Yu.V.Tikhonov, A.M.Pimenov, R.T.Toguzov, T.Grune, W.Siems, H.Schmidt and G.Gerber, Biomed.Biochim.Acta 49:125 (1990).

6. A.M.Pimenov, Yu.V.Tikhonov and R.T.Toguzov, Experimental Oncol. 11:33 (1989).

7. Yu.V.Tikhonov, A.M.Pimenov, S.A.Uzhevko and R.T.Toguzov, J.Chromatogr. 520:419 (1990).

8. A.M.Pimenov, Yu.V.Tikhonov and R.T.Toguzov, J.Chromatogr. 365:221 (1986).

9. R.T.Toguzov, Yu.V.Tikhonov, A.M.Pimenov, V.Yu.Prokudin, W.Dubiel, M.Ziegler and G.Gerber, J.Chromatogr. 434:447 (1988).

10. D.A.Carson, E.Lakow, D.Wasson and N.Kamatani, Immunol. Today 2:234 (1981).

11. R.T.Toguzov, V.F.Demin, V.Yu.Prokudin, Yu.V.Tikhonov and A.M.Pimenov, Biomed.Biochim.Acta 48:279 (1989).

12. M.L.Clark and J.R.Senior, Gastroenterology 56:887 (1969).

REFERENCE VALUES OF OROTIC ACID, URACIL AND PSEUDOURIDINE IN URINE

Satoru Ohba**, Kiyoshi Kidouchi*, Chie Nakamura*, Toshiyuki Katoh*, Masanori Kobayashi** and Yoshiro Wada**

*Department of Pediatrics, Nagoya City Higashi General Hospital
1-2-23 Wakamizu, Chikusa-ku, Nagoya 464, JAPAN
**Department of Pediatrics, Nagoya City University Medical School
1 Kawasumi, Mizuho-ku, Nagoya 467, JAPAN

INTRODUCTION

Measuring urinary orotic acid and uracil which are intermediates in pyrimidine biosynthesis, is important for screening inborn errors of metabolism, such as urea cycle[1] and pyrimidine metabolism disorders[2,3]. Pseudouridine is a degradation substance from transfer RNA and its measurement might be useful as a marker for cancer and cancer therapy evaluation[4]. Until now, there have been no reports of reference values classified by age. We are presenting the orotic acid, uracil and pseudouridine reference values measured by using high performance liquid chromatography (HPLC) with an automated column switching system[5].

MATERIALS AND METHODS

Single voided urine samples were collected from healthy control subjects (newborns, n=14; children aged 1-15, n=129; adults, n=25), and from a patient with partial ornithine transcarbamylase (OTC) deficiency and his mother. These urine samples were frozen and stored at -20℃. Immediately prior to analysis, each sample was passed through a 0.45 μm Centricut filter to remove particle matter. For the quantitative determination of orotic acid, uracil and pseudo-

TABLE I REFERENCE VALUES OF OROTIC ACID, URACIL AND PSEUDOURIDINE IN URINE CLASSIFIED BY AGE

age		orotic acid	uracil	pseudouridine
neonate 0d.	(n=14)	17.1± 8.7	6.1± 4.3	852.3±396.8
neonate 5d.	(n=14)	22.3± 9.9	17.3± 14.4	1039.7±387.9
0 y.	(n=16)	16.1± 8.3	134.4± 89.1	906.6±235.5
1 y.	(n= 7)	28.4± 9.4	307.3± 93.2	859.2±148.0
2 y.	(n=16)	18.1± 9.7	173.0± 92.9	697.7±108.2
3 y.	(n= 9)	13.7± 3.9	155.7± 68.3	612.0±103.1
4 y.	(n= 7)	12.0± 4.0	127.8±108.8	548.5±123.9
5 y.	(n= 8)	10.3± 5.6	105.8± 51.6	511.6±125.3
6 y.	(n= 6)	9.7± 6.7	96.4± 29.5	458.5± 37.9
7-15 y.	(n=60)	8.8± 4.9	91.9± 67.5	330.0±109.4
adult	(n=25)	4.9± 1.8	56.2± 26.1	205.1± 31.3

(mean±S.D., μmol/g creatinine)

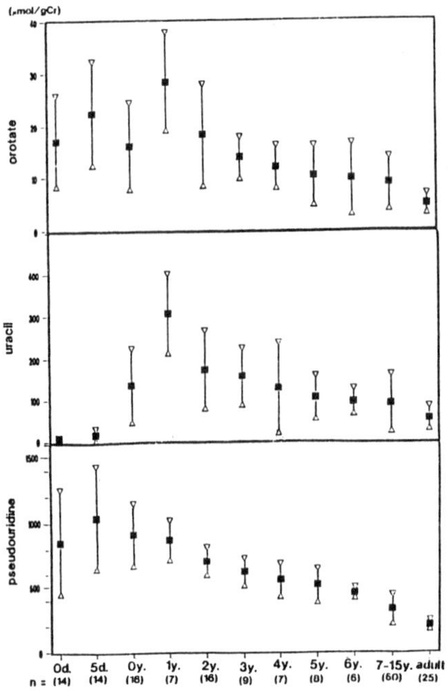

Fig. 1. Reference values of urinary orotic acid, uracil and pseudouridine classified by age (represent mean ± S. D.)

uridine in urine, high performance liquid chromatography (HPLC) was performed as described previously[5]. The urinary creatinine level was measured by Jaffe's method by using auto analyzer.

RESULTS AND DISCUSSION

TABLE and Fig. 1 show reference values of orotic acid, uracil and pseudouridine in urine classified by age. We found that the respective values varied greatly according to age. The urinary excretion patterns of orotic acid and uracil were significantly different from pseudouridine, and their maximum value occurred at the age of one year. Orotic acid values in our study were slightly lower compared to those reported by the colorimetric procedure[6] which measured total orotic acid (orotic acid plus orotidine). Recently, creatinine-related uracil excretion was reported as being unrelated to age[7]. However, our study showed that it was lowest in the newborn period. Further study for uracil reference values including the newborn period is necessary.

Fig. 2 shows sequential alterations of urinary orotic acid and uracil values during non-hyperammonemia in a patient with partial OTC deficiency(a) and in his mother(b). Orotic acid values were almost within the reference value range, except for periodic high values. However, uracil values were always high (>+2 S. D.) during the study period. Fig. 3 shows urea cycle and pyrimidine synthesis. In urea cycle disorders such as OTC deficiency, accumulated mitochondrial carbamyl phosphate diffuses into the cytosole, where pyrimidine biosynthesis is stimulated. Generally, carrier detection of OTC deficiency requires a protein loading test followed by a urinary orotate measurement. Recently, Hauser ER et al reported the allopurinol test, by which they clarified increased pyrimidine biosynthesis through urinary orotidine measurement after allopurinol loading[8]. Now, we found that the patient ex-

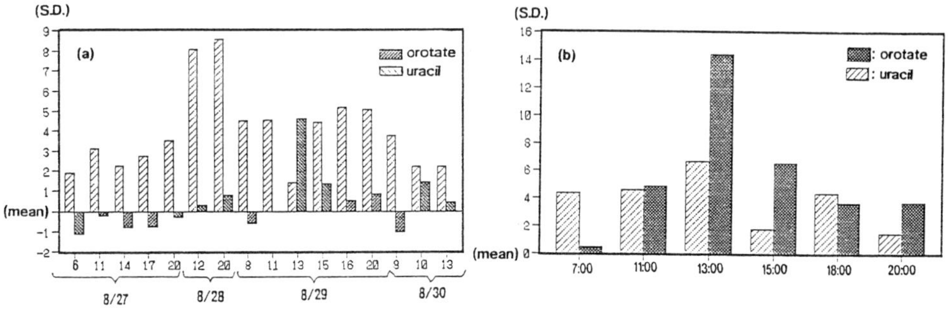

Fig. 2. Sequential alterations of urinary orotic acid and uracil values during non-hyperammonemia in a patient with partial OTC deficiency(a), and in his mother(b), showing the deviation from the reference values(mean) according to age

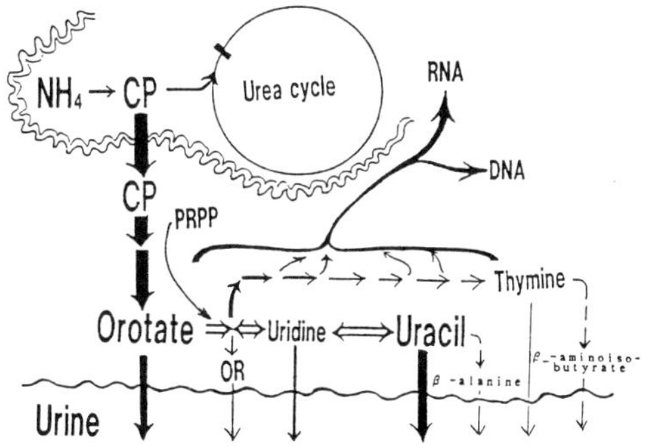

Fig. 3. Urea cycle and pyrimidine biosynthesis, with the relationships among intramitochondrial and cytosolic CP, and urinary excretion of its metabolites (CP:carbamyl phosphate, OR: orotidine, PRPP:phosphoribosyl pyrophosphate)

creted a high value of uracil during the period of normal orotic acid excretion, without protein or drug loading. It may be possible to screen for OTC deficiency (including partial OTC deficiency) by measuring urinary uracil, without loading test, and with the added benefits of increased safety and convenience.

REFERENCES

1. A. H. van Gennip, E. J. van Bree-brom, J. Grift, P. K. De Bree and S. K. Wadman, Urinary purines and pyrimidines in patients with hyperammonemia of varous origins. Clin. Chem. Acta, 104: 227-239 (1980)
2. C. M. Huguley, J. A. Bain, S. L. Rivers and R. B. Scoggins, Refractory megaloblastic anemia associated with excretion of orotic acid. Blood, 14: 615-634 (1959)
3. S. K. Wadman, R. Berger, M. Duran, P. K. De Bree, S. A. Stoker-de Vries, F. A. Beemer, J. J. Weots-Binnerts, T. J. Penders and J. K. van der Woude, Dihydropyrimidine dehydrogenase deficiency leading to thymine-uraciluria.

An inborn error of pyrimidine metabolism. J. Inher. Metab. Dis., 8 Suppl. 2: 113-114 (1985)
4. K. Itoh, M. Mizugaki and N. Ishida, Detection of elevated amounts of urinary pseudouridine in cancer patients by use of a monoclonal antibody. Clin. Chem. Acta, 181: 305-316 (1989)
5. S. Ohba, K. Kidouchi, T. Katoh, T. Kibe, M. Kobayashi and Y. Wada, Automated determination of orotic acid, uracil and pseudouridine in urine by HPLC with column switching J. Chromatogr. in press
6. M. L. Harris and V. G. Oberhorzer, Conditions affecting the colorimetry of orotic acid and orotidine in urine. Clin. Chem., 26: 473-479 (1980)
7. B. Assmann and H. J. Haas, Determination of creatinine-related urinary uracil excretion in children by HPLC. J. Chromatogr., 525: 277-285 (1990)
8. E. R. Hauser, J. E. Finkelstein, D. Valle and S. W. Brusilow, Allopurinol-induced orotidinuria. A test for mutations at the OTC locus in women. N. Engl. J. Med. 322: 1641-5 (1990)

AUTOMATED QUANTITATIVE ANALYSIS FOR OROTIDINE AND URIDINE / THYMINE IN URINE BY HIGH-PERFORMANCE LIQUID CHROMATOGRAPHY WITH COLUMN SWITCHING

Kiyoshi Kidouchi*, Chie Nakamura*, Toshiyuki Katoh*,
Tetsuya Kibe**, Satoru Ohba**, Yoshiro Wada**

*Department of Pediatrics, Nagoya City Higashi General Hospital
1-2- 23 Wakamizu, Chikusa-ku, Nagoya 464 (Japan)
**Department of Pediatrics, Nagoya City University Medical School
1 Kawasumi, Mizuho-ku, Nagoya 467 (Japan)

INTRODUCTION

The measurement of urinary orotidine, uridine and thymine which are intermediates in pyrimidine biosynthesis, is important for screening and diagnosis inborn errors of metabolism, such as urea cycle[1] and pyrimidine metabolism disorders[2], as well as the measurement of urinary orotic acid and uracil[1,3]. Several high-performance liquid chromatography (HPLC) methods have been reported by a number of researchers. However, these HPLC methods for uridine and thymine analysis require rather complicated sample preparation in order to obtain accurate results[1]. Although, a simple method for measuring orotidine and orotic acid were recently reported[4], it is not capable of evaluating near-normal levels of orotidine .

We have established a simple and accurate quantitative method for urinary orotic acid, pseudouridine and uracil by using an HPLC with a dual column[3]. Based on this technique, we present here an accurate quantitative method requiring no sample preparation apart from filtration for urinary orotidine, and uridine / thymine. By using this method, preliminary reference values, and urinary pyrimidine excretion status in a patient with ornithine transcarbamylase (OTC) deficiency are evaluated.

EXPERIMENTAL

Chemicals and solutions

Analytical grade sulfuric acid and acetonitrile were purchased from Wako (Tokyo, Japan). Deionized water was passed through a Milli-Q Labo (Nihon Millipore Kogyo, Yonezawa, Japan). Orotidine, uridine and thymine were purchased from Sigma Chemical Co. (St. Louis, MO, U. S. A.). Other reference standards were analytical grade from Sigma.

Urine Samples

Single voided urine samples were collected from healthy control subjects (newborns, n=15; children aged 1-6, n=10; adults, n=12), and from a patient with OTC deficiency(liver OTC activity was 0% of control values) and his younger brother. These urine samples were frozen and stored at -20°C. Immediately prior to analysis, each sample was passed through a 0.45 μm Centricut filter (Kurabou, Osaka, Japan) to remove particle matter.

TABLE I

RECOVERY(n=5) AND COEFFICIENT OF VARIATION FOR URIDINE, THYMINE AND OROTIDINE ADDED TO A URINE SAMPLE

added ST	recovery(%)			coefficient of variation(%)		
(nmol/ml)	OROTI	URIDI	THYMI	OROTI	URIDI	THYMI
5.0	104.3	104.3		1.9	1.3	
12.5	106.0	106.0	104.8	1.2	1.4	1.1
25.0	101.7	103.2	97.2	0.7	0.9	0.7
50.0	106.3	104.3	104.6	0.5	1.0	0.9
100.0	99.8	99.8	101.9	0.6	0.8	0.8

ST:Standards, OROTI:Orotidine, URIDI:Uridine, THYMI:Thymine

HPLC Apparatus

Two Eyela PLC-5 liquid chromatograph (Tokyo Rikakikai, Tokyo, Japan), with pumps and detectors were used. A computerized system controller SC-15 (Tokyo Rikakikai) consisting of an electric valve, gradient system, and auto sampler KSP-600 (Kyowa Seimitsu, Tokyo, Japan) was used. An on-line Erma Degasser ERC-3611 (Erma, Tokyo, Japan) was used for eluent delivery, and an integrator CR4A Chromatopack (Shimadzu corporation, Kyoto, Japan) was used for data analysis.

HPLC procedure

A precisely measured 5 and 20 μl aliquot of each sample was applied to the first column (a reversed-phase ODS-C18 column: MIC GEL ODS-1MU, 150x4.6 mmI.D., particle size 5 μm, Mitsubishi Kasei Corporation, Tokyo, Japan) at 30°C. The flow-rate was 0.8 ml/min.

For orotidine analysis; one fraction from the ODS column was delivered to the second column (anion-exchange column: MCI GEL CA08F SO_4^- form 250 x 4.6 mmI.D., particle size 7-8 μm, Mitsubishi Kasei Corporation, Tokyo, Japan), by an automated column switching system[3]. The ODS column was eluted with 5mM H_2SO_4 (pH 2.3)for 13 min., and was then washed with 50% of acetonitrile /water for 5 min. Finally, it was equilibrated with 5mM H_2SO_4 for 10 min. The anion-exchange column at 60°C was eluted isocratically with 20mM H_2SO_4/0.4M Na_2SO_4 (pH 2.3). The flow-rate was 0.8 ml/min. The two columns were connected for a period of 4 ～ 5 min.(4 min.: just before the orotidine was eluted; 5 min.: just after the orotidine was eluted) using an electric switching valve. The eluate from the anion-exchange column was continuously monitored at 254nm and 280nm. External standards were analyzed at a rate of once per ten urine samples. Retention times and peak heights were recorded using the CR4A Chromatopack integrator.

For uridine and thymine measurement; one fraction for a period of 10 ～

TABLE II

QUANTIFICATION RATIOS AT 254nm TO 280nm OF OROTIDINE, URIDINE AND THYMINE PEAKS IN URINE SAMPLES

Pyrimidine	Urine Sample (Number)	Concentration (mmol/L)	Injected Volume (μl)	280nm/254nm(%) (Mean ± S.D.)
Orotidine	32	4.6 ～ 152.1	20	99.3 ± 4.1
Uridine	6	2.5 ～ 32.5	20	101.6 ± 26.5
	17	4.7 ～ 1064.6	5	100.9 ± 4.1
Thymine	9	1.5 ～ 17.1	20	101.7 ± 7.9

13 min. (10 min. : just before the uridine was eluted; 13 min. : just after the thymine eluted) from the ODS column was delivered to the second column (cation-exchange column: MCI GEL CK08EH H^+form 300x8 mmI.D., particle size 9 μm, Mitsubishi Kasei Corporation, Tokyo, Japan). The cation-exchange column was eluted isocratically at 30°C with 5mM H_2SO_4, and the flow-rate was 0.8 ml/min. The eluate from the cation-exchange column was continuously monitored at 254nm and 280nm.

RESULT and DISCUSSION

Separation of the orotidine, uridine and thymine as well as orotic acid, pseudouridine, uracil and uric acid group, was achieved by using the first column (ODS-C18). Orotidine and uridine / thymine were eluted for a period of 4 to 5 min. and 10 to 13 min. respectively. The orotidine was then applied to the second column (anion-exchange column) by the column switching system, and were well separated. The chromatographic profile of the urine samples from a healthy adult and a patient with hereditary orotic aciduria, had one major peak corresponding to the standard mixture of orotidine(retention time and absorption ratio of the 280/254nm). The uridine and thymine were also well separated by using the cation-exchange column.

Retention times (mean±S.D., in series) of orotidine(n=34), uridine(n=20) and thymine(n=20) were 22.54±0.05, 18.54±0.04 and 31.58 ±0.05 min., respectively. There were no significant variations over a short period (analysis time for 50 samples).

Calibration curves for orotidine, uridine and thymine were obtained by processing aliquots of an aqueous standard mixture solution at different concentrations (1, 5, 10, 50, 100, 125, 250, 500, 1000 nmol/ml). The relationships between standard concentrations and peak heights were linear, in the concentration range from 1 to 1000 nmol/ml for orotidine and uridine, and from 5 to 1000 nmol/ml for thymine. The correlation coefficients for the three compounds under study (r, obtained from five measurements) were 0.9999, 0.9997 and 0.9995, respectively. Detection limits were 5 picomol(orotidine and uridine) and 10 picomol(thymine) per 5 μl injected.

TABLE III

URINARY OROTIDINE, URIDINE AND THYMINE IN CONTROLS AND IN A PATIENT WITH OTC DEFICIENCY/HIS MOTHER, AND OROTIC ACIDURIA

Group or Subject	Age	n	Orotidine Mean ± S.D. (Range)	Uridine Mean ± S.D. (Range)	Thymine Mean ± S.D. (Range) (mol/g creatinine)
Control					
Newbons	0&5days	29	*3.5 ± 4.7		
			*(0.3 ～ 22.7)	(n.d ～32.6)	*(n.d ～ 27.6)
Children	1-6y	10	*8.2 ± 5.0		
			*(3.1 ～ 18.3)	(n.d ～ 39.1)	*(n.d ～ 35.8)
Adults	18y<	12	*3.4 ± 1.4		
			*(1.4 ～ 5.9)	(n.d. ～10.1)	*(n.d ～ 8.3)
OTC					
A.Y.	1M	11	24.7 ± 8.7		
			(8.1 ～ 37.9)	(n.d ～ 58.3)	(n.d. ～ 8.4)
M.Y.	2Y	7	13 .8± 6.7	373.3 ± 246.6	
			(5.0 ～ 24.2)	77.0 ～ 768.7)	(n.d. ～ 21.1)
(Hyperammonemia)		1	14.7	1765.4	6.6
Orotic Aciduria	6Y	2	161.6	5186.0	
uridine theray			(92.7 ～ 230.5)	(1747.8, 8624.2)	(1.7, 11.3)

* injected volume: 20 μl

Table I shows the recovery of standard compounds added to urine. Orotidine and uridine recoveries were between 99.8% and 106.8%. Thymine recovery was between 97.2% and 104.8% at an added concentration from 12.5 nmol/ml to 100 nmol/ml. Repeatability (coefficient of variation, %) was between 0.5 and 1.9% of peak heights.

The absorption ratio at 254nm to 280nm of each urine sample peak was identical to the standard for orotidine, uridine and thymine, and the ratio of their values measured by both uv absorption(%, mean±S.D.) were 99.3±4.1(n=32, sample volume: 20 ul), 100.9±4.1(n=5, sample volume: 5ul), 101.7±7.9(n=20, sample volume: 20ul) respectively(Table II). When 20ul urine samples in which the uridine value is low (as in healthy control urine) were analyzed, the ratio of uridine varied slightly due to the presence of an interfering substance which is usually more absorbent at 280nm than at 254nm. However, calculated at 254nm, the value was within the normal range given in previous reports[1,2]. Moreover, for high uridine concentrations detected in 5ul urine samples, the ratio of uridine was stable as shown table II. This showed that automated sample extraction using the ODS column, followed by elimination of ionic substances and separation of the three compounds using ion-exchange columns were satisfactory.

Preliminary reference values for the urine of healthy newborns, children and adults are given in Table III. Although orotidine values in our study were slightly lower compared to the reported by HPLC methods[1], uridine and thymine values are similar to those given in previous reports[1,2]. Orotidine values in the urine of a non-hyperammonemic patient with OTC deficiency sometimes overlapped with control ranges. However, contrasted to Van Gennip's patients[5], the mean values were slightly high compared to healthy children. During hyperammonemic period orotidine excretion in urine did not increase, as much as that of the orotic acid[1,5] and the uridine excretion. The orotic acid and uridine excretions were as high as in a patient with hereditary orotic aciduria on a uridine treatment program[Table III].

It has been hypothesized that an increased flux in the pyrimidine biosynthetic pathway occurs in patients with asymptomatic OTC deficiency and its heterozygous carriers. Allopurinol-induced orotidinuria was used as a test for carrier detection[6]. As observed in Van Gennip's patients[1]. We have reported that the patient with partial OTC deficiency usually excreted a high value of uracil, during the period of normal orotic acid excretion and normal serum ammonia level[3]. Therefore, the evaluation of the increased flux in the pyrimidine biosynthetic pathway by urinaly orotidine, orotic acid and uracil analysis may be reliable method for the screening of OTC deficiency with normal ammonia levels, as well as for screening for carriers.

This method and our previously established method[3] should prove useful in the screening of congenital metabolic diseases such as pyrimidine metabolism disorders, inborn errors in the urea cycle, and others.

REFERENCES

1 A. H. Van Gennip, E. J. Van Bree-blom, J. Grift, P. K. De Bree and S. K. Wadman, Clin. Chem. Acta, 104: 227 (1980).

2 J. A. J. M. Bakkeren, R. A. De Abreu, R. C. A. Sengers, F. J. M. Gabreels, J. M. Mass and W. O. Renier, Clini Chem. Acta 140: 247 (1984).

3 S. Ohba, K. Kidouchi, T. Katoh, T. Kibe, M. Kobayashi and Y. Wada, J. Chromatogr., in press, (1991).

4 S. W. Brusilow and E. Hauser, J. Chromatogr., 493: 368 (1989).

5 A. H. Van Gennip, J. Grift, P. K. De Bree, G. J. M. Zegers, J. W. Stoop and S. K. Wadman, Clin. Chem. Acta, 93: 419 (1979).

6 E. R. Hauser, J. E. Finkelstein, D. Valle and S. W. Brusilow, N. Engl. J. Med., 322: 1642 (1990).

HPLC ASSAY OF URIDINE MONOPHOSPHATE SYNTHASE (UMPS) IN CHORIONIC VILLUS SAMPLES (CVS) AND ERYTHROCYTES (RBC)

Lynette D. Fairbanks, John A. Duley, Ailsa J. Shores, H. Anne Simmonds

Purine Research Laboratory, Clinical Science Laboratories
UMDS Guy's Hospital

INTRODUCTION

The activities of the enzymes which catalyse the last two steps of the pyrimidine de novo synthetic pathway - orotic acid phosphoribosyltransferase (OPRT) and orotidine-5'-monophosphate decarboxylase (ODC) - are contained in a single bifunctional protein, UMP synthase (UMPS) (1). Patients with the autosomal recessive disorder, hereditary oroticaciduria, have coordinate deficiencies of both these activities (2). Previous assays for the human enzyme have measured CO_2 release from [^{14}C]-orotic acid (2). This paper reports results using an HPLC method employing an ion-pair reversed phase system, developed to analyse the products of UMPS activity in lysates of RBC and CVS.

MATERIALS AND METHODS

Chemicals and solvents: Radiolabelled 6-^{14}C orotic acid was from New England Nuclear UK. The pyrimidine standards and other chemicals used were from Sigma UK and HPLC grade solvents from Rathburn Chemicals UK. Heparinised blood from healthy volunteers was processed as described previously (5). CVS was either termination material or biopsy taken at 9-14 weeks under ultra-sound guidance.

UMP synthase assay in lysates of RBCs and CVS. The method of sample preparation used for RBCs has been published (5). CVS samples were separated from blood or decidua and the villi homogenised in buffer (25mM tris, 1mM DTT, 1mM EDTA, 1mM PMSF and 0.1% BSA) followed by freeze/thawing (one cycle), and centrifugation at 9,000g.
Assay for UMPS: 25ul of lysed RBC or 50ul of CVS were incubated at 37°C with 0.5uCi ^{14}C orotic acid (58.5uM) in phosphate buffer pH 7.4 containing 0.75mM PPRP, 0.8mM $MgCl_2$. Incubations with XMP (1.5mM) were pre-incubated for 15 minutes before addition of substrate. Incubations were terminated at 1 hour with 40% TCA, centrifuged at 9,000g and neutralised with diethyl ether to a pH above 5.0.
UMP synthase assay in intact RBCs and CVS: A frond of villus was also incubated in 18mM P_i with radiolabelled orotic acid, as described for erythocytes (4)

HPLC: Sensitivity was increased by using ^{14}C labelled substrate and an in-line radio-detector was coupled to the HPLC. Intact cell studies were anaylsed by anion exchange HPLC (4) and lysates by both the anion exchange (4) and ion-pair systems (5) as follows:
Ion-pair system (system 1) consisted of an ODS5 C18 column (120 X 4.5mm) with an isocratic buffer system ,40mM ammonium acetate plus 5mM tetrabutyl ammonium sulphate, adjusted to pH 2.75.
Anion exchange system (system 2) consisted of an APS hypersil column (250mm x 4.6mm) with a linear gradient from a 100% buffer A to 60% buffer B in 20 minutes with a 5 minute equilibrium delay. Buffer A was 5mM KH_2PO_4 (pH 2.65) and buffer B 0.5M KH_2PO_4 + 1.0M KCL (pH 3.4).

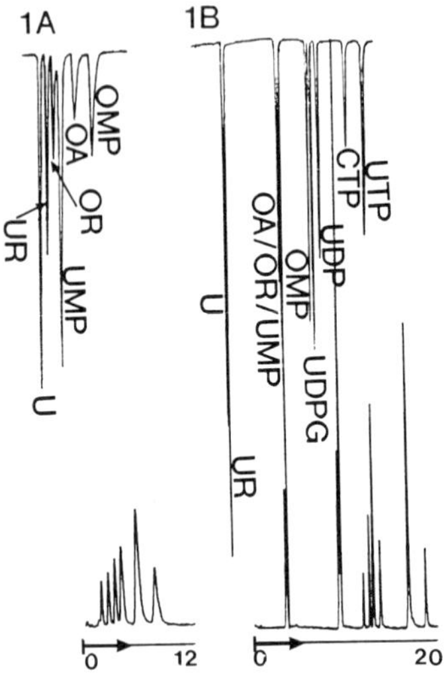

Figure 1 shows the u.v trace at 254 nm (upper), 280nm (lower) at 0.5 AUFS, chart speed following injection of a mixture of U, UR, OR, OA, OMP, UMP, UDP, UTP, CTP and UDPG, on :1A) system 1 and 1B) system 2.

RESULTS

1. Separation of a standard mixture by HPLC.
The best resolution of OA, OR, UR, U, UMP, and OMP was obtained in 12 minutes using the isocratic ion-pair system, but the di- and tri-ribonucleotide derivates were not eluted using this system (Figure 1A). The anion exchange system gave a good separation of the di- and tri-ribonucleotides also found as assay products, but similar retention times of OR, OA and UMP (Figure 1B) . Both systems were required to separate all the assay products.

2. Activity of OPRT and ODC in RBC or CVS lysates.
Incorporation into the different products in the RBC lysate assay is illustrated in Figure 2. For the RBC lysate assay OMP, UMP, UDP and UTP were the principal products with some UR (Figure 2A & 2B). With XMP (an ODC inhibitor) in the assay, the only product formed was OMP (Figure 2C). In the CVS studies the principal products were OMP, UMP and UR (Figure 2D & 2E). Incubations with XMP resulted in OMP being the major product (Figure 2F).

The mean UMP synthase activity was 0.362 nmol/mgHb/h in RBCs, and 16.11 nmol/mg protein/h in CVS. These results are similar to those reported using other methods (1,3).

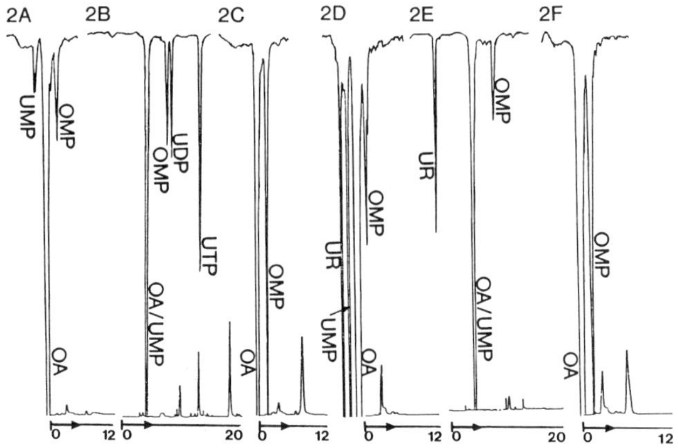

Figure 2 compares the radiotrace (upper) with the u.v. trace (lower) recorded at 254nm and 1.0 AUFS, following injection of a 75ul of RBC lysate assay on system 1 (2A); system 2 (2B); 2C) and + XMP on system 1 (2C); CVS lysate assay on system 1 (2D); system 2 (2E); and + XMP, system 1 (2F).

3. Activity of OPRT and ODC in intact RBCs and CVS

Figure 3 demonstrates the incorporation of radiolabel into the different nucleotide, nucleoside and base pools in RBCs (Figure 3A) and CVS (Figure 3B) under conditions favouring PP-ribose-P synthesis (18mM Pi). In RBCs a mean of approximately 50% of the product was found in UTP, 18% in UDPG, 18% in uridine, 10% in UDP and 2% in CTP. Using intact CVS, orotic acid was incorporated into UTP, UDPG and CTP, the percentage being much lower.

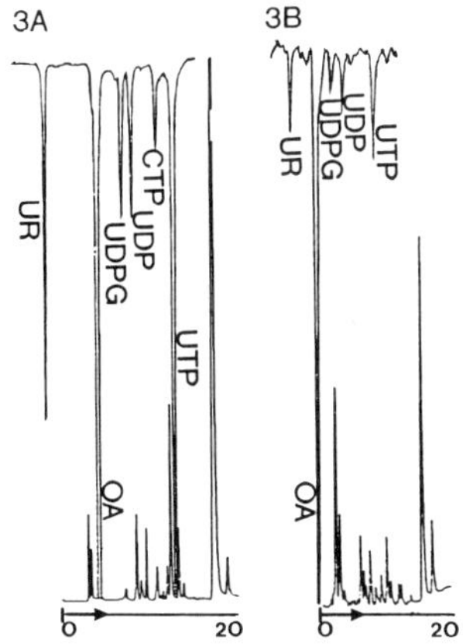

Figure 3 compares the radiodetector trace (upper) with the u.v. trace at 254nm 1.0 AUFS (lower), following injection of 75ul of : 3A) intact RBCs on system 2 ; 3B) intact CVS on system 2

DISCUSSION

HPLC demonstrated that numerous radiolabelled products were formed using lysates of either erthrocytes or CVS in the UMPS assay. The prinicpal products were OMP, UMP, UDP, UTP with some UR which could not all be separated using either of the anion or reversed-phase systems. Accurate identification was obtained using both systems: the anion exchange system, which separated OMP,UDP,UTP but not OA, OR and UMP, followed by the reversed phase ion-pair system, which separated UR, UMP and OMP, but from which the di- and tri-nucleotides did not elute. Incorporation of the ODC inhibitor, XMP, into the assay enabled rapid separation of the ^{14}C radiolabelled substrate and products of the assay in both lysed human RBCs or CVS lysates using only the reversed phase ion-pair HPLC coupled with in-line radiodetection. The production of OMP by OPRT is the rate limiting step. Consequently the activity of this enzyme should reflect the activity of the UMP synthase complex as a whole and this was found to be the case.

A substrate cycle has been suggested in Erlich ascites cells involving the interconversion of orotic acid - OMP - orotidine - orotic acid (6). Rapid degradation of the product has been identified as a potential pitfall in the assay of HGPRT in fibroblast and CVS lysates due to the much higher 5'-nucleotidase and PNP activity. Our results do not suggest substrate cycling could be a problem. No significant OR was found in any of the uninhibited lysate assays and in the assays incorporating XMP the product OMP was found in the same amount as the sum of products found (OMP, UMP, UDP and UTP) in the uninhibited assay. XMP also inhibits the enzyme 5'nucleotidase. These findings, consequently, would exclude the presence of such a cycle in the human RBC, under these conditions.

In summary, lysate assays showed that orotic acid is converted to UMP and is also further metabolised to UTP in RBC but in CVS is degraded to UR. HPLC has also proved useful in demonstrating significant incorporation of orotic acid into UTP, CTP and UDPG in the intact RBC as well as the CVS, indicating the existence of activity of the enzymes involved and their integrity.

References

1. Suttle DP, Becroft DMW, Webster DR. Hereditary orotic aciduria and other disorders of pyrimidine metabolism. Chapter 43. In:Scriver CR, Beaudet AL, Sly WS, Valle D, eds. The Metabolic Basis of Inherited Disease. New York:McGraw-Hill, 1989; 6th edition:1095-1129
2. Livingstone LR. Jones ME. The purification and preliminary characterisation of UMP synthase from human placenta J Biol Chem 1987;262:15726-15733.
3. Perignon JL, Durandy A, Peter MO, Freycon F, Durmez Y, Griscelli C. Early prenatal diagnosis of inherited severe immunodeficiencies linked to enzyme deficiencies. J Pediatr 1987;111:595-598.
4. Morris, G.S., Simmonds, H.A. & Davies, P.M. (1986) Use of biological fluids for the rapid diagnosis of potentially lethal inherited disorders of purine and pyrimidine metabolism. Biomed Chromatog 1, 109-118.
5. Fairbanks LD, Simmonds HA, Webster DR. Use of intact erythrocytes in the diagnosis of inherited purine and pyrimidine disorders. J Inher Metab Dis 1987;10:174-186.
6. Traut TW. Uridine-5'-phosphate synthase:evidence for substrate cycling involving this bifunctional protein. Arch Biochem Biophys 1989;268:108-115.

Acknowledgements: These studies were supported by the Regional Medical Development Fund, the Special Trustees and Friends of Guy's Hospital and the Arthritis and Rheumatism Council

A RP-HPLC METHOD FOR THE MEASUREMENT OF GUANINE, OTHER PURINE BASES AND NUCLEOSIDES

C.P.Quaratino, E.Messina, P.Bertuglia, G.Spoto,
G.Sciarra, G.Bucci, and A.Giacomello

Institute of Pediatrics, University of Rome
Institutes of Biochemical Sciences and
Human Physiopathology,
University of Chieti, and
Hospital of Castel Di Sangro-Italy

INTRODUCTION

A method for the analysis of purine bases and their nucleosides by RP-HPLC has been developed using a procedure similar to that reported in Reference 1. A chromatogram of standard compounds is shown in Figure 1. The method has been employed for the evaluation of purine metabolites in plasma from subjects admitted to the neonatal intensive care unit of the University of Rome (A) and for the study of allopurinol metabolism in human heart (B).

A) PURINE METABOLITES IN PLASMA FROM INFANTS
The main clinical characteristics of the 11 patients studied are summarized in table 1.

All patients needed ventilatory assistance. Intermittent arterial samples were drawn for blood gas analyses; $tcPO_2$ and $tcPCO_2$ were also monitored. When possible aliquots (0.5-1.0 ml) of the blood samples were rapidly centrifuged (within 5 min) and plasma was stored at -70 ℃ until assayed for purine metabolites (31 samples). Since uric acid has a retention time very close to that of hypoxanthine and guanine, samples were preincubated with uricase-Boehringer (10 ug/5ul) 2 hours at room temperature. Proteins were removed by heat treatment (5 min at 90 ℃). After centrifugation, aliquots (5-20 ul) of the supernatants were injected in the column and eluted as described in Figure 1. Complex chromatograms were obtained. The identity and purity of the peaks were established by the retention time, peak shape and treatment of each sample with specific enzyme preparations (xanthine oxidase, purine nucleoside phosphorylase and guanase from Boehringer) (1). Quantitative measurements were carried out by comparison using standard solutions of known concentrations. The peak with the retention time of xanthine not always completely disappeared after xanthine oxidase treatment. Clinical studies and animal experiments have suggested that hypoxanthine might be a suitable marker for asphyxia (2-4). In this preliminary study the relationship between hypoxanthine levels and blood gas measurements is reported. In each case the highest hypoxanthine levels in plasma were obtained in association with the presence of acidosis (pH <7.3) but with a variable PO_2 (PO_2>95 mm Hg in

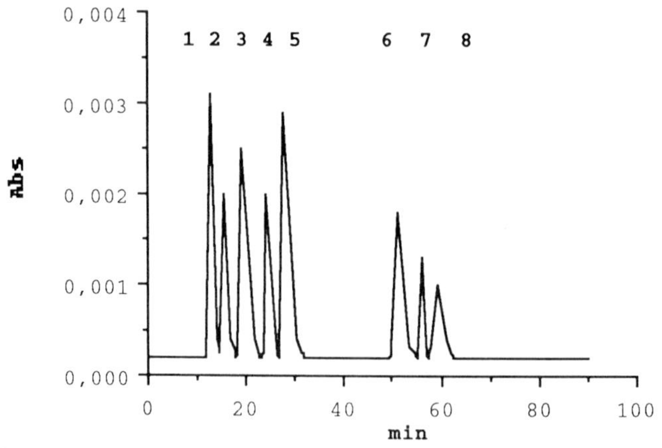

Fig.1. Chromatogram of standard compounds (3X0-150 pmol). Peaks: 1=guanine, 2=hypoxanthine, 3=xanthine, 4=oxypurinol, 5=allopurinol, 6=adenosine, 7=guanosine, 8=inosine.
Conditions: Supelco column (250x2.1 mm I.D.), Supelcosil LC-18 (3 µm) with precolumn; mobile phase, 0.2 ml/min, 0.04mol/l phosphoric acid at pH 3.0 with KOH and with 1% (v/v) methanol at room temperature. The detector was set at 254 nm.

Table 1. Characteristics of patients

Newborn Infants				
Case	B.W. (g)	G.A. (wk)	APGAR S. (1'/5')	Main Clinical Problems
1	1630	31	8/7	Prematurity
2	2950	38	?	Perinatal Asphyxia
3	1200	32	?	Prematurity
4	1200	30	5/7	Perinatal Asphyxia, RDS, Prematurity
5	3240	35	6/8	IDM, RDS, Polycythemia
6	1390	30	7/8	Esophageal Atresia, Apneic Spells
7	1760	31	3/6	Prematurity, Perinatal, Asphyxia, RDS
8	1300	31	7/9	Prematurity, RDS, Prematurity,RDS, Perinatal Asphyxia

Older Infants and Toddlers			
Case	W (Kg)	A. (months)	Main Clinical Problems
10	7.0	15	ICIP, Cardiac Arrest, Seizures
11	11.7	29	Seizures, Prolonged Apnea Crisis

RDS=Respiratory Distress Syndrome; IDM=Infant of Diabetic Mother; ICIP=Idiopathic Chronic Interstitial Pneumonia

5 cases, $PO_2<50$ mmHg in 3 cases and $69 < PO_2 < 86$ in 3 cases). This observation is in good agreement with literature data usually showing correlations of arterial hypoxanthine levels with decreased pH and increased base deficit, but not with decreased PO_2 (4). As a typical example, the results obtained in case 8 are reported in Figure 2.

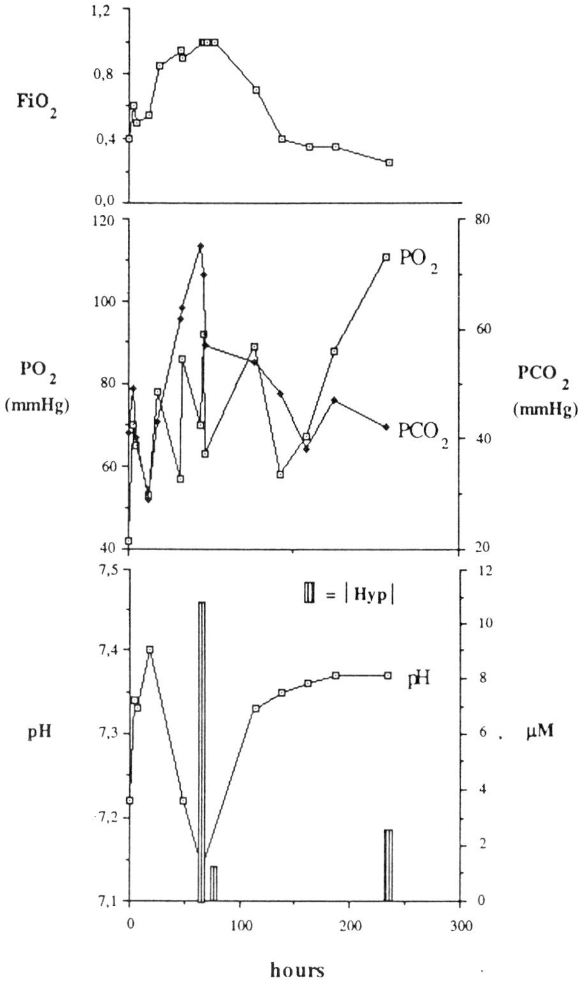

Fig.2. Results obtained in case 8. FiO_2, PCO_2, PO_2 and pH are plotted against the postnatal age in hours. Plasma Hypoxanthine levels were measured at 66.0, 76.8 and 234.5 hours. The highest hypoxanthine value corresponds to the lowest pH value. Hypoxanthine concentration in the last sample is higher than that observed in the second one in which a lower pH value was obtained. This result could be , at least in part, explained by the higher PO_2 in the third sample (see text).

Many factors may influence plasma hypoxanthine levels (3),for example purine uptake and/or release by human erytrocytes. The rate of IMP synthesis from exogenous hypoxanthine in mature human red blood cells markedly increases with decreasing pH and the oxygen partial pressure in the incubation mixture (5). At pH values lower than 7.3 the rate of hypoxanthine uptake markedly decreases with increasing oxygen partial pressure. Although hemoglobin F has a lower affinity for 2,3-DPG than does hemoglobin A, similar results were obtained using red cells from cord blood (6). These in vitro obtained data could, at least in part, explain the higher hypoxanthine levels observed with a high PO_2.

B) ALLOPURINOL METABOLISM IN HUMAN HEART

Activated species of oxygen generated by the enzyme xanthine oxidase have been postulated to cause cellular injury during reperfusion of ischemia (7). The role of xanthine oxidase in the pathogenesis of ischemic myocardial injury, however is controversial. Biochemical assays have not detected significant xanthine oxidase activity in homogenates of human myocardium (8). However immunohistochemical techniques have been reported to demonstrate the enzyme's presence in the capillary endothelium of the human heart. (9). Furthermore, it was reported that the human heart produced uric acid during transient myocardial ischemia induced by coronary angioplasty (10) and that uric acid concentration is higher in the coronary sinus than in aortic blood (11). Allopurinol is both an inhibitor of, and a substrate for, xanthine oxidase. In man its oxydation to oxypurinol, its main metabolite, by xanthine oxidase is rapid (12). In three patients undergoing open heart surgery using cardiopulmonary bypass, 9 times 800-1200 ml aliquots of a cold cardioplegic infusate containing from 0 to 100 mg/l of allopurinol were injected through the aortic cannula and aspirated after 20 min from the right atrium. When the aspirates, concentrated up to 13 times, were analyzed for purine metabolites using the HPLC method described in the present paper, no trace of oxypurinol could be detected, suggesting that the activity of xanthine oxidase and/or of other enzymes that convert allopurinol to oxypurinol is low in human heart (13).

REFERENCES

1) R.J. Simmonds and R.A.Harkness,J.Chrom.226:369-381(1981)
2) O.D.Saugstad,Pediatr.Res. 9:158-161 (1975)
3) V.Ruth, K.O.Raivio,Pediatr.Res.18:355-358(1984)
4) J.Pietz,N.Guttenberg and L.Gluk. Ostetrics and Gynecology 72:762-766(1988)
5) C.Salerno,G.Gerber,A.Giacomello,Adv.Exp.Med.and Biol.195B:75-77 (1986)
6) C.Salerno, E.Messina, Personal communication
7) J.M.McCord,N.Engl.J.Med.312:159-163 (1985)
8) C.M.Grum,K.P.Gallagher,M.M.Kirsh and M.Shlafer. J.Mol.Cell.Cardiol.21:263-267 (1989)
9) E.D.Jarasch,G.Bruder,H.W.Heid.Acta Physiol.Scand.548,39-46 (1986)
10)T.Huizer et al. J.Mol.Cell.Cardiol.21,691-695(1989)
11)G.Ronca et al.Adv.Exp.Med.Biol.253A:387-391(1988)
12)G.Elion et al.Biochem.Pharmac.15:863-880(1986)
13) G.Calafiore,F.Cuccurullo, A.Giacomello,C.P.Quaratino, G.Spoto in press

SOME ASPECTS OF PURINE NUCLEOTIDE METABOLISM IN HUMAN LYMPHOCYTES BEFORE AND AFTER INFECTION WITH HIV-1 VIRUS: NUCLEOTIDE CONTENT

Roberto Pagani, Antonella Tabucchi, Filippo Carlucci Roberto Leoncini, Elena Consolmagno, *Massimo Molinelli and **Patrizia Valerio

Inst. of Biochemistry and Enzymology, *Inst. of Pediatrics, **Inst. of Dentistry - University of Siena

INTRODUCTION

The evaluation of intracellular metabolism of purine nucleotides is important in many conditions, during, for example, proliferation, or maturation of the cell -i.e. leukemia and SCID (= severe congenital immunodeficiency)- and can be studied following different procedures:

1) evaluation of purine nucleotide content (1);

2) incorporation of labelled precursors of the purines, e.g. ^{14}C-glycine or ^{14}C-formate with kinetics based on biomathematical models (2);

3) assay of each individual enzymatic activity, involved in the various pathways of purine nucleotide metabolism, the de novo synthesis, the salvage pathway, and the purine nucleotide catabolism (2).

The analysis of all these parameters for 24 enzymes would be too cumbersome and could not be carried out simultaneously on a single sample of lymphocytes. On the other hand, the determination of purine nucleotide content is easy and routine giving precious informations on purine nucleotide metabolism.

Every time a cell is infected by replicating virus, it is reasonable assume that an alteration of purine nucleotide metabolism in the cell will occur, as a consequence of the alterations in nucleic acid metabolism. The overall importance of this problem should be obvious, and we underline that it has not yet been considered.

As problem is also relevant for the HIV-1 virus infection, we decided to study some aspects of purine nucleotide metabolism both in normal and in HIV-1 virus infected lymphocytes, and specifically, their purine nucleotide content.

We have elaborated a procedure taken from the literature for determination of the most important purine nucleotides in these cells: we present the first data on normal PBL, before and after infection with HIV-1 virus.

MATERIALS AND METHODS

We prepared the lymphocytes from heparinized peripheral blood of healthy donors using 17 subjects (13 males and 4 females), aged from 28 to 53 years. Blood was taken at 8.00 a.m. from fasting subjects. 15 ml of blood were normally adeguate for all determinations.

The procedure was essentially that of Böyum (3) and is briefly reported here.

Purine and Pyrimidine Metabolism in Man VII, Part B
Edited by R.A. Harkness *et al.*, Plenum Press, New York, 1991

The lymphocytes were isolated from whole blood using the Lymphocyte Separation Medium (L.S.M.) of Flow, as previously described (3).

The cells were counted in a Delcon cell counter, staining with Trypan Blue to assay the percentage of viable cells, which was always >95%. No contamination by erythrocytes or platelets was evident.

The cells were centrifuged at 1,000xg, 0.4 N $HClO_4$ (200 µl/10x10^6 cells) was added to the pellet, this was left for 15 min at 0°C, again centrifuged at 8,000xg for 10 min.

The acid-soluble fraction containing the free nucleotides was neutralized with K_2CO_3 (K-perchlorate was eliminated by centrifugation) and submitted to HPLC.

Purine nucleotides were determined through HPLC using a Beckman System Gold instrument equipped with a Whatman Partisil 10 SAX column (250 x 46 mm) and a precolumn with the same packing. 100 µl (equivalent to 5x10^6 cells) was injected, overloading the injection loop with 130 µl.

De Korte's linear gradient was used to separate the nucleotides (1), which were identified through retention time, the coelution with internal standards, and assay with specific enzymes. The concentration of the compounds was calculated by calibration with standard solutions.

We calculated the most important ratio between the different nucleotides, namely:

a) ATP/ADP ratio

b) GTP/GDP ratio

c) energy charge for adenylate (EC-AMP)

d) energy charge for guanylate (EC-GMP)

These parameters are indicated by the literature as an index of energy and viability of the cells (1,2,4).

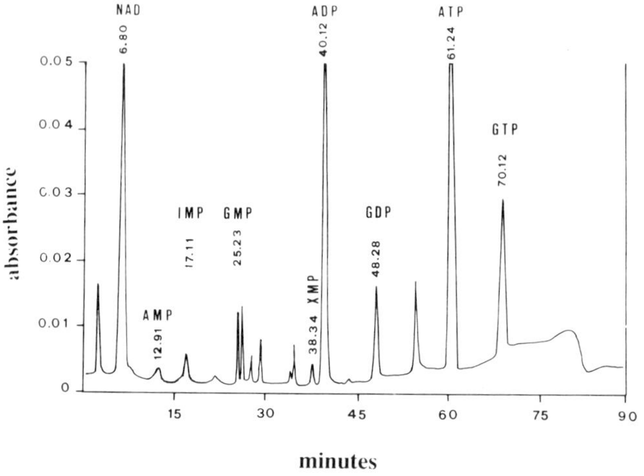

Fig. 1. Typical UV profile of purine nucleotide in pheripheral blood lymphocytic extract from healthy subject.

RESULTS AND DISCUSSION

A typical chromatogram obtained from normal subjects is reported in Figure 1. Here it is evident that a good separation of all the most important nucleotides (NAD-AMP-XMP-IMP-GMP-ADP-GDP-ATP-GTP) is possible.

Table 1 reports the average of our determinations in healthy subjects. From this table it is evident that the NAD content in the lymphocytes is high: a very low content and, at times, undetectable content of mononucleotides (AMP-IMP-XMP-GMP) can be noted in some cases, as well as a higher content in ADP, GDP and GTP and a very high content in ATP. In Table 1 we show the different ratios.

Figure 2 presents the individual values of some nucleotides obtained from both females and males, and their marked variability. The results obtained, the pattern of the nucleotides, the order of magnitude in purine nucleotides, are substantially coincident with the few data reported in the literature (2,4).

The various parameters also agree with the literature, and, the values of EC coincide with those of other Authors, in Ø,85 (4).

Table 1. Purine nucleotide content in human peripheral blood lymphocytes

NAD	AMP	IMP	GMP	XMP	ADP	GDP	ATP	GTP
179.46 ±66.18	47.8 ±13.74	49.11 ±12.69	12.15 ±2.45	14.87 ±8.14	440.58 ±89.42	60.94 ±11.78	948.29 ±120.22	176.82 ±18.15

The values, expressed as pmol/10^6 cells, represent the mean ± S.E. of seventeen cases.

E.C. AMP	0.82	ATP/ADP	2.26
E.C. GMP	0.85	GTP/GDP	2.99
		A/G	5.72

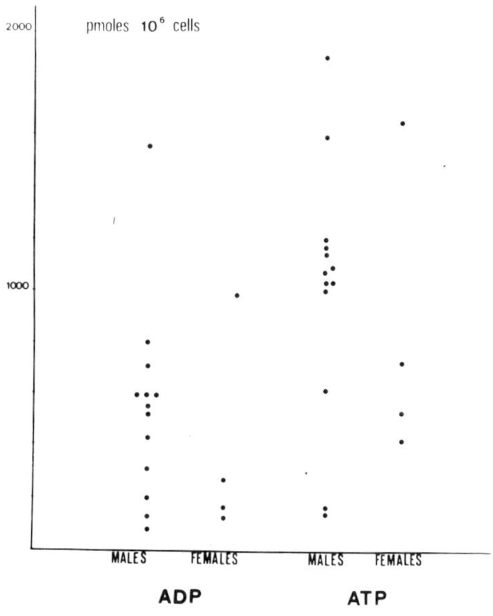

Fig. 2. ADP and ATP content in healthy subjects: single cases are reported. The values are expressed as pmol/1Ø^6 cells.

We call attention to two important aspects of our research:
1) the low lymphocytes content in purine nucleotide monophosphates when compared with the other values; it could be due to the marked activity of 5'-nucleotidase, the enzyme which regulates their intracellular content, according to the observations of other Authors (5).
2) the nucleotide concentration varied considerably, as shown by the high SE values and the high dispersion in Figure 2.

The problem is whether this is due to single factors (such as age, sex, diet, other parameters) which must be further investigated, or to a variable state of "activation" of the lymphocytes, in the subjects, due to stimulation by different antigens, which would necessary involve important differences in the nucleotide content.

Our research suggests the need of standardization of the purine nucleotide content in the lymphocytes, and proposes a basis for its comparative evaluation under different conditions.

Our procedure can also be applied to lymphoblastoid cell lines, whether or not infected with HIV-1 virus, such as A3.01 cell line (human lymphoblastoid line $CD4^+$), 8E51 (line A3.01 transfected with reverse transcriptase deficient HIV-1); H9 cell line (continuos lymphoblastoid line); H9/HTLV-III cell line (H9 permanently infected with transcriptase positive HIV-1); PBL from normal donors in course of experimental infection.

Some preliminary results of this study are reported in the following communication of this section.

This work was financied by a contribution of the Ministero della Sanità, Istituto Superiore di Sanità, AIDS Project 1990, Rome - Italy

REFERENCES

1. D.De Korte, W.A.Haverkort, A.H.Van Gennip and D.Roos, Nucleotide profile of normal human blood cells determinations by high performance liquid chromatography, Anal.Biochem., 147: 197 (1985)
2. Y.M.T.Marjinen, D.De Korte, W.A.Haverkort, E.J.S.Breejen, A.H. Gennip and D.Roos, Studies on the incorporation of precursors into purine and pyrimidine nucleotides via "de novo" and salvage pathways in normal lymphocytes and in lymphoblastic cell lines, Bioch. Bioph. Acta, 1012: 148 (1989)
3. A.Böyum, Isolation of leukocytes from blood and bone marrow. Scand.J.Clin.Lab.Invest., 97: 21 (1968)
4. G.J.Peters, A.R.De Abreu, A.Oosternhof and J.H.Veerkampf, Concentration of nucleotides and deoxynucleotides in peripheral and phytohemagglutinin stimulated mammalian lymphocytes, Bioch. Bioph. Acta, 795: 7 (1983)
5. A.S.Sun, J.F.Holland, T.Ohnuma and M.Slankard-Chahinian, 5'-nucleotidase activity in permanent lymphoid cell lines. Implication for cell proliferation and aging in vitro, Bioch. Bioph. Acta, 714: 530 (1982)

PRENATAL DIAGNOSIS OF LESCH-NYHAN SYNDROME BY PURINE ANALYSIS OF AMNIOTIC FLUID AND CORDOCENTESIS

Felícitas A. Mateos, Juan G. Puig, Teresa H. Ramos, Manuel L. Jiménez,
Nuria M. Romera and Antonio G. González

Divisions of Internal Medicine, Clinical Biochemistry and Gynaecology
"La Paz" Hospital, Universidad Autónoma, Madrid, Spain

INTRODUCTION

Lesch-Nyhan syndrome is an X-linked disorder caused by a deficiency of the enzyme hypoxanthine guanine phosphoribosyltranferase (HGPRT, E.C.2.4.2.8) that catalyzes the conversion of hypoxanthine and guanine to their respective nucleotides (1,2). Lack of this enzymatic activity promotes an increase of hypoxanthine, xanthine and uric acid in biological fluids and an enhancement of adenine phosphoribosiltransferase (APRT, E.C.2.4.2.7) in cells (2). Because of Lesch-Nyhan syndrome (LNS) severity, early prenatal diagnosis may be requested to provide genetic advice. The prenatal diagnosis of LNS has been performed by enzyme assay or DNA analysis in amniocytes or chorionic villus sampling (3,4). Cordocentesis permits the obtaining of fetal blood for enzyme assay although its usefulness in the antenatal diagnosis of LNS has not been reported. Furthermore, there are no studies on the concentration of oxypurines in amniotic fluid in pregnancies with fetuses at risk of LNS. In this study we made a prenatal diagnosis of the LNS by cordocentesis in the 21st gestational week. In addition, we report the results of oxypurine levels in amniotic fluid.

CASE REPORT

A pregnant woman, aged 26, was known to be a carrier of the LNS, and requested antenatal diagnosis in her 20th gestational week. The sex of the fetus (male) had been determined during the 18th gestational week by chomosomic study. She had a cousin and two second cousins that suffered the Lesch-Nyhan syndrome. Two brothers died as a consequence of an undiagnosed illness with the clinical characteristics of the Lesch-Nyhan syndrome (Fig.1). On the 20th gestational week, an umbilical cord puncture was performed by the transabdominal way. Two ml of blood and 2 ml of amniotic fluid were obtained. In a blood drop we performed the Nierhaus and Betke reaction. The sample belonged to the mother. We measured uric acic, hypoxanthine and xanthine concentrations in the amniotic fluid simultaneously obtained, and were found to be increased as compared to normal values. Four days later, a second cordocentesis was performed and the blood sample obtained belonged to the fetus. HGPRT activity was undetectable and APRT was increased. The diagnosis of LNS was established. At the request of the mother, the pregnancy was terminated, and 2 ml of fetal blood by cardiac puncture were taken.

METHODS

Creatinine, uric acid, hypoxanthine and xanthine concentrations were measured in the amniotic fluid obtained in the 20th gestational week. The results were compared with those obtained in 14 amniotic fluid from normal pregnancies and similar gestational age. Uric acid and creatinine were quantified in a multichannel autoanalyzer (Hitachi 737. Boehringer Manhein). Hypoxanthine and xanthine concentrations were measured by high performance liquid chromatography (HPLC) (5). HGPRT and APRT enzyme activities in pre and post abortion fetal erythrocytes and in the maternal blood were quantified by HPLC (6,7). The results were compared with those obtained in 50 normal controls and in 11 patients with HGPRT deficiency.

RESULTS

Uric acid, hypoxanthine and xanthine concentrations were around 2, 2.8 and 3 times higher, respectively, in the amniotic fluid of our pregnant woman than in 14 control amniotic fluids from normal pregnancies (Table I). HGPRT activity in fetal erythrocytes, obtained by cordocentesis and post abortion, was undetectable. APRT activities in the same samples were increased. HGPRT and APRT activities in the pregnant woman were similar to those obtained in 50 normal controls.

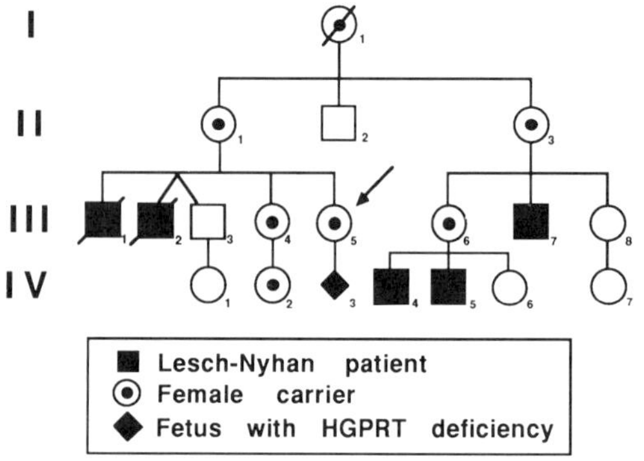

Figure 1. Pedigree of a family with the Lesch-Nyhan syndrome.

DISCUSSION

We report the prenatal diagnosis of HGPRT deficiency by means of cordocentesis and oxypurines quantification in amniotic fluid. The antenatal diagnosis of Lesch-Nyhan syndrome is usually performed in amniotic cell cultures obtained between the 15th and 17th gestational weeks (3,4). With this procedure the results can be provided within the 18th to 20th gestational week, and the risk of fetal loss is small (0.5-1.5%). However, in some cases, the amniotic cell cultures may be contaminated with maternal cells causing a fallacious mosaicism (8). The villus chorionic biopsy permits an earlier diagnosis (8-12th gestational week) (3) but the possibility of fetal loss is higher (2.3-3.4%) and the sample can be mixed with maternal cells thus resulting in an inaccurate diagnosis (8,9).

Table 1. Amniotic fluid purine concentrations in normal women and in a female carrier of the Lesch-Nyhan syndrome.

	Normal Amniotic Fluid (n=14)	Problem Amniotic Fluid
Gestational week	17±3	20
Uric acid (μmol/L)	1547±357	2975
Hypoxanthine (μmol/L)	0.6±0.5	4.5
Xanthine (μmol/L)	1.0±0.4	5.2
Uric acid (μmol/g creatinine)	20.8±4.9	39.2
Hipoxanthine (μmol/g creatinine)	74±49	600
Xanthine (μmol/g creatinine)	135±54	680

Values are the mean±SD.

Umbilical cord puncturing is a relatively new diagnosis procedure that can be performed in the 18-20th gestational week. The chance of fetal death with this technique is around 1.9% (10). This procedure is accurate because of the possibility of identifiying the origin of the sample *in situ* by acid eluction of fetal haemoglobin. However, puncturing the umbilical cord in early stages of pregnancy can be extremely difficult.

In our patient we attempted cordocentesis in the 20th gestational week. Although we could not get fetal blood, we obtained amniotic fluid and postulated that the analysis of oxypurines could provide information about the HGPRT enzyme activity of the fetus. In the first half of a normal pregnancy, the amniotic fluid composition is similar to the fetal extracelular fluid and its study has been used for the diagnosis of inherited diseases (11). An increase of purine nucleotide degradation is readily recognized by an increase of nucleosides and purine bases in biological fluids (12). Few studies have examined the normal concentration of amniotic fluid oxypurines in the first gestational trimester (13). However, some authors have demonstrated that in fetal hypoxia there is an increase of hypoxanthine, xanthine and uric acid levels in cord blood and in amniotic fluid (14,15). In HGPRT deficiency lack of hypoxanthine reutilization and acceleration of *de novo* purine synthesis explains the increase of oxypurines in body fluids (2). The amniotic fluid examined by us showed hypoxanthine, xanthine and uric acid concentrations higher than two standard deviations above the mean observed in 14 control amniotic fluids from normal pregnancies with a similar gestational age. This finding in a fetus later confirmed to be deficient for HGPRT activity suggests that amniotic fluid analysis in early stages of pregnancy can be useful for the prenatal diagnosis of the LNS. The possibility of contamination of amniotic cell cultures and chorionic villus biopsy by maternal cells (16,17) renders analysis of oxypurine concentrations in amniotic fluid a new technique for the early diagnosis of LNS. Because of the small volume of the sample needed and the short time required for oxypurine analysis (roughly, 2 hours), this method could be regarded as a valuable alternative in the early diagnosis of inherited purine metabolic diseases.

ACKNOWLEDGEMENTS

We are indebted to Javier Díaz and Mrs. Mª Paz Canencia for valuable technical assistance and to Erik Lundin for help in the preparation of the manuscript. This work was supported by grants from Caja de Madrid and Fondo de Investigaciones Sanitarias de la Seguridad Social (FISS, 90/0575), Spain.

REFERENCES

1. M. Lesch and W.L. Nyhan. A familial disorder of uric acid metabolism and central nervous system function. Am J Med 36:561-570 (1964).
2. J.T. Stout and C.T. Caskey. Hypoxanthine phosphoribosyltransferase deficiency: the Lesch-Nyhan syndrome and gouty arthritis. In: C.H.R. Scriver, A.L. Baudet, W.S. Sly and D. Valle, eds. The metabolic basis of inherited disease. New Yoork: McGraw-Hill. 1007-1028 (1989).
3. D.A. Gibbs, I.R. McFayden, M.A. Crawfurd, et al. First trimester prenatal diagnosis of Lesch Nyhan syndrome. Lancet 2:1180-1183 (1984).
4. D.A. Gibbs, B.M.C. Headhouse and W.E. Watts. Family studies of the Lesch Nyhan syndrome: the use of a restriction fragment length polymorphism (RPLF) closely linked to the disease gene for carrier state and prenatal diagnosis. J Inherited Metab Dis 9:44-57 (1986).
5. F.A. Mateos, J.G. Puig, M.L. Jiménez, et al. Hereditary xanthinuria. Evidence for enhanced hypoxanthine salvage. J Clin Invest 79:847-852 (1987).
6. H.J. Rylance, R.C. Wallace and G. Nuki. Hypoxanthine-guanine phosphoribosyltransferase assay using high performance liquid chromatography. Clin Chim Acta 127:159-165 (1982).
7. H.J. Rylance, R.C. Wallace and G. Nuki. Adenine phosphoribosyltransferase: assay using high performance liquid chromatography. Clin Chem Acta 148:267-272 (1985).
8. Canadian Collaborative CVS-Amniocentesis Clinical Trial Group. Multicentre randomized clinical trial of chorion villus sampling and amniocentesis. Lancet 1:1-6 (1989).
9. G.G. Rhoads, L.G. Jackson, S.E. Schelesselman, et al. The safety and efficacy of chorionic villus sampling for early prenatal diagnosis of cytogenetic abnormalities. N Engl J Med 320:609-617 (1989).
10. F. Daffos, M. Capella-Pavlovsky and F.A. Forestier. A new procedure for fetal blood sampling in utero: preliminary results of 53 cases. Am J Obtet Gynaecol 146:985-987 (1983).
11. A.E. Seeds. Current concepts of amniotic fluid dynamics. Am J Obstet Gynecol 138:575-586 (1980).
12. I.H. Fox. Adenosine triphosphate degradation in specific disease. J Lab Clin Med 106:101-110 (1985).
13. R.A. Harkness. Hypoxanthine, xanthine and uridine in body fluids, indicators of depletion. J Chromatogr Biom Appl 429:255-278 (1989).
14. O.D. Saugstad. Hypoxanthine as an indicator of hypoxia. Its role in health and disease through free radical production. Pediatr Res 23:143-150 (1988).
15. F.A. Mateos, J.G. Puig T.H. Ramos, et al. Erythrocyte ATP (iATP) as an indicator of neonatal hypoxia. Adv Exp Med Biol 253A: 345-352 (1989).
16. P. Aula, K. Matilla, O. Piroinen, et al. First trimester prenatal diagnosis of aspartylglucosaminuria. Prenat Diag 9:617-620 (1989).
17. P.A. Benn and L.Y.F. Hsu. Maternal cell contamination of amniotic fluid cell cultures. Results of a US nationwide survey. Am J Med Genet 15:297-305 (1983).

GENE THERAPY IN MAN AND MICE: ADENOSINE DEAMINASE DEFICIENCY, ORNITHINE TRANSCARBAMYLASE DEFICIENCY, AND DUCHENNE MUSCULAR DYSTROPHY

Markus Grompe[1], Kohnosuke Mitani[1,2], Cheng-Chi Lee[1], Stephen N. Jones[1,2], and C. Thomas Caskey[1,2]

[1]Institute for Molecular Genetics and
[2]Howard Hughes Medical Institute
Baylor College of Medicine, Houston, TX

INTRODUCTION

Gene therapy is defined as the delivery and expression of a functional gene into somatic tissues of patients (or animals) endogenously deficient in this gene. Depending on the natural tissue pattern of expression and the sites most affected by disease processes, different anatomic sites may be the targets for gene transduction. To date, retrovirus-mediated gene transfer into hemopoietic stem cells is the most advanced of gene therapy systems and the most promising in terms of clinical applications in the near future. There are, however, additional efforts under way in our laboratory to develop efficient means of transducing genes into the liver, intestine, skeletal and cardiac muscles.

ADENOSINE DEAMINASE (ADA) DEFICIENCY

ADA deficiency is a rare autosomal recessive disorder accounting for approximately 15% of the cases of severe combined immunodeficiency disease (SCID). ADA catalyzes the irreversible deamination of adenosine and deoxyadenosine to inosine and deoxyinosine, respectively. In the absence of ADA, deoxyadenosine accumulates in many tissues, especially those of the lymphoid system which constitutively express high amounts of

Purine and Pyrimidine Metabolism in Man VII, Part B
Edited by R.A. Harkness *et al.*, Plenum Press, New York, 1991

the enzyme. It has been suggested that the accumulation of deoxyadenosine and its metabolites, particularly deoxyATP, inhibits DNA synthesis resulting in profound T cell and subsequent B cell dysfunction. Although HLA-identical bone marrow transplantation has been curative for ADA deficiency, less than 30% of the patients have an HLA-matched sibling available and the results have not been encouraging. Treatment with injections of bovine ADA protein conjugated to polyethylene glycol (PEG) has been shown to reverse the biochemical abnormalities and restore some T lymphocyte function, however, total immune reconstitution has not been demonstrated. Recently, a trial has begun to treat ADA-deficient SCID patients who are on PEG-ADA treatment with *ex vivo*-expanded autologous peripheral T cells after retroviral-mediated gene transfer of the human ADA gene. Both PEG-ADA injections and administration of retroviral vector-modified peripheral T cells provide only temporary treatment of the ADA enzyme defect. In contrast, transfer of the human ADA gene into hematopoietic stem cells would lead to permanent correction of the defect, making ADA deficiency an ideal candidate for somatic gene therapy of bone marrow cells. Our group is working on gene transfer into pluripotent stem cells which, in theory, requires only one a few infusions, and which results in expression of ADA in all hematopoietic cell lineages.

As a vehicle for gene transfer, we have been using a retroviral vector which has no selectable marker and which transcribes human ADA complementary DNA (cDNA) from the viral LTR promoter. Low density mononuclear bone marrow (BM) cells from normal allogenic donors and an ADA-deficient patient were infected *in vitro* by cocultivation with the amphotropic virus-producing cell line in the presence of recombinant human interleukins 3 and 6. In order to estimate the efficiency of gene transfer into hematopoietic progenitors, a fraction of the infected BM was cultivated in a short-term colony-forming assay. DNA was prepared from individual colonies and a polymerase chain reaction (PCR) performed to detect the ADA provirus. To identify more primitive hematopoietic progenitors, infected BM was maintained in myeloid long term culture (LTC) on irradiated human primary stroma cells. Immediately after gene transfer, the infection efficiency of clonogenic progenitors was 90% in normal BM. After nine weeks in LTC the ADA provirus was present, on average, in 34% of clonogenic cells. Since pre-existing clonogenic cells are not maintained in LTC for longer than four to five weeks, the results suggest successful gene transfer into primitive hematopoietic progenitors, possibly representing pluripotent stem cells.

In order to detect expression of the retroviral vector in infected cells, we have developed a highly sensitive and specific strategy using the PCR. In this scheme, the antisense PCR primer consists of the 3' end of the R region in the virus LTR and a poly(dT) stretch. After reverse transcription of mRNA with oligo(dT), this antisense primer, with a sense primer, amplifies only cDNA from mRNA expressed from the retroviral vector but does not amplify provirus DNA, although cDNA and provirus DNA have a similar structure. We have been applying this method to determine the expression of ADA-coding retroviral vectors in infected human hematopoietic progenitors. In one experiment, one out of thirteen colonies (8%) at week six of LTC had a detectable level of retroviral transcript, suggesting transduction of the ADA gene into primitive hematopoietic progenitors resulting in exogenous ADA production. The transduced ADA activity was measured by a microassay in the ADA-deficient BM at weeks zero and two of LTC. The ADA activity increased to the level of normal erythroid cells following retroviral gene transfer.

Enriched stem cell populations are now being used in these studies. Successful transduction of ADA activity into these enriched cell populations will allow optimization of gene transfer methodology for eventual clinical applications.

ORNITHINE TRANSCARBAMYLASE (OTC) DEFICIENCY

Ornithine transcarbamylase (OTC) deficiency is an X-linked urea cycle disorder, which in its severe form presents with hyperammonemic coma in the newborn period. Despite improved dietary and pharmacologic treatments, the prognosis in this disorder is bad and most patients are either mentally retarded or die early in life. Since the OTC enzyme is expressed primarily in liver, transplantation of this organ has been performed in a number of patients and has been successful in a few cases. However, limited donor organ supplies and immunologic barriers make this a less than ideal treatment and OTC deficiency is therefore an excellent candidate disorder for gene replacement therapy targeted at the liver. In addition, research in OTC deficiency is facilitated by the availability of two murine models of the disease, the *spf* and *spf-ash* mice.

To test the feasibility of therapy by gene transfer and to define the sequence elements required for expression of the OTC gene, we microinjected *spf* oocytes with a construct consisting of the human OTC gene under the transcriptional control of 800 base pairs of 5' flanking sequence from the murine gene. This experiment was successful and led to full correction of the phenotype by expression of human OTC in the intestine, thus demonstrating that OTC gene therapy could also be directed at small bowel mucosal cells (Jones *et al.*, 1990).

More recently we have developed retroviral vectors capable of transducing the human OTC cDNA into primary hepatocytes from both *spf* and *spf-ash* mice. The most promising vector, ΔN2OTC, was created by deleting the neomycin resistance gene from the vector N2 and replacing it with 1200 base pairs of OTC cDNA. This construct has produced high titers ($>10^6$/ml) in ecotropic (gp+E86) as well as amphotropic (Am12) packaging cells, and has been used to infect primary hepatocytes of *spf* and *spf-ash* mice. Hepatocytes were isolated by perfusing the liver with collagenase and then plated in hormonally defined media at low density. The cells were infected with viral supernatant at 48 hours, trypsinized at 72 hours and replated at high density for long-term culture. This protocol permitted optimal infection as well as maintenance of the hepatocyte phenotype. Cells were harvested for DNA, RNA, enzyme assay and immunological staining at 4 days, 10 days and 18 days post-infection. The copy number of the provirus in the population of infected hepatocytes was estimated by Southern blot and found to vary between one and five in multiple experiments, indicating a very high infection efficiency. The amount of human OTC mRNA produced by the infected cells was studied by Northern blot and RNA slot blots. Hybridization conditions were developed that lead to the specific detection of only human sequences, despite the high degree of homology between the murine and human genes. The amount of human RNA produced varied between 20% and 200% of that found in a normal adult human liver control, and was easily detectable by Northern blot analysis. Enzyme levels were two- to four-fold higher in infected *spf* hepatocytes at 10 days and five- to ten-fold higher at 18 days post-infection. However, overall OTC activity was down-regulated in the cultured cells and the enzymatic correction achieved did not reach the OTC enzymatic levels of wild-type livers. The long-term cultured cells were assayed for two hepatocyte specific markers (albumin and N-acetyl glutamine synthetase) at 18 days and found to consist of >90% hepatocytes, indicating that the correction was actually achieved in these cells and not endothelial cells or fibroblasts. The high viability of the infected cells after trypsinization and the maintenance of hepatocyte markers indicate that they could be used in *in vivo* transplantation experiments in the future.

DUCHENNE MUSCULAR DYSTROPHY (DMD)

Duchenne muscular dystrophy (DMD) is an X-linked progressive myopathy due to a defect in the dystrophin locus. The dystrophin gene produces a 14kb messenger RNA (mRNA) that codes for a large cytoskeletal membrane protein. We have recently reported the synthesis of a 14kb full-length complementary DNA (cDNA) from the mouse muscle dystrophin mRNA, and the expression of this cDNA in COS cells (Lee *et al.*, 1991). The recombinant dystrophin was indistinguishable from mouse muscle dystrophin by western analysis using anti-dystrophin antibodies and was determined by an immunofluorescent technique to be localized in the cell membrane.

An animal model for DMD is the *mdx* mouse which has a point mutation in the dystrophin mRNA transcript resulting in a deficiency of the protein. The *mdx* mouse provides an ideal animal model to test the ability of the cloned dystrophin minigene to correct dystrophin deficiency. In order to drive the DMD minigene in an animal system, we have constructed a vector based on the muscle creatine kinase promoter (MCK) which has previously been shown to exhibit tissue-specific expression in skeletal and cardiac muscle. The DMD minigene driven by this MCK promoter was injected into normal mouse embryos and a male animal carrying a single copy of the transgene was generated. This animal was then bred with an *mdx* female and analysis was carried out on the male progeny from this breeding. Western analysis using anti-dystrophin antibodies shows that there is recombinant dystrophin expression in skeletal and cardiac muscle. Using more sensitive polymerase chain reaction techniques, it was determined that the primary expression of the transgene is in the skeletal and cardiac muscle with much lower levels in spleen, kidney, brain, liver and smooth muscle. The recombinant dystrophin was essentially the same size as the native muscle protein. Immunofluorescent studies using anti-dystrophin antibodies revealed that the recombinant dystrophin was located at the muscle membrane protein, as is the case with the native dystrophin. Studies are currently in progress to determine whether the expression of the recombinant dystrophin has corrected the *mdx* defect.

ACKNOWLEDGMENTS

The assistance of Belinda Rossiter in preparing this manuscript is appreciated. Funding was received from United States Public Health Service grant DK42696, the Muscular Dystrophy Association, and the Howard Hughes Medical Institute. MG is an Association of Medical School Pediatric Department Chairman, Inc. Pediatric Scientist Training Program Fellow supported by National Institutes of Health grant #00850, and CTC is a Howard Hughes Medical Institute Investigator.

REFERENCES

Jones, S.N., Grompe, M., Munir, M.I., Veres, G., Craigen, W.J., and Caskey, C.T., 1990, Ectopic correction of ornithine transcarbamylase deficiency in sparse fur mice. J. Biol. Chem. 265:14684.

Lee, C.C., Pearlman, J.A., Chamberlain, J.S., and Caskey, C.T., 1991, Expression of recombinant dystrophin and its localization to the cell membrane. Nature 349:334.

REGULATION OF THE HUMAN ADENOSINE DEAMINASE GENE BY FIRST INTRON SEQUENCES: A T-CELL ENHANCER

Dan Wiginton, Bruce Aronow, Richard Silbiger, Steven Potter, and John Hutton

Division of Basic Science Research, Children's Hospital Medical Center, University of Cincinnati
Cincinnati, Ohio, USA

INTRODUCTION

Adenosine deaminase (ADA) is expressed ubiquitously in mammalian cells and tissues but the levels vary greatly according to tissue and species. In humans the thymus exhibits levels of ADA up to 100-fold higher than most other tissues. In the large first intron of the human ADA gene, our laboratory previously identified a complex cis regulatory region 4-10 kb downstream of the first exon (Aronow et al., 1989). This region exhibits an array of DNase I hypersensitive sites, some of which are tissue specific. Inclusion of intronic fragments encompassing this region in hybrid transgene constructions conveyed generalized expression of the transgene in all tissues and very high level transgene expression in thymus. Transfection-transient assay experiments with plasmids containing hybrid gene constructions indicated that this intronic region contains a lymphoid specific enhancer that functions in a potent manner in T cells that express high levels of ADA. Studies indicated that this putative enhancer lay within a 1.3 kb intronic fragment that encompasses hypersensitive sites II and III. In the studies described here, we have continued to utilize hybrid gene constructions that contain the chloramphenicol acetyltransferase (CAT) coding sequence (as a reporter gene) and human ADA gene fragments to promote expression. These constructions have been used in transfection studies to characterize the enhancer.

METHODS

All plasmids constructed for the experiments described here are derivatives of the promoterless vector pSV0-CAT (Gorman et al.,1982a). The preparation of the plasmids pADA-CAT 4.0, pADA-CAT 4/12, and pADA-CAT 4/I1.3 have been described previously (Aronow et al.,1989). The plasmids with intronic deletions were produced by utilizing convenient restriction enzyme sites or by deletion with Bal-31 exonuclease (Maniatis et al.,1982). The DEAE-dextran transfection procedure was modeled after a protocol supplied to us by Dr. Jeffrey Leiden (Gottesdiener et al.,1988). The endogenous ADA activity and growth conditions for the human lymphoid cell lines used for transfection studies has been reported previously (Aronow et al., 1989). DNase I footprinting was patterned after the methods used in Dr. Gary Felsenfeld's laboratory (Emerson et al., 1985). Nuclear extracts utilized for footprinting were prepared by the method of Dignam et al.,1983. Production of cell extracts, CAT enzyme activity measurements, and protein determinations in these extracts were described in Aronow et al., 1989.

RESULTS

Deletion analysis for enhancer activity in the human ADA gene first intron

The plasmid pADA CAT 4/12 was previously constructed to contain ~16 kb of the human ADA gene. This included ~4 kb of 5' flanking DNA, the non-coding portion of the first exon, and ~12 kb of the first intron. The coding portion of the first exon has been replaced by a DNA cartridge

(~1.6 kb) containing the chloramphenicol acetyl transferase (CAT) coding sequence as well as sequences specific for splicing and polyadenylation of the mRNA. Specific restriction enzyme sites were exploited to produce a series of seven plasmids derived from pADA CAT 4/12 with various segments of the intronic DNA deleted.

The plasmid series was analyzed for enhancer function by transfection-transient assay for CAT activity in human lymphoid cell lines. The cell lines used and their previously determined level of endogenous ADA activity were: MOLT 4 (T cell) 1650, CCRF-CEM (T cell) 270, and Raji (B cell) 14 nmol / min / mg protein. High levels of enhancer activity were found only when the plasmids were transfected into MOLT 4 and only from those plasmids which had intronic fragments containing DNase I hypersensitive site (HS) III. Low levels of enhancer activity associated with HS III were detected in the CEM line but not in Raji. Intronic fragments containing one or more of the other five hypersensitive sites previously identified in the first intron imparted no detectable enhancer activity to the plasmid construction in the absence of HS III.

Bal-31 deletion analysis of the HS III enhancer domain

Our previous studies had indicated that an enhancer lay within a 1.3 kb fragment of intronic DNA that encompassed HS II and HS III. In our present studies Bal-31 exonuclease was used to produce deletions (5',3' or both) in this 1.3 kb intronic fragment of the parental plasmid, pADA CAT 4/I1.3. The resulting plasmids, with deletions, were utilized to map and characterize the the enhancer (Figure 1). To accomplish this the plasmids in the series were assayed in the MOLT 4 cell line by transfection - transient assay. The enhancer mapped to a 175 bp fragment that colocalized with HS III. Deletion into this fragment defined three distinct segments: a central core sequence of 25-30 bp that is essential and two segments that flank the core on either side and complement the core in distinct ways (see Figure 1). DNase I protection analysis of a DNA fragment encompassing the enhancer domain was carried out with MOLT 4 nuclear extract and the indicated "footprints" were obtained.

Analysis of the enhancer core sequence

The core sequence of the enhancer was analyzed by utilizing short synthetic DNA sequences.Two oligomers were synthesized (32mers) and hybridized to produce a double-stranded oligomer with complimentary XhoI and SalI "sticky ends". This oligomer was designed to contain a 28 bp sequence corresponding to the enhancer core identified by Bal-31 deletion analysis. XhoI and SalI "sticky ends" are homologous and compatible for ligation. This was exploited by multimerizing the oligo and ligating the multimer into a SalI site engineered into pADA CAT 4.0 (Aronow et al.,1989) just downstream of the CAT cartridge. One to four copies (in tandem) were inserted into the SalI site. The resulting plasmids were analyzed for enhancer activity by transfecting them into the lymphoid cell lines indicated in Table 1 and assaying for transient CAT activity expression. CAT activity was normalized relative to the expression level of the parental plasmid. The results in Table 1 indicate that a single copy of the core sequence has low enhancer activity. However multiple copies of the core have potent enhancer activity (up to 1000-fold for tetramers) in MOLT 4 cells. The enhancer effect was very greatly diminished in CEM and very low or non-existent in Raji. Sequencing was performed to determine relative orientation of the monomers and the subunits in the multimers. Differences in the absolute and relative orientations of the core sequences appear to have little if any effect on the relative enhancement of expression by either the monomers or the multimers.

Analysis of intronic fragments for enhancer characteristics

A wide variety of plasmid constructions were prepared and tested by transfection-transient assay to evaluate the characteristics of the putative enhancer. The enhancer was found to have the "classical" enhancer characteristics of independence from location, distance, and orientation. We have previously shown (Aronow et al.,1989) that a large 12 kb intronic fragment which contains all six of the HS regions functions equally well as an enhancer in either orientation when placed downstream of the promoter and CAT cartridge. Here we report that a 1.3 kb intronic fragment containing HS II and HS III which was previously shown to function as an enhancer downstream (pADA CAT 4/I1.3 in Aronow et al.,1989) also functions upstream in either orientation. The enhancer was found to function equally well at distances from the basal promoter of from 300 bp up to 9 kb. Full enhancer function was found to require only the "basal" promoter of the ADA gene (comprised of the first 150 bp 5' of the major transcriptional start site) although the enhancer functioned quite well in constructions with ~4 kb fragments of 5'-flanking DNA as promoters.

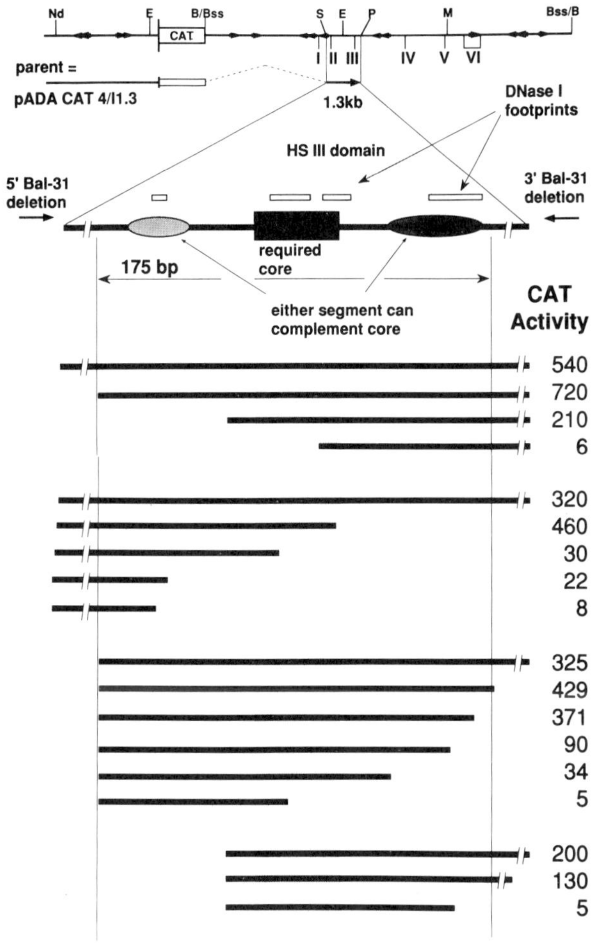

Figure 1 - Bal-31 deletion analysis of the enhancer domain. At the top is shown the hybrid gene segment of pADA CAT 4/12 which includes ~4 kb of 5' flanking DNA and first exon sequence and ~12 kb of the first intron from the human ADA gene. The coding portion of the first exon has been replaced by the CAT cartridge. Arrows indicate human ALU repetitive sequences and Roman numerals indicate identified DNase I hypersensitive sites.

Table 1 Core Enhancer Analysis (Oligo Multimers)

	# of constructs analyzed	Relative CAT Activity*		
		Molt 4	CEM	Raji
Parental	1	1	1	1
+Monomer	2	7	2	2
+Dimer	4	51 ± 17	4 ± 2	4
+Trimer	1	340	9	4
+Tetramer	5	850 ± 330	12 ± 3	3 ± 1

*Activity normalized against parental value

Effect of the enhancer fragments on heterologous promoters

Plasmid constructions were produced to test the effect of the ADA enhancer on other promoters.The heterologous promoters tested were from the HSV thymidine kinase (TK) gene, the mouse metallothionein I (MMT) gene, the SV40 early region promoter (SV2), and the Rous sarcoma virus (RSV) LTR. The function of the enhancer on heterologous promoters was tested in MOLT 4 cells. The TK promoter is a 253 bp fragment that contains a TATA box, a CAAT box, and two Sp1 sites (McKnight et al., 1981). The MMT promoter is a 336 bp fragment that contains a TATA box and a number of metal regulatory elements (Glanville et al.,1981). The SV2 promoter is a 323 bp fragment as described for pSV2-CAT in Gorman,1985 and the RSV promoter is a 524 bp fragment (including the enhancer) as described for pRSV-CAT in Gorman et al.,1982b. Pairs of plasmid constructions were prepared with each of these promoters driving the CAT cartridge, with and without an enhancer fragment inserted downstream. The enhancer fragment utilized was a 330 bp piece of DNA that completely encompasses the enhancer domain as defined by deletion analysis. The enhancer increased expression from the TK, MMT, and SV2 promoters by factors of 6-fold, 37-fold, and 68-fold respectively.The TK construction had the highest expression among the enhancerless plasmids. Expression from the RSV LTR was not increased by the enhancer and appeared to be moderately inhibited (33%). This is probably due to the presence of the strong RSV enhancer, since the the constructions with or without the ADA enhancer were both expressed at very high levels in MOLT 4 cells.

REFERENCES

Aronow, B., Lattier, D., Silbiger, R., Dusing, M., Hutton, J., Jones, G., Stock, J., McNeish, J., Potter, S., Witte, D., and Wiginton, D., 1989, Evidence for a complex regulatory array in the first intron of the human adenosine deaminase gene, Genes Dev. 3: 1384.

Dignam, J.D., Lebovitz, R.M., and Roeder, R.G., 1983, Accurate transcription initiation by RNA polymerase II in a soluble extract from isolated mammalian nuclei, Nucl. Acids Res. 11: 1475.

Emerson, B.M., Lewis, C.D., and Felsenfeld, G., 1985, Interaction of specific nuclear factors with the nuclease-hypersensitive region of the chicken adult beta-globin gene: Nature of the binding domain, Cell 41: 21.

Glanville, N., Durnam, D.M., and Palmiter, R.D., 1981, Structure of mouse metallothionein-I gene and its mRNA, Nature 292: 267.

Gorman, C., Moffat, L.F., and Howard, B.H., 1982a, Recombinant genes which express chloramphenicol acetytransferase in mammalian cells, Mol. Cell. Biol. 2: 1044.

Gorman, C.M., Merlino, G.T., Willingham, M.C., Pastan, I., and Howard, B.H., 1982b, The Rous sarcoma virus long terminal repeat is a strong promoter when introduced into a variety of eukaryotic cells by DNA-mediated transfection, Proc. Natl. Acad. Sci. USA 79: 6777.

Gorman, C., 1985, High Efficiency Gene Transfer into Mammalian Cells, in "DNA Cloning- Vol. 2", P.M. Glover, ed., IRL Press, Washington, D.C.

Gottesdiener, K.M., Karpinski, B.A., Lindsten, T., Strominger,.J.L., Jones, N.H., Thompson, C.B., and Leiden, J.M., 1988, Isolation and structural characterization of the human 4F2 heavy-chain gene, an inducible gene involved in T-lymphocyte activation, Mol. Cell. Biol. 8: 3809.

Maniatis, T., Fritsch, E. F., and Sambrook, J., "Molecular Cloning: A Laboratory Manual", Cold Spring Harbor Laboratory, Cold Spring, N.Y.

McKnight, S.L., Gavis, E.R., and Kingsbury, R., 1981, Analysis of transcriptional regulatory signals of the HSV thymidine kinase gene: Identification of an upstream control region, Cell 25: 385.

EXPRESSION OF THE APRT GENE IN AN ADENOVIRUS VECTOR SYSTEM AS A MODEL FOR STUDYING GENE THERAPY

Qing Wang, Vincent Konan and Milton W. Taylor

Department of Biology, Indiana University, Bloomington, IN 47405

INTRODUCTION

Most protocols for gene therapy which correct metabolic defects involve the use of retrovirus vectors to introduce normal cellular genes into mutant somatic tissue (1-2). Although retroviruses have many advantages in gene therapy, such vectors containing endogenous promoters readily undergo deletion and rearrangement as well as recombination, which makes them unstable and potentially pathogenic (3-4). Human adenoviruses have been studied as possible helper independent gene transfer and expression vectors (5-8). The *aprt* gene is a good model system for studying the parameters of gene therapy and gene transfer. It is constitutively expressed in all cell types (9), its deficiency does not affect cell morphology or cell growth in culture; it is small (2.8 kb), and conversion from *aprt-* to *aprt+* is easily selected. Many *aprt-* mutant cell lines are available (10) and can be used as target cells for in vitro studies.

In this paper, we report the construction of two recombinant adenovirus-5 vectors containing the *aprt* gene (including introns), lacking its native promoter but linked to a retroviral (Moloney murine sarcoma virus) promoter sequence (11). The recombinant viruses containing the *aprt* gene in opposite orientations were used to infect LAT cells, a semi-permissive mouse fibroblast L cell line defective in APRT and TK. The recombinant virus directs the synthesis of APRT following infection in both orientations. However the levels of APRT differ depending on the orientation of the *aprt* gene. The production of antisense RNA may be the reason for the lower level of APRT activity in Ad5/aprt1 (opposite orientation from E3 promoter). Integration of Ad5/aprt1 into the mouse cell chromosome occurred at a low frequency with loss of most of the viral sequences. In such transductants APRT activity is fully functional. Our work would indicate that adenovirus-human gene constructs might be a suitable vectors system for human gene therapy.

Purine and Pyrimidine Metabolism in Man VII, Part B
Edited by R.A. Harkness *et al*., Plenum Press, New York, 1991

RESULTS AND DISCUSSION

The recombinant viruses Ad5/aprt1 and Ad5/aprt2 were constructed as described in Fig. 1. The resulting plasmid pBR-10L contains the *aprt* in leftward orientation and pBR-10R contains a rightward insertion of *aprt*. These two plasmids were used to transfect 293 cells with Eco R1 digested Ad 5 DNA. The infectious recombinant viruses from the resultant plaques were analyzed for the presence of a complete insert in both orientations by restriction analysis.

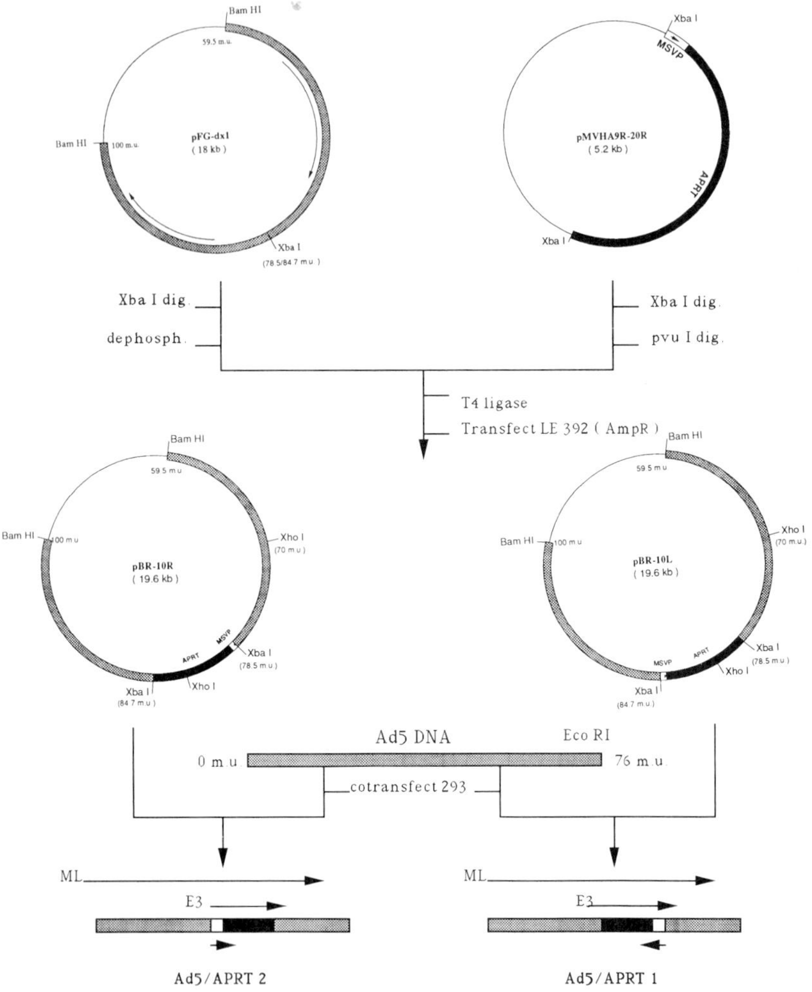

Fig. 1. Construction of recombinant viruses: Standard cloning techniques were used to construct plasmids pBR-10L and pBR-10R. The plasmid pMVHA9R-20R contains an Xba 1-Xba 1 fragment including 180 bps of the MSV promoter sequence and the CHO *aprt* structural gene from 5 bps upstream of the ATG start site to 200 bps downstream of the poly A signal site (11). This Xbal fragment was excised and cloned into the unique Xba 1 site of pFG-dx1 which contains a Bam H1 fragment of adenovirus-5 genome from 59.5 to 100 map units with a deletion of the Xba 1 D fragment.

In cells infected with Ad5/aprt2, ^{3}H-adenine incorporation was detected 10 hrs postinfection, increased rapidly until 24-36 hrs and continued to increase through the experiment (Fig. 2a). Whereas in cells infected with Ad5/aprt1, ^{3}H-adenine incorporation was not detected until 24 hrs. That this incorporation of ^{3}H-adenine into nucleic acids in vivo is due to the activity of APRT in infected LAT cells was further confirmed by an in vitro assay of APRT activity in the cell extracts (Fig. 2b). The APRT activity in Ad5/aprt1 infected cells is only 5-13% of that in Ad5/aprt2 infected cells and 10-25% of wild type cell level. This observation

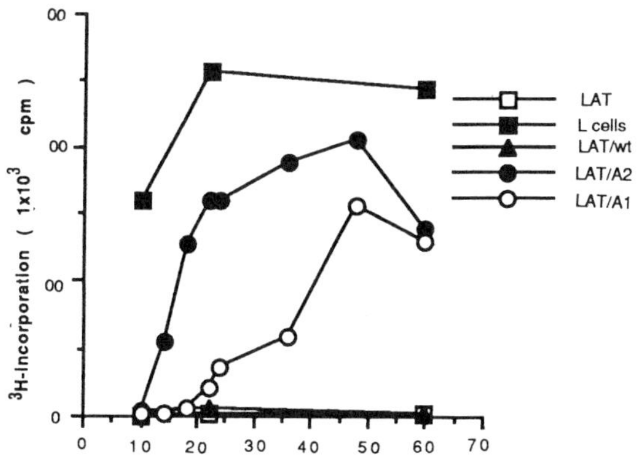

Fig. 2a. Time course of 3H-adenine incorporation in LAT cells after viral infection.

1×10^6 cells were infected at an m.o.i of 10 with Ad5/aprt1 and Ad5/aprt2 and incubated at 37 °C. Uninfected cells and cells infected with wt virus were used as negative controls. The *wt* L cells were assayed as a positive control. At different time intervals, cells were pulsed with ^{3}H-adenine at 1uCi/ml for 2 hr. The nucleic acids were precipitated by 10% TCA and the level of ^{3}H incorporation were determined.

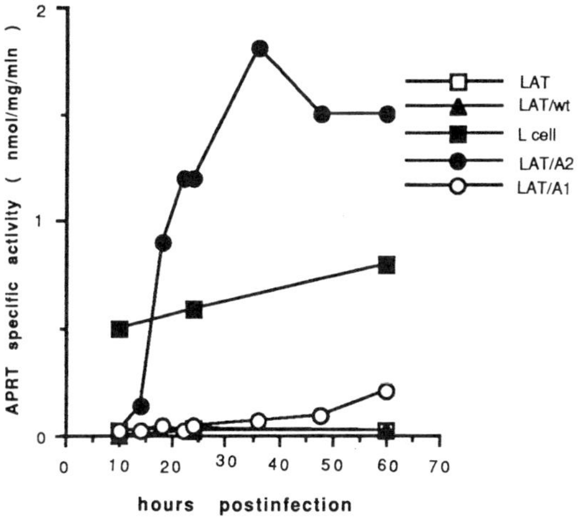

Fig. 2b. Time course of CHO APRT activity expressed in recombinant viruses infected cells.

1×10^6 cells were infected with Ad5/aprt1 and Ad5/aprt2 at an m.o.i of 10 and incubated at 37 °C. *wt* L cells and LAT without viral infection and with *wt* virus infection were assayed as control. The cells were collected at different times after infection. APRT activities in the lysate were analyzed. The specific activity is defined as the conversion of 1 nmol of ^{3}H-AMP per minute per mg protein.

is consistent with other studies of adenoviral vectors (7, 8), in which a foreign gene cloned in the opposite orientation relative to the viral E3 or ML promoter is expressed at a very low level. The apparent discrepancy between the level of adenine incorporation into DNA, and the specific activity of APRT in cells infected with Ad5/aprt1 may reflect the level of enzyme required to synthesize saturating level of AMP that are incorporated into DNA.

The transcription of *aprt* was studied by Northern analysis using a 2.1 kb Eco R5-Xba 1 fragment containing the CHO *aprt* coding sequences. The results (Fig. 3) demonstrate that the *aprt* gene is efficiently transcribed irrespective of orientation. The APRT probe detected two major species of RNA of 4.2 kb and 10 kb and a minor 1.5 kb species in Ad5/aprt1 infected cells. Four different species of mRNA of 1.2 kb. 3.6 kb, 4.2 kb and 10 kb could be detected in Ad5/aprt2 infected cells. The heterogeneity of the specific APRT mRNA species may reflect multiple initiation, termination and splicing events. Our RNA studies (data not shown) also indicated that antisense RNA might be synthesized in Ad/aprt1. In order to quantitate this antisense mRNA and to determine which transcript of the three species detected by the double strand APRT probe is indeed antisense mRNA, a single stranded sense DNA probe was rehybridized to the RNA following removal of the double stranded DNA probe. The most abundant 4.2 kb transcript synthesized in Ad5/aprt1 infected cells hybridizes to the sense strand DNA probe. The synthesis of the antisense RNA might be driven from viral ML promoter is also indicated from our previous experiments (data not shown). The lower level of APRT activity found in Ad5/aprt1 could be explained by inhibition of translation of APRT mRNA due to RNA-RNA hybridization to antisense RNA. If this was so, a blockage of the ML promoter should result in an increase in APRT activity. Cytosine arabinoside (Ara-C) blocks viral DNA synthesis, as well as the transcription from ML promoter (12). In contrast to the situation with Ad5/aprt2 in which APRT activity is decreased by Ara-C at all times after infection, there is a slight increase in APRT activity in the case of Ad5/aprt1.

To obtain stable transductants, LAT cells were infected at an moi of 10, and placed in selective medium. 15 days after infection, colonies were detectable at a frequency of approximately 1×10^{-6}. The *aprt* gene was stably integrated in LAT cells, since these clones

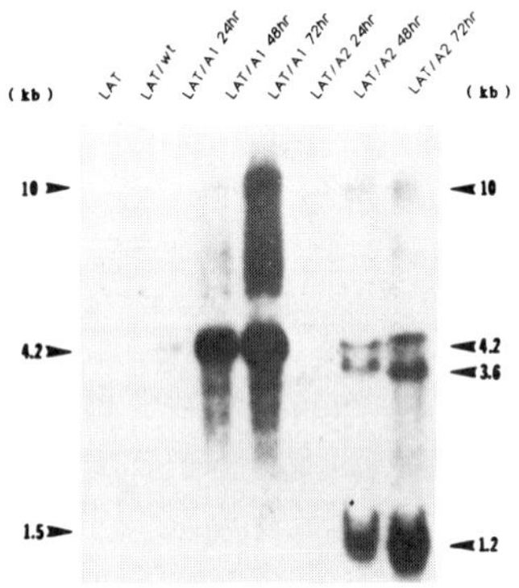

Fig. 3. Northern analysis of APRT transcripts in recombinant viruses infected cells.

Total cellular RNA was isolated at 24, 48 and 72 hr postinfection (15 m.o.i.). 15 ug of RNA samples were fractionated on 1% formaldehyde agarose gel and transferred to nitrocellulose filter. Specific APRT transcripts were identified by hybridizing to a ^{32}P-labeled 2.1 kb Eco R5 - Xba 1 *aprt* fragment.

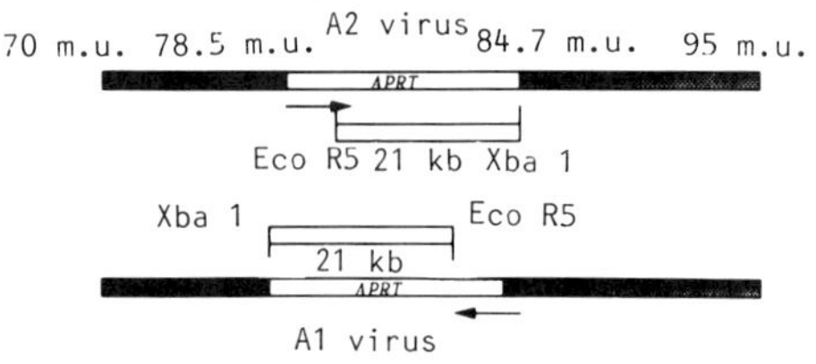

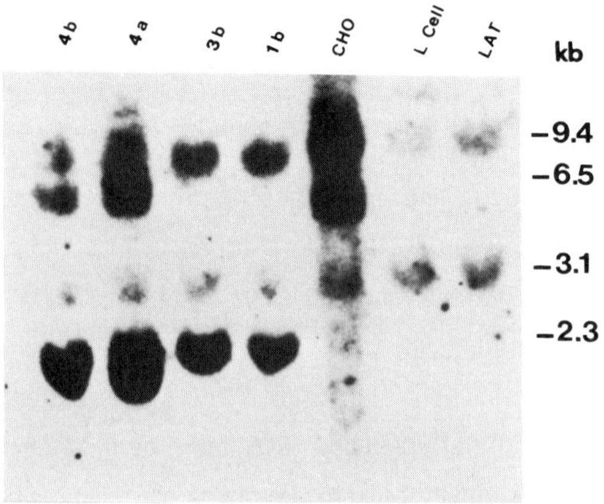

Fig. 4. Southern blot analysis of the CHO aprt gene in transductants.

20 ug of DNA from the control cells (LAT, L cell and CHO) and the transductants (1B, 3B, 4A and 4B)was digested with EcoR1 and run on a 0.8% agarose gel. Following transfer to a nitrocellulose filter, the DNA was hybridized to a radio labeled 2.7 kb BamH1-BamH1 fragment of CHO *aprt* probe.

could be grown in the absence of selective pressure for 6 months and still maintain their *aprt*+ phenotype. These results clearly show that adenovirus-5 can be used as a vector to transduce the CHO *aprt* gene.

To determine whether the CHO aprt gene had been integrated into the mouse LAT cells by homologous or heterologous recombination, and if the entire viral genome was integrated along with the *aprt* gene. Southern analysis was performed. When the DNA was digested with EcoR1 and probed with a 2.7 kb CHO aprt fragment (Fig. 4), besides the expected 2.3 kb band, two additional bands (9.4 kb in 1B and 3B; 6.5 kb in 4A and 4B) were observed. These findings suggest that integration is probably the result of a single double crossover event between viral and mouse genome near the *aprt* locus. When the DNA was cut with appropriate enzymes and hybridized to viral sequences upstream and downstream of the aprt gene, no viral sequences were detected. These results indicated that the four clones arose independently and most probably arose by a cross-over event in viral sequences on either side and near the CHO *aprt* gene, thus excluding most of the viral sequences.

These results support the concept that adenovirus can be made into a generalized transducing vector. Further work is now being done using a deletion mutant of Ad/aprt virus and aprt- CHO cells to test whether the adeno/aprt vector can be targeted for homologous recombination.

REFERENCES

1. Anderson, W.F., Prospects for human gene therapy. Science 226:401-409 (1984).
2. Temin, H.M., Retrovirus vectors: promise and reality. Science 246:983 (1989).
3. Bandyopadhyay, P.K. and Temin, H.M., Expression of complete chicken thymidine kinase gene inserted in a retrovirus vector. Mol. Cell. Biol. 4:749-754 (1984).
4. Emerman, M. and Temin, H.M., High-frequency deletion in recovered retrovirus vectors containing exogenous DNA with promoters. J. Virol. 50:42-49 (1984).
5. Yamada, M., Lewis, J.A. and Grodzicker, T., Overproduction of the protein product of a nonselected foreign gene carried by an adenovirus vector. Proc. Natl. Acad. Sci. USA 82:3567-3571 (1985).
6. Haj-Ahmad, Y., and Graham, F.L., Development of helper-independent human adenovirus vector and its use in the transfer of the herpes simplex virus thymidine kinase gene. J. Virol. 57:267-274 (1986).
7. Johnson D.C., Ghosh-Choudhury, G., Smiley, J.R., Fallis, L and Graham, F., Abundant expression of herpes simplex virus glycoprotein gB using an adenovirus vector. Virology 164:1-14 (1988).
8. Schneider, M., Graham, F.L. and Prevec, L., Expression of the glycoprotein of vesicular stomatitis virus by infectious adenovirus vectors. J. Gen. Virol. 70:417-427 (1989).

9. Taylor, M.W. et al., The APRT system, In molecular cell genetics, Gottesman, M.M., Ed., John Wiley, New York, 311-332 (1985).
10. Simon, A.E. and Taylor, M.W., High frequency mutation of the adenine phosphoribosyl transferase locus in CHO cells is due to deletion of the gene. Proc. Natl. Acad. Sci. USA 80:810-814 (1985).
11. Tang, D.C. and Taylor, M.W., Transcriptional activation of the adenine phosphoribosyl transferase promoter by an upstream butyrate-induced moloney murine sarcoma virus enhancer-promoter element. J. Virol. 64:2907-2911 (1990).
12. Flint, S.J., Regulation of adenovirus mRNA formation. Adv. Vir. Res. 31:169-228 (1986).

TRANSCRIPTIONAL REGULATION OF RIBONUCLEOTIDE REDUCTASE

Daniel A Albert[1] and Enrique Rozengurt

Imperial Cancer Research Fund, Lincolns' Inn Fields
London, WC2A 3PX
[1]Department of Medicine, Univesity of Chicago, Box 74
Chicago, IL 60637, USA

INTRODUCTION

The Swiss 3T3 cell system can be used to explore the molecular mechanism of proliferation. These fibroblasts can be arrested in a quiescent homogeneous G_o/G_1 state by serum deprivation and then induced to proliferate by a variety of mitogenic growth factors (1). An important endeavour in determining the basis of growth factor action is to elucidate the signal transduction pathways involved that result in the mitogenic response. These studies have led to the formulation of a model that proposes the existence of multiple growth factor-activated signalling pathways that synergistically lead to a mitogenic response (2). These pathways depicted in Figure 1 include one mediated by activation of protein kinase C, another mediated by cAMP-dependent protein kinase (A) and others that are, as yet, less clearly defined, which mediate responses to polypeptide growth factors that activate receptor tyrosine kinases (3). Activation of PKC specifically results in the rapid increase in the phosphorylation of an acidic cellular protein that migrates with an apparent M_r of 80000 (80K) (4). By contrast, the cAMP-mediated mitogenic response is characterised by the rapid increase in the phosphorylation of a M_r = 58000 cellular protein identified as vimentin (5). More distal effects include activation of the cellular proto-oncogenes c-fos and c-myc (6). Consistent with these physiologic events is the observation that these two genes have response elements in their 5' upstream control elements that respond to either a PKC signal (AP-1 site) or a cAMP signal (CREB site). Even more distal events are under current investigation.

Amongst the most essential elements of the proliferative response is the activation of the apparatus necessary for DNA synthesis. Multiple complex processes including DNA conformational changes, large increases in the pools of deoxynucleotide precursors and the synthesis of a number of enzymes that participate in DNA synthesis are inherent to this process. One critical enzyme is ribonucleotide diphosphate reductase which catalyzes the reduction of ribose diphosphates to deoxyribose diphosphates and is the only in vivo source of deoxynucleotides for DNA synthesis (7). This enzyme appears to be regulated in part by allosteric effector mechanisms and in part by de novo synthesis (8) and possibly by post-translational modification (9). The enzyme is composed of two subunits M1 and M2 that are regulated independently (10).

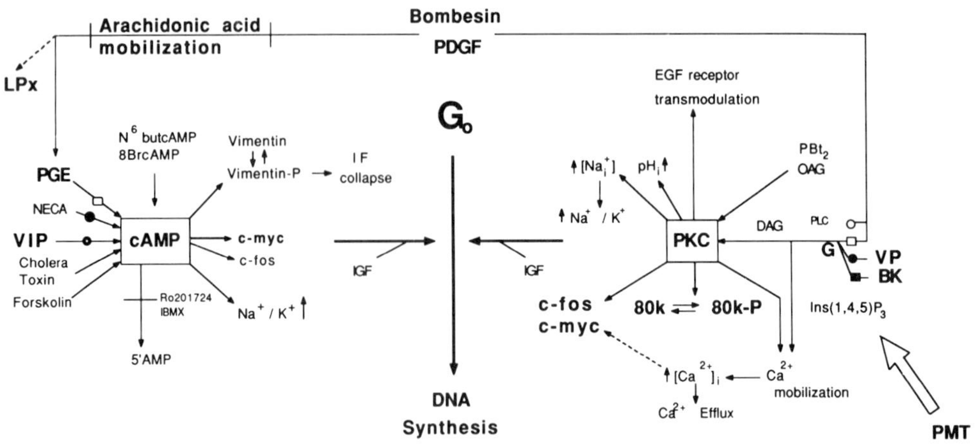

Fig. 1.

To further investigate the regulation of ribonucleotide diphosphate reductase during cellular growth, we have utilized the 3T3 system to explore the control of gene expression of the M1 and M2 subunits as cells are activated by mitogenic stimuli. The data presented here shows that the genes coding for the M1 and M2 subunits of ribonucleotide reductase are expressed in late G_1 phase and suggest that the increase in their expression occurs in a coordinate fashion.

MATERIALS AND METHODS

Cell Culture

Swiss 3T3 fibroblasts were maintained in Dulbecco's modified Eagle's medium (DMEM) in the presence of fetal calf serum (10%), penicillin (100 U/ml) and streptomycin (100 ng/ml) in a humidified atmosphere containing 10% CO_2 at 37°C. Experimental cultures were plated at 6 x 10^5 cells/90 mm dish (Nunc Petri) and used 6 days later when they were confluent, quiescent and the media was depleted of serum growth factors.

Assays of DNA Synthesis

DNA synthesis was determined by incubating cells, washed free of serum, in medium containing [^{3}H]thymidine (1 µCi/ml; 1 µM) and additions as indicated. The incorporation of radioactivity into TCA precipitable material was measured as previously described (11).

Cytofluorometric Analyses

Cells were detached by treatment with trypsin (0.025%) and 4 mM EDTA, suspended in DME containing 10% bovine serum, centrifuged at 1000 rpm for 5 min, washed with phosphate-buffered saline, then resuspended in staining solution containing propidium iodide (5%) Na citrate (1.0 mg/ml) and triton X-100 (0.1%). Stained nuclei were analyzed on a Fluorescence Activated Cell Sorter (FACS Star 4) (Becton Dickinson, CA) after a 15 min incubation.

Northern Blot Analysis

Quiescent cultures of Swiss 3T3 cells were stimulated with bovine fetal calf serum then washed twice in cold (4°) PBS then lysed with 4 M guanidine isothiocyanate. Total RNA was isolated by centrifugation through a cesium chloride gradient. RNA samples (20 μg/lane) were fractionated on a 1% agarose/6% formaldehyde gel by electrophoresis then transferred to a Hybond-N (or N plus, Amersham) nylon membrane by capillary action then fixed by incubation at 80°C for 2 h. Northern blots were prehybridized for 4 to 6 h at 42°C in a solution containing 50% formamide, 5x Denhardt's, 5x SSC, 0.5% SDS and 500 μg/ml denatured salmon sperm DNA. Hybridization were performed by the addition of 4.0 to 3.0 x 10^6 cpm/ml of ^{32}P-labelled DNA probe to the prehybridization mix and incubating overnight (approximately 18h). Blots were washed free of unbound probe by sequential exposure to 2 x SSC 0.1% SDS at 42°C for 30', then 0.1 x SSC 0.1% SDS 55°-60°C for 30'. Following autoradiography at -70°C bound DNA probes were removed by incubating blots in 0.1 x SSC 0.1% SDS at 100°C for 15'. Blots could then be rehybridized to a different probe. Autoradiographic bands were quantitated by scanning densitometry on a LKB ultrascan XL densitometer.

Materials

Probes used for this study include cDNA for the M1 and M2 subunits of ribonucleotide reduction (courtesy of Ingrid Caras, Genentech and Lars Thelander, University of Umea, respectively), the plasmid FM564 containing 1.2 kb PST-1 glyceraldehyde 3 phosphate dehydrogenase (human) cDNA (gift of C.Williams and L. Lim, Institute of Neurology, London, UK) and the pSV-c-*myc*1 plasmid containing the 2.8 kb HindIII XbaI murine c-*myc* insert. Probes were labelled with [^{32}P]dCTP by random priming (Amersham, Multiprime) (12).

RESULTS

Quiescent cultures of Swiss 3T3 cells were transferred to medium containing 10% fresh serum and assayed for M1 and M2 mRNA levels by Northern hybridization at various times after stimulation. Parallel cultures were used to determine DNA replication by [^{3}H]thymidine incorporation and cytofluorimetry. The results presented in Figure 2 show there was little expression of the M2 gene in quiescent cells in contrast to M1 which had a low but discernible degree of expression. Correspondingly, the increase in the level of M2 mRNA ranged from 20 to 50 fold after 18 hours of stimulation. This magnitude of induction appears to be similar to that of other genes that are sensitive to growth stimulation including the c-*myc* message. The induction of M1 (5-10 fold) is less pronounced than that of M2 and peaked later (24 h vs 18 h for M2). The induction of message for M1 and M2 precedes entry into S phase which begins about 16 h after addition of serum as judged by either [^{3}H]thymidine incorporation or cytofluorimetry.

DISCUSSION

The induction of gene expression after stimulation of cell proliferation follows a defined sequence of events. The expression of the cellular proto-oncogene fos peaks 30 min after stimulation whereas the c-*myc* mRNA levels reach a maximum about 3 hours after serum stimulation in Swiss 3T3 cells (6). In contrast, the genes that code for

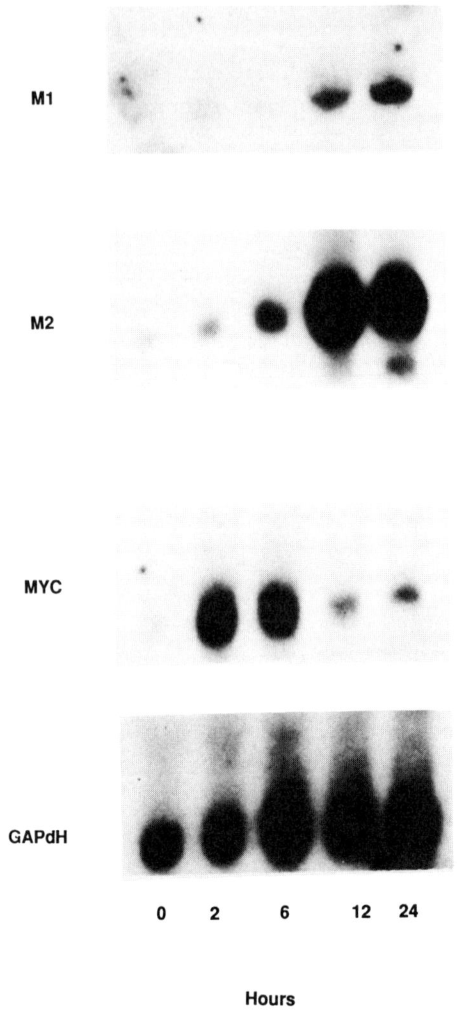

Fig. 2

the M1 and M2 subunits of mouse ribonucleotide reductase are induced about 6 to 8 h after serum addition.

The enzyme ribonucleotide reductase is essential for DNA synthesis. In quiescent cells its activity is very low to undetectable but it increases dramatically following serum stimulation. This increase appears to correlate more closely with an increase in the M2 rather than M1 subunit. Thus, it is surprising that serum stimulation of quiescent 3T3 cells results in a substantial induction of the expression of the genes for both the M1 and M2 subunits of ribonucleotide reductase. Furthermore, it is surprising that the timing of induction of expression suggest relatively co-ordinated expression since the M1 and M2 subunits appear to be regulated independently at the protein level. M1 protein is long lived with a half-life of about 18 h whereas M2 has a short half-life of about 3 h. M2 protein appears to increase 5 to 10 fold during the progression from G_1 to S phase whereas M1 remains relatively constant (2-fold increase or less). One of the most important generalizations that has emerged from studies concerning the regulation of cell proliferation is that this process is triggered by distinct pathways that act in a synergistic manner. Nevertheless, these pathways must converge prior to the stimulation of DNA synthesis. The induction of the machinery required for DNA replication, including ribonucleotide reductase, is an obvious point of signal convergence. Alternatively, the induction of these proteins may be downstream to other, earlier, events.

In further studies it will be important to characterize the signal transduction pathways that induce expression of the genes that code for ribonucleotide reductase.

REFERENCES

1. Rozengurt, E. Early signals in the mitogenic response. Science 234: 161-166, 1986.
2. Rozengurt, E., Erusalimsky, J., Mehmet, H., Morris, C., Nånberg, E. and Sinnett-Smith, J. Signal transduction in mitogenesis: further evidence for multiple pathways. Cold Spring Harbor Symp. Quant. Biol. 53: 945-954, 1988.
3. Rozengurt, E. Neuropeptides as cellular growth factors: role of multiple signalling pathways. Eur. J. Clin. Invest. 21: 123-134, 1991.
4. Rozengurt, E., Rodriguez-Pena, A. and Smith, K.A. Phorbol esters phospholipase C and growth factors rapidly stimulate the phosphorylation of a Mr 80,000 protein in intact quiescent 3T3 cells. Proc. Natl. Acad. Sci. USA. 80: 7244-7248.
5. Escribano, J. and Rozengurt, E. Cyclic AMP increasing agents rapidly stimulate vimentin phosphorylation in quiescent cultures of Swiss 3T3 cells. J. Cell Physiol. 137: 223-234, 1988.
6. Mehmet, H., Sinnett-Smith, J.W., Moore, J.P., Evan, G.I. and Rozengurt, E. Differential induction of c-fos and c-myc by cyclic AMP in Swiss 3T3 cells: significance for the mitogenic response. Oncogene Res. 3: 281-286, 1988.
7. Reichard, P. Interactions between deoxyribonucleotide and DNA synthesis. Ann. Rev. Biochem. 57: 349-374, 1988.
8. Albert, D.A. and Gudas, L.J. Ribonucleotide reductase and deoxyribonucleoside triphosphate metabolism during the cell cycle of S49 wild type and mutant T lymphoma cells. J. Biol. Chem. 260: 679-686, 1985.
9. Albert, D.A. and Nodzenski, E. The M2 subunit of ribonucleotide reductase is a target of cyclic AMP dependent protein kinase. J. Cell. Physiol. 130: 262-269, 1987.

10. Eriksson, S. and Martin, D.W., Jr. Ribonucleotide reductase in cultured mouse lymphoma cells. Cell cycle dependent variation in the subunit protein M2. J. Biol. Chem. 256: 9436-9440, 1987.
11. Dicker, P. and Rozengurt, E. Phorbol esters and vasopressin stimulate DNA synthesis by a common mechanism. Nature 287: 607-612, 1980.
12. Feinberg, A.P. and Vogelstein, B. A technique for radiolabelling DNA restriction endonuclease fragments to high specific activity. Anal. Biochem. 132: 6-13, 1988.

MUTATIONAL BASIS OF ADENINE PHOSPHORIBOSYLTRANSFERASE DEFICIENCY

Amrik Sahota, Ju Chen, Peter J. Stambrook*, and
Jay A. Tischfield

Department of Medical and Molecular Genetics, Indiana
University School of Medicine, Indianapolis, IN 46202;
*Department of Anatomy and Cell Biology, University of
Cincinnati College of Medicine, Cincinnati, OH 45267

INTRODUCTION

Adenine phosphoribosyltransferase (APRT, EC 2.4.2.7) catalyzes the synthesis of AMP from adenine and 5-phosphoribosyl-1-pyrophosphate. In APRT deficiency (McKusick 102600), adenine is oxidized by xanthine oxidase to the highly insoluble and nephrotoxic derivative, 2,8-dihydroxyadenine. The accumulation of this compound in the kidney can lead to stone formation and eventual renal failure (Simmonds et al. 1989).

Two different types of APRT deficiency have been described. Type II deficiency (complete enzyme deficiency *in vivo* but partial deficiency in cell extracts) has been found only in Japan. Type II patients are homozygotes or compound heterozygotes for the *APRT*J* missense mutation in exon 5 (Kamatani et al. 1987; Hidaka et al. 1988; Kamatani et al. 1990; Sahota et al. 1990). Type I APRT deficiency (complete deficiency *in vivo* or *in vitro*), on the other hand, has been found in patients from many different countries, including Japan. Type I patients are homozygotes or compound heterozygotes for a variety of null alleles collectively designated *APRT*Q0* (Fujimori et al. 1985; Hidaka et al. 1987). Four different *APRT*Q0* mutant alleles have been identified to date (Hidaka et al. 1987; Hidaka et al. 1988; Sahota et al. 1990; Mimori et al. 1991). In this report we describe the identification of 11 additional *Q0* mutant alleles.

METHODS

DNA was isolated from blood or from transformed lymphoblasts and a 2.4 kb fragment spanning the entire coding region and the flanking sequences of the *APRT* gene was amplified by the polymerase chain reaction (PCR), using the Perkin-Elmer Cetus thermal cycler and the GeneAmp kit (Chen et al. 1991). Primers NC720 (sense sequence) and C719 (antisense sequence), located at positions -161 to -140 and 2242 to 2221, respectively, of our

Purine and Pyrimidine Metabolism in Man VII, Part B
Edited by R.A. Harkness *et al.*, Plenum Press, New York, 1991

wild-type sequence were used for amplification. (The A of the ATG start codon is designated base 1.) The PCR reaction mixture was electrophoresed on an agarose gel and the 2.4 kb fragment was purified from the gel. The PCR-amplified DNA was then ligated with M13mp18 RF DNA which had been digested with *Hinc*II. Following *E. coli* JM109 transformation, single stranded DNA from three to five positive plaques from each patient was sequenced completely (Chen et al. 1991). Selected regions of the PCR-amplified DNA from each patient were also sequenced directly. Nucleic acid sequences were analyzed using the Microgenie (V5.0) software from Beckman.

RESULTS AND DISCUSSION

We have investigated the molecular basis of APRT deficiency in 17 non-Japanese patients (excluding siblings) and 8 Japanese patients (Table 1). One of the Japanese patients had Type I APRT deficiency and the other seven had Type II deficiency. Fifteen different mutations in the *APRT* gene have been identified, and we have seen 14 of these in the patients we have studied (Table 2). Six of the mutations were located in exon 3, and two at the exon 4-intron 4 splice donor site. Two of the exon 3 mutations were of the nonsense type.

Table 1. Country of origin of patients.

Country	No.	Comments
Austria	1	
Bermuda	1	
Britain	1	
Canada	2	Sibs
France	1	
Germany	2	Sibs
Greece	1	
Hungary	1	
Iceland	5	
Iraq	1	
Japan	8	
Pakistan	1	
USA	2	One heterozygote

A common missense mutation in exon 3 (asp_{65} -> val) was identified in six patients (five from Iceland and one from Britain) (Table 3). A second common mutation (a T insertion at the exon 4-intron 4 splice donor site) was found in twin brothers from Germany, a patient from Austria, and in a heterozygote from the United States. This mutation has been previously seen in one allele from two Belgian brothers (Hidaka et al. 1987). A G -> T transversion instead of an insertion at this junction was found in an Iraqi patient. A patient from Pakistan was homozygous for a nonsense mutation in exon 3 (arg_{87} -> end). The other non-Japanese patients had a variety of mutations, mainly single base changes and small deletions.

Table 2. *APRT* mutations.

Mutation	Change	Location	Comments
Insertion	T	Exon 2	Frameshift after Ile61
Deletion	GGCCCCA	Exon 3	Frameshift after Pro93
	AC	Exon 3	Frameshift after Pro95
	TTC	Exon 5	Phe173 or Phe174 deletion*
Missense	ATG -> GTG	Exon 1	Met1 -> Val
	GAC -> GTC	Exon 3	Asp65 -> Val
	CGA -> CAA	Exon 3	Arg67 -> Gln
	CTG -> CCG	Exon 4	Leu110 -> Pro
	ATT -> TTT	Exon 4	Ile112 -> Phe
	ATG -> ACG	Exon 5	Met136 -> Thr
	TGC -> CGC	Exon 5	Cys153 -> Arg
Nonsense	CGA -> TGA	Exon 3	Arg87 -> End
	TGG -> TGA	Exon 3	Trp98 -> End
Splice donor	TG:GTAA -> TG:GTTAA	Intr 4	Exon 4 skipped
	TG:GTAA -> TG:TTAA	Intr 4	Exon 4 skipped

* We have not seen this mutation (Hidaka et al. 1987) in the patients we have studied. It is included here for the sake of completeness.

A Type I patient from Japan was homozygous for an exon 3 nonsense mutation (trp98 -> end) (Sahota et al. 1990). This mutation has also been found in several other Japanese patients (Mimori et al. 1991). The same mutation was found in one allele from a Type II compound heterozygote (Sahota et al. 1990). The other allele from this patient had the previously described exon 5 missense mutation (met136 -> thr) typical of Type II patients (Hidaka et al. 1988). The remaining Japanese patients were homozygous for the exon 5 mutation.

Table 3. Common mutations in the *APRT* gene.

GAC -> GTC	Asp65 -> Val	Exon 3	Type I	Iceland, Britain
TGG -> TGA	Trp98 -> End	Exon 3	Type I	Japan
ATG -> ACG	Met136 -> Thr	Exon 5	Type II	Japan
TG:GTAA -> TG:GTTAA		Intr 4	Type I	Austria, USA, Germany, Belgium

The findings presented here demonstrate that: (i) a small number of different mutations is likely to be responsible for the majority of cases of APRT deficiency; (ii) exon 3 and the exon 4-intron 4 junction are hot spots for mutation; and (iii) APRT activity is not essential *in vivo*.

SUMMARY

The mutational basis of APRT deficiency was studied in non-Japanese and Japanese patients. Fifteen different mutations have been identified altogether. Of these 4 were common, 6 were located in exon 3, and two at the exon 4-intron 4 junction. The common mutations were a missense mutation in exon 3 (asp65 -> val) and a T insertion at the exon 4-intron 4 junction in non-Japanese patients, a nonsense mutation in exon 3 (trp98 -> end) in Type I Japanese patients, and an exon 5 missense mutation (met136 -> thr) in Type II patients. The other mutations in Type I patients consisted mainly of single base changes and small deletions.

ACKNOWLEDGMENTS

We thank all the patients and their families for donating blood samples, and Mann Hans for help with word processing. This work was supported by NIH grants DK38185 and CA36897.

REFERENCES

Chen J, Sahota A, Stambrook PJ, Tischfield JA (1991) Polymerase chain reaction amplification and sequence analysis of human mutant adenine phosphoribosyltransferase genes: The nature and frequency of errors caused by *Taq* DNA polymerase. Mutat Res 249:169-176

Fujimori S, Akaoka I, Sakamoto K, Yamanaka H, Nishioka K, Kamatani N (1985) Common characteristics of mutant adenine phosphoribosyltransferase from four separate Japanese families with 2,8-dihydroxyadenine urolithiasis associated with partial enzyme deficiencies. Hum Genet 71:171-176

Hidaka Y, Palella TD, O'Toole TE, Tarle' SA, Kelley WN (1987) Human adenine phosphoribosyltransferase: Identification of allelic mutations as a cause of complete deficiency of the enzyme. J Clin Invest 80:1409-1415

Hidaka Y, Tarle' SA, Fujimori S, Kamatani N, Kelley WN, Palella TD (1988) Human adenine phosphoribosyltransferse deficiency: Demonstration of a single mutant allele common to the Japanese. J Clin Invest 81:945-950

Kamatani N, Terai C, Kuroshima S, Nishioka K, Mikanagi K (1987) Genetic and clinical studies on 19 families with adenine phosphoribosyltransferase deficiencies. Hum Genet 75:163-168

Kamatani N, Kuroshima S, Yamanaka H, Nakashe S, Take H, Hakoda M (1990) Identification of a compound heterozgote for adenine phosphoribosyltransferase deficiency (*APRT*J/APRT*Q0*) leading to 2,8-dihydroxyadenine urolithiasis. Hum Genet 85:500-504

Mimori A, Hidaka Y, Wu VC, Tarle' SA, Kamatani N, Kelley WN, Palella TD (1991) A mutant allele common to the Type I adenine phoshoribosyltransferase deficiency in Japanese subjects. Am J Hum Genet 48:103-107

Sahota A, Chen J, Asako K, Takeuchi H, Stambrook PJ, Tischfield JA (1990) Identification of a common nonsense mutation in Japanese patients with Type I adenine phosphoribosyltransferase deficiency. Nuc Acids Res 18:5915-5916

Simmonds HA, Sahota AS, Van Acker KJ (1989) Adenine phosphoribosyltransferase deficiency and 2,8-dihydroxyadenine lithiasis. In: Scriver CR, Beaudet AL, Sly WS, and Valle D (eds) The Metabolic basis of inherited disease, 6th ed, McGraw-Hill, New York, pp 1029-1044

ANALYSIS OF THE PROMOTER REGION OF THE CHO APRT GENE

Bin Ru She and Milton W. Taylor

Department Of Biology,
Indiana University, Bloomington IN. 47405

Introduction

APRT (adenosine phosphoribosyltransferase) catalyzes the condensation of adenine with PRPP to form adenosine monophosphate and pyrophosphate. The enzyme is constitutively expressed in all tissues examined. The CHO APRT gene does not have a TATA or CAAT box in its 5' region; like most housekeeping genes the 5' region is GC-rich and contains three consensus Sp1-binding sequences. A deletion of upstream region of nucleotide -89 (relative to the transcription initiation sites) does not affect gene expression, while an additional 29 base pair deletion decreases expression by two-fold. To further study the cis-acting elements essential for gene expression we constructed mutations in the 5' region by linker-scanning techniques. We report here that (i) mutants with linker-scanning mutations between the nucleotides -33 and +22 retained 40% activity of the wild type gene; (ii) mutations in the region from nucleotides -100 to -88, and from -48 to -39 resulted in two-fold increased expression; and (iii) mutations at the transcription start sites did not affect gene expression.

In order to study the role of Spl in regulating APRT gene expression, gel-shift assays and DNase-1 footprinting analysis were performed using purified Spl. The results showed that Spl bound to two of the three consensus Spl-binding sequences. Coincidentally, mutations at those two regions slightly decreased expression in one case but increased expression two-fold in the other, while mutations at the third (unbound) consensus Spl sequence did not affect gene expression.

Materials And Methods

Cell culture and transfection CHO APRT$^-$ cells were maintained in F-12 medium supplemented with 10% calf serum. Cells were seeded at a density of 10^6 cells per 100mm dish. After 48 hours of incubation, cells were transfected according to the methods of Sussman and Milman[3] Each dish of cells was transfected with 2 ug of DNA being tested and 2 ug of pSV2CAT (as an internal control). 48 to 72 hours after the transfection the medium was removed and the cells were washed twice with PBS. 1 ml of TEN (40 mM Tris-Cl pH7.5, 1mM EDTA and 150 mM NaCl) was added to each dish and cells were collected by scraping. After precipitation the cell pellet was resuspended in 150 ul of 0.25 M Tris-Cl (pH7.5), broken by freeze/thaw and centrifuged. The supernatant was saved for APRT and CAT assays.

Enzyme assay The APRT assay was done following the methods described by Park et al[2], and the CAT assay was done according to Gorman et al[4].

Purine and Pyrimidine Metabolism in Man VII, Part B
Edited by R.A. Harkness *et al.*, Plenum Press, New York, 1991

```
                -90      -80       -70       -60       -50       -40       -30       -20       -10        +1      +10
wt          CCCGCCCCGTCCCGCCCCCTCCCGGCCTCTCCTCGTGCTGGATCGCTCCCTAAGGACGCCCCGCTCCAGAAGCCCCACCTACCAAGGACGCCCCACCCTTGTTCCCGGA-100%
LS-97/-89   CCgaagatcTtCCGCCCCCTCCCGGCCTCTCCTCGTGCTGGATCGCTCCCTAAGGACGCCCCGCTCCAGAAGCCCCACCTACCAAGGACGCCCCACCCTTGTTCCCGGA--73 %
LS-90/-78   CCCGCCCCGgaagatCgatcttCGGCCTCTCCTCGTGCTGGATCGCTCCCTAAGGACGCCCCGCTCCAGAAGCCCCACCTACCAAGGACGCCCCACCCTTGTTCCCGGA--193%
LS-75/-67   CCCGCCCCGTCCCGCCCCCTCCCGaagatcttcCGTGCTGGATCGCTCCCTAAGGACGCCCCGCTCCAGAAGCCCCACCTACCAAGGACGCCCCACCCTTGTTCCCGGA--100%
LS-66/-57   CCCGCCCCGTCCCGCCCCCTCCCGGCCTCTCCTgaaGaTcttcCGCTCCCTAAGGACGCCCCGCTCCAGAAGCCCCACCTACCAAGGACGCCCCACCCTTGTTCCCGGA--87 %
LS-59/-51   CCCGCCCCGTCCCGCCCCCTCCCGGCCTCTCCTCGTGCTGaAgatCgatCTtcGGACGCCCCGCTCCAGAAGCCCCACCTACCAAGGACGCCCCACCCTTGTTCCCGGA--105%
LS-54/-44   CCCGCCCCGTCCCGCCCCCTCCCGGCCTCTCCTCGTGCTGGATCGCTgaagAtcttCGCCCCGCTCCAGAAGCCCCACCTACCAAGGACGCCCCACCCTTGTTCCCGGA---86 %
LS-48/-40   CCCGCCCCGTCCCGCCCCCTCCCGGCCTCTCCTCGTGCTGGATCGCTCCCTgAaGAtcttCCGCTCCAGAAGCCCCACCTACCAAGGACGCCCCACCCTTGTTCCCGGA--222%
LS-39/-30   CCCGCCCCGTCCCGCCCCCTCCCGGCCTCTCCTCGTGCTGGATCGCTCCCTAAGGACGCCgaagatCttcAGCCCCACCTACCAAGGACGCCCCACCCTTGTTCCCGGA---96 %
LS-33/-15   CCCGCCCCGTCCCGCCCCCTCCCGGCCTCTCCTCGTGCTGGATCGCTCCCTAAGGACGCCCCGCTCgAagtcttCgAagatCttcGGACGCCCCACCCTTGTTCCCGGA----14 %
LS-25/-16   CCCGCCCCGTCCCGCCCCCTCCCGGCCTCTCCTCGTGCTGGATCGCTCCCTAAGGACGCCCCGCTCCAGAAGCCgaAgaTcttcAGGACGCCCCACCCTTGTTCCCGGA---42 %
LS-24/-15   CCCGCCCCGTCCCGCCCCCTCCCGGCCTCTCCTCGTGCTGGATCGCTCCCTAAGGACGCCCCGCTCCAGAAGCCCgAagatCttcGGACGCCCCACCCTTGTTCCCGGA---56 %
LS-20/-12   CCCGCCCCGTCCCGCCCCCTCCCGGCCTCTCCTCGTGCTGGATCGCTCCCTAAGGACGCCCCGCTCCAGAAGCCCCACCgAagAtcttCGCCCCACCCTTGTTCCCGGA---21 %
LS-16/-8    CCCGCCCCGTCCCGCCCCCTCCCGGCCTCTCCTCGTGCTGGATCGCTCCCTAAGGACGCCCCGCTCCAGAAGCCCCACCTACCgAaGAtcttCCACCCTTGTTCCCGGA---54 %
LS-8/+5     CCCGCCCCGTCCCGCCCCCTCCCGGCCTCTCCTCGTGCTGGATCGCTCCCTAAGGACGCCCCGCTCCAGAAGCCCCACCTACCAAGGACGCgaagatCgatcTtCCGGA---52 %

                 -30       -20       -10       +1        +10       +20       +30       +40       +50       +60      +50
                                               ↓↓
wt          CCGCTCCAGAAGCCCCACCTACCAAGGACGCCCCACCCTTGTTCCCGGACTGGTATGACCCCAGCCTGCTGACATCCCTCCGCCCTTTCTCGTGCACGCGGCTATGGCG-100%
LS-3/+6     CCGCTCCAGAAGCCCCACCTACCAAGGACGCCCCACgaagaTcttCGGACTGGTATGACCCCAGCCTGCTGACATCCCTCCGCCCTTTCTCGTGCACGCGGCTATGGCG-100%
LS-2/+8     CCGCTCCAGAAGCCCCACCTACCAAGGACGCCCCACCgaaGaTCttcGACTGGTATGACCCCAGCCTGCTGACATCCCTCCGCCCTTTCTCGTGCACGCGGCTATGGCG--65 %
LS-1/+9     CCGCTCCAGAAGCCCCACCTACCAAGGACGCCCCACCCgaagatCttcACTGGTATGACCCCAGCCTGCTGACATCCCTCCGCCCTTTCTCGTGCACGCGGCTATGGCG--38 %
LS+9/+7     CCGCTCCAGAAGCCCCACCTACCAAGGACGCCCCACCCTTGTTCCCGaAgatcTtcGACCCCAGCCTGCTGACATCCCTCCGCCCTTTCTCGTGCACGCGGCTATGGCG--42 %
LS+10/+19   CCGCTCCAGAAGCCCCACCTACCAAGGACGCCCCACCCTTGTTCCCGGgaaGaTcTtcCCCCAGCCTGCTGACATCCCTCCGCCCTTTCTCGTGCACGCGGCTATGGCG--45 %
LS+15/+22   CCGCTCCAGAAGCCCCACCTACCAAGGACGCCCCACCCTTGTTCCCGGACTGGaAgatCttCAGCCTGCTGACATCCCTCCGCCCTTTCTCGTGCACGCGGCTATGGCG--57 %
LS+22/+31   CCGCTCCAGAAGCCCCACCTACCAAGGACGCCCCACCCTTGTTCCCGGACTGGTATGACCgaAGatcttcGACATCCCTCCGCCCTTTCTCGTGCACGCGGCTATGGCG-148%
LS+30/+31   CCGCTCCAGAAGCCCCACCTACCAAGGACGCCCCACCCTTGTTCCCGGACTGGTATGACCCCAGCCTGaaGACATCCCTCCGCCCTTTCTCGTGCACGCGGCTATGGCG-86 %
LS+30/+37   CCGCTCCAGAAGCCCCACCTACCAAGGACGCCCCACCCTTGTTCCCGGACTGGTATGACCCCAGCCTGaaGAtcTtCCTCCGCCCTTTCTCGTGCACGCGGCTATGGCG--94 %
LS+30/+43   CCGCTCCAGAAGCCCCACCTACCAAGGACGCCCCACCCTTGTTCCCGGACTGGTATGACCCCAGCCTGgaagatcgatcttcCCCTTTCTCGTGCACGCGGCTATGGCG--50 %
LS+40/+49   CCGCTCCAGAAGCCCCACCTACCAAGGACGCCCCACCCTTGTTCCCGGACTGGTATGACCCCAGCCTGCTGACATCCCgaaGatCTTcCTCGTGCACGCGGCTATGGCG-103%
LS+46/+53   CCGCTCCAGAAGCCCCACCTACCAAGGACGCCCCACCCTTGTTCCCGGACTGGTATGACCCCAGCCTGCTGACATCCCTCCGCCgaagaTCtTcCACGCGGCTATGGCG--58 %
LS+58/+67   CCGCTCCAGAAGCCCCACCTACCAAGGACGCCCCACCCTTGTTCCCGGACTGGTATGACCCCAGCCTGCTGACATCCCTCCGCCCTTTCTCGTGCAgaaGatctTcGCG---4 %
△-33/+6     CCGCTC ---------------------------------------- CGGACTGGTATGACCCCAGCCTGCTGACATCCCTCCGCCCTTTCTCGTGCACGCGGCTATGGCG-28 %
△-3/+19     CCGCTCCAGAAGCCCCACCTACCAAGGACGCCCCAC ------------------------ CCCCAGCCTGCTGACATCCCTCCGCCCTTTCTCGTGCACGCGGCTATGGCG--56%
△-33/+19    CCGCTC ------------------------------------------------------ CCCCAGCCTGCTGACATCCCTCCGCCCTTTCTCGTGCACGCGGCTATGGCG-40%
```

Figure1. The linker-scanning and deletion mutants and their APRT activities. Plasmids carrying mutated APRT genes were cotransfected with pSV2CAT to CHO APRT- cells. APRT and CAT activities were then measured. CAT activity was used as an internal control to correlate the APRT activity. 100% denotes the expression of a wild type APRT gene. The arrows indicate the transcription initiation sites of the wild type APRT gene. The boxed regions are consensus Sp1 binding sequences. The nucleotides printed in darkened little letters are mutated nucleotides. The dashes indicate deleted regions. The underlined ATG is the translation initiation codon.

Gel-shift and DNase1-footprinting assays. The 403 bp BamHI-HinfI fragment of CHO APRT gene was isolated and end labeled by T4 DNA polymerase. The labeled fragment was isolated and used as the probe for both gel-shift and footprinting assays (Fig.2, A). For gel-shift assay $2x10^4$ cpm of probe, 4 ng of Sp1 (purchased from Promega) and the competitor DNA were mixed in Sp1 binding buffer and incubated in ice for 10 min. Free probe and Sp1-bound probe were resolved in 3.5% polyacrylamide gel with TB buffer (40 mM Tris-base, borate). The DNase1-footprinting assay was performed following the methods described by Briggs et al.[5] except that the DNase1 digestion was performed in ice for 50 seconds with the concentration of DNase1 at 1 ug/ml.

Results and Discussion

To assay the expression of APRT mutants, 2 ug of the mutant DNA on puc19 and 2ug of pSV2CAT were cotransfected into CHO APRT- cells. Both APRT and CAT activities were measured using in vitro enzyme assays 48-72 hours after transfection. The CAT activity was used as an internal control to measure efficiency of transfection, and correlate APRT activity.

Mutations at the transcription initiation sites. The transcription of CHO APRT gene initiates predominantly at two sites, 64 and 65 bp upstream of the translation initiation codon, and are denoted as +1 and -1 respectively. A mutant, LS -3/+6, in which both initiation sites are mutated (T → G and T → A; Fig. 1) retained 100% expression of the wild type gene, indicating the flexibility of the selection of transcription initiaion nucleotides. We are currently defining the initiation sites in this mutant.

Mutations on the translation initiation codon. The original translation initiation codon, AUG, was changed to UUC in LS +58/+67. This mutant had very low, but detectable (4% of the wild type, Fig.1) APRT activity. There are only two additonal AUG codons in the 5' untranslated region and 1st exon of APRT mRNA. Translation initiates at either AUG codon will result in a frameshift and totally destroy the enzyme function. The residual enzyme activity could come from a candidate GUG codon. It has been shown that in a eukaryotic system the GUG codon can serve as the initiation codon at low efficiency. Several GUG codons were found in the vicinity of and in-frame with the original AUG codon.

Mutations in the region between nucleotides -33 and +22. A series of mutants (LS -33/-15 to LS +15/+22), which were mutated in the region from nucleotides -33 to +22, lost 40-86% of gene expression (Fig. 1). This region, including transcription start sites and the flanking regions, probably is the binding site for RNA polymerase and other auxiliary factors. A preliminary DNase1-footprinting study using crude nuclear extracts of CHO cells showed protection from DNase1-digestion in this region (data not shown). To further test the importance of this region in gene regulation we constructed a mutant with the deletion of this region (from nucleotides -33 to +19) and assayed the gene expression. To our surprise, this mutant still possessed 40% expression of the wild type gene (Fig.1), raising the possibility that either the RNA polymerase does not bind to the immediate vicinity of the transcription initiation site or the selection of the binding site is not very strict, thus in the absence of the original binding site the polymerase can bind to other cryptic binding sites. The analysis of the transcription initiation sites of this deletion mutant is also underway.

Mutations in the regions between nucleotides -90 and -78 and between nucleotides -48 and-40. Two mutants, LS -90/-78 and LS -48/-40, which were mutated in the regions between nucleotides -90 and-78 and nucleotides -48 and-40 repectively expressed a twofold increase in APRT activity compared to the wild type gene. This result suggests the presence of negative regulators.

The interaction between Sp1 and the APRT promoter Sp1 has been shown to activate transcription in both tissue specific and housekeeping genes. The activating

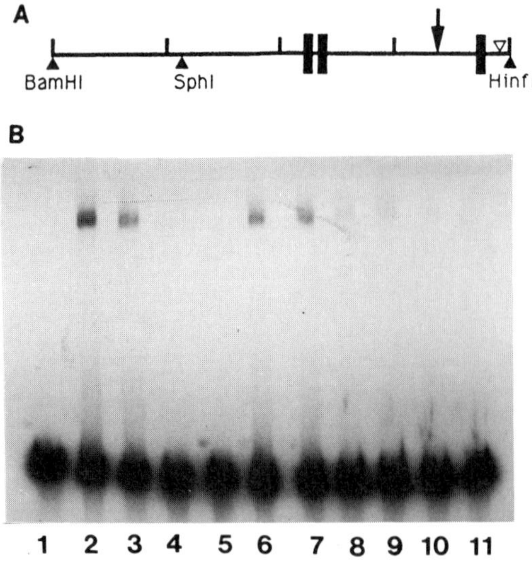

Figure2. The gel-shift assay of Sp1-binding on APRT promoter
The 290 bp SphI-HinfI fragment was used as probe for gel-shift and DNase1-footprint assays. (A) The solid bars represent the consensus Sp1-binding sequences. The arrows indicate the transcription initiation sites while the white triangle indicates the translation initiation codon. (B) The probe (1.7ng) was mixed with 4 ng Sp1 and competitor DNA as indicated in Sp1-binding buffer and resolved in 3.5% polyacrylamide gel with Tris-borate buffer. lane1: probe, lane2: probe+ Sp1, lane3: probe+ Sp1 + 4ng unlabeled probe, lane4: probe + Sp1+ 40 ng unlabeled probe, lane5: probe +Sp1+ 160ng unlabeled probe, lane6-lane11: probe+ Sp1+ salmon sperm DNA at the following amount-- lane6: 40ng, lane7: 160ng, lane8: 0.5ug, lane9: 1.0 ug, lane10: 1.5 ug, lane11: 2.0 ug.

mechanism remains uncertain. The consensus Sp1-binding core sequence has been shown to be CCGCCC or GGGCGG while the sequences flanking the core sequence appear to affect the binding affinity. The 5' region of CHO APRT gene contains three consensus Sp1-binding sequences (CCCGCC). Two of these are located upstream to the transcription start sites. The other is down stream to the start sites. To study the interaction between Sp1 and the APRT promoter we have performed gel-shift and DNase1-footprinting assays. The probe used was a 290 bp SphI-HinfI fragment containing all three Sp1-binding sites (Fig. 2, A). 1.7 ng of labeled probe was incubated with 4 ng of Sp1 as well as different competitor DNA. The results showed that although 40 ng of unlabeled probe competed out the binding of Sp1 to the labeled probe, 2 ug of salmon sperm DNA was required to compete with the binding to the same extent, indicating the specific binding of Sp1 to the probe (Fig. 2, B). Furthermore DNase1-footprinting assays have been done to clarify the specific binding sites. We found that under the assay conditions Sp1 did not bind to the consensus Sp1-binding sequence downstream to the transcription start sites. However it did bind to the two upstream binding sequences (data not shown). Coincidentally, mutations in the downstream binding sequence did not affect the gene expression in vivo (Fig. 1, LS +40/+49) but the mutations on the two upstream binding sites increased or slightly decreased gene expression (Fig. 1, LS -90/-78 and LS -97/89). The result suggests that the downstream binding site has very low affinity for Sp1 (compared with the upstream binding site). We do not exclude the possibility that the binding pattern may be different in vivo. Sp1 seems to interact with other transacting factors in order to function, hence only when the other transacting factors regulating the APRT gene are known can we clarify the actual roles of Sp1 in the regulation of APRT gene.

The linker-scanning and deletion analysis data presented suggest that the expression of the APRT gene might be tightly controlled by multiple negative and positive regulatory elements to ensure the constitutive but limited expression. We are currently analyzing the quantities and the 5' ends of the mRNA produced by different mutants. A major challenge will be the identification of the transacting regulatory factors.

References

1. M. Kozak, Mol. Cell. Biol. 9:5073 (1989).
2. J. Park and M.W. Taylor, mol. Cell. Biol. 8:2536 (1988).
3. D.J.Sussman and G. Milman, Mol. Cell. Biol.4:1641 (1984).

4. C.M. Gorman, L.F. Moffat and B.H. Howard, Mol. Cell. Biol. 2:1044 (1982).
5. M.R. Briggs, J.T. Kadonaga, S.P. Bell and R. Tjian, Science234:47 (1986).
6. D. Gidoni, W.S. Dynan and R. Tjian, Nature 312:409 (1984).
7. J.T. Kadonaga, K.A. Jones and R. Tjian, TIBS11:20 (1986).

A SPLICE MUTATION AT THE ADENINE PHOSPHORIBOSYLTRANSFERASE LOCUS DETECTED IN A GERMAN FAMILY

Birgit S. Gathof[1], Amrik Sahota[2], Ursula Gresser[1], Ju Chen[2], Peter S. Stambrook[3], Jay A. Tischfield[2], Nepomuk Zöllner[1]

[1]Medizinische Poliklinik, Pettenkoferstr. 8a, D-8000 München 2, [2]Department of Molecular and Medical Genetics, University of Indianapolis, Indiana, USA, [3]Department of Anatomy and Cell Biology, Cincinnati College of Medicine, Cincinnati, USA

INTRODUCTION

Approximately 40 cases of type I adenine phosphoribosyltransferase (APRT) deficiency (complete deficiency) have been described. About 20 different mutations have been identified in patients with this disorder (Hidaka et al. 1987, Chen et al. 1991, Sahota et al. 1991). In Germany, five other cases of type I APRT deficiency have been found, two of Turkish and three of unknown ethnic origin. Our group has described a type I Caucasian family from southern Germany (Zöllner and Gresser 1990). The aim of the present study was to investigate the molecular nature of the mutation in this family (Gathof et al. 1991a).

PATIENTS AND METHODS

The family examined consists of twin sons and their non-consanguineous parents. Monozygosity of twins has been confirmed by forensic hematologic tests (Gathof et al. 1991b). According to enzyme activities the twins have complete APRT deficiency, whereas the parents are heterozygotes.

Genomic DNA was isolated from blood of the parents and the twins. The APRT gene was amplified by polymerase chain reaction (PCR), subcloned into M13 and sequenced (Chen et al. 1991). For RFLP analysis, PCR and genomic DNA were digested with Taq I and Sph I, respectively.

RESULTS

The father was heterozygous for the Taq I RFLP and the mother for the Sph I RFLP. The twins were homozygous for the more common RFLP for both of these enzymes. Thus, the patients are homozygous, rather than hemizygous, for APRT deficiency.

Two clones from the father were sequenced completely. The sequence of both clones was identical to the wild-type sequence (Chen et al. 1991) with the exception of a single T insertion between bases 1831-1832 or 1832-1833. (Because the insertion produces adjacent thymidines, its exact position is indeterminable.) The same insertion was found in nine clones from the twins (7 from one and 2 from the other). These findings strongely suggest, that the patients are homozygous for this mutation.

DISCUSSION

The same mutation was found in another Caucasian patient and in one allele from an APRT heterozygote Caucasian (Sahota et al. unpublished). This mutation has been described previously in one of the alleles from two Belgian brothers who are compound heterozygotes for APRT deficiency (Hidaka et al. 1987). These patients are not known to be related to the family investigated in this study.

The mutation alters the exon four - intron four splice donor site from TG:GTAA to TG:GTTAA (insertion between 1831-1832) or TG:GTTAA (insertion between 1832-1833) and results in an A-to-T transversion at the third base downstream from the cleavage site (:). A purine base at this position is essential for normal splicing (Carothers et al. 1990). Aberrant splicing at this site causes the deletion of exon four from mRNA (Hidaka et al. 1987).

CONCLUSION

We have identified a splice donor mutation in APRT-deficient twin brothers from southern Germany. The same mutation has been found in patients from Belgium and in two other Caucasians. Thus, this mutation appears to be a common cause of APRT deficiency in the Caucasian population.

REFERENCES

Carothers AM, Urlaub G, Mucha J, Harvey RB, Chasin LA, Grunberger D (1990) Splicing mutations in the CHO DHFR gene preferentially induced by (+/-)-3 alpha, 4 beta-dihydroxy-1 alpha, 2 alpha-epoxy-1,2,3,4-tetrahydrobenzo[c]phenanthrene. Proc Natl Acad Sci USA 87: 5464-5468

Chen J, Sahota A, Stambrook P, Tischfield JA (1991) Polymerase chain reaction amplification and sequence analysis of human mutant adenine phosphoribosyltransferase gene: The nature and frequency of errors caused by Taq DNA polymerase. Mutat Res (in press)

Gathof BS, Sahota A, Chen J, Gresser U, Stambrook PS, Tischfield JA, Zöllner N (1991a) Identification of a common splice donor mutation at the APRT locus in a German family. 7th International / 3rd European Joint Symposium on Purine and Pyrimidine Metabolism in Man, Bournemouth. Int J Pur Pyr Res 2, Suppl 1: abs 48

Gathof BS, Sahota A, Gresser U, Chen J, Stambrook PJ, Tischfield JA, Zöllner N (1991b) Identification of a splice mutation at the adenine phosphoribosyltransferase locus in a German family. Klin Wochenschr (in press)

Hidaka Y, Pallela T, O'Toole TE, Tarlé SA, Kelley WN (1987) Human adenine phosphoribosyltransferase. Identification of allelic mutations at the nucleotide level as a cause of complete deficiency of the enzyme. J Clin Invest 80: 1409-1415

Sahota A, Chen J, Stambrook PJ, Tischfield JA (1991) Mutational basis of adenine phosphoribosyltransferase deficency. 7th International / 3rd European Joint Symposium on Purine and Pyrimidine Metabolism in Man, Bournemouth. Int J Pur Pyr Res 2, Suppl 1: abs 159

Zöllner N und Gresser U (1990) Nephrolithiasis in twins with APRT-deficiency. Stones as a marker of an inborn error of metabolism. Bildgebung / Imaging 57: 64-66

GERMLINE AND SOMATIC MUTATIONS LEADING TO ADENINE PHOSPHORIBOSYLTRANSFERASE (APRT) DEFICIENCY

Masayuki Hakoda, Naoyuki Kamatani, Sanae Ohtsuka and Sadao Kashiwazaki

Institute of Rheumatology, Tokyo Women's Medical College
NS BLDG, 2-4-1 Nishishinjuku, Shinjuku-ku, Tokyo, Japan

INTRODUCTION

Both germline and somatic mutations can lead to the deficiency of adenine phosphoribosyltransferase (APRT) in the human body. When the mutations occur in the germline, the mutational alleles are transmitted to the descendants and can cause a disease in the homozygotes. Thus, individuals having, in the germline, the defective APRT alleles in double dose develop 2,8-dihydroxyadenine (DHA) urolithiasis, and renal failure in severe cases (1). Germline APRT deficiency, therefore, is of significant medical interest.

Somatic mutations, however, do not cause any diseases but lead to the deficiency of APRT in progeny cells. These somatic cells deficient in APRT can be detected efficiently using a somatic cell genetics technique (2). Mechanisms of in vivo somatic mutations is of scientific interest since they are known to be closely associated with in vivo tumorigenesis and antibody diversity (3). We will show that in vivo somatic mutation is of medical interest also because heterozygosity for APRT deficiency can be diagnosed by the detection of somatic mutants. APRT is a rare genetic locus in which both germline and in vivo somatic mutations can be studied. Therefore, comparison of APRT gene mutations between germline and somatic cells is of great importance, and is likely to lead to a better understanding of mutations in general in the human body.

METHODS

Diagnosis of APRT deficiency: Diagnosis of homozygous APRT deficiency was done by our previously described T-cell method (4). Thus, when viable T-cells were resistant to 6-methylpurine and 2,6-diaminopurine (DAP), the subject was diagnosed as a homozygote. Diagnosis of heterozygotes were done by the somatic mutation method as described previously (2, 5). Thus, viable T-cells were cloned in the presence of DAP, and the subject was diagnosed as heterozygote when DAP resistant clones were obtained at frequencies around 10^{-5}.

Identification of a germline mutation: Polymerase chain reaction (PCR) technique was used to amplify a part of the germline APRT sequence. The amplified DNA was dot-blotted onto nylon membranes and hybridized with allele-specific oligonucleotide (ASO) probes. One probe had the 19-bp sequence of mutational human APRT gene including the *APRT*J* mutation site in exon 5, while the other had the normal 19-bp sequence of the same region (6). The conditions for the

Purine and Pyrimidine Metabolism in Man VII, Part B
Edited by R.A. Harkness *et al.*, Plenum Press, New York, 1991

labelling of the ASO probes with radioactive ATP, those for the hybridization of the probes and those for the washing were as described (6).
Southern blot analysis: Genomic DNA was extracted and digested with various restriction endonucleases, and submitted to the electrophoresis in agarose gel. The separated DNA fragments were blotted onto nylon membranes. The membranes were hybridized with the radiolabelled APRT cDNA probe (7). The membranes were washed and then exposed to X-ray films . After 1-3 weeks, the films were developed.
Selection of DAP-resistant T-cells: Perpheral blood mononuclear cells obtained were cultured in the presence of 0.7 μg/ml phytohemagglutinin, 0.5 ng/ml recombinant human interleukin 2, 10^4 cells/well of X-irradiated Raji cells, and 100 μM DAP in 96-well microtiter plates (2). Normal T-cells fail to proliferate in these conditions, but APRT-deficient cells do grow. The DAP-resistant T-cell clones thus otained could grow into enough cell numbers for the extraction of DNA.

RESULTS AND DISCUSSION

Genotypes of APRT deficient patients: 39 separate families with homozygous APRT deficiency were tested. By the PCR-ASO hybridization method as described in the Methods, the genotype of each family was determined. Table 1 shows the results (6).

Table 1. Genotypes of families with APRT deficiency

Genotype	type	No. of families
*APRT*J/APRT*J*	type II	24
*APRT*J/APRT*Q0*	type II	6
*APRT*Q0/APRT*Q0*	type I	9

Data from reference (6)

Table 2. Statistics of alleles with APRT deficiency

Allele	No.	Percentage
*APRT*J*	54	69%
*APRT*Q0*	24	31%

Data from reference (6)

The homozygotes were completely deficient in APRT activities in 9 families (type I deficiency), while in the remaining 30 families, the homozygotes were only partially deficient (type II) (6). Comparison between the genotypes and deficiency types in the families has demonstrated that all patients with a genotype *APRT*Q0/APRT*Q0* were of type I, while all with genotypes of *APRT*J/APRT*J* and *APRT*J/APRT*Q0* were of type II (6).
Southern blot analysis: There are two restriction fragment length polymorphism (RFLP) sites which are highly polymorphic among Japanese. One of them is detected by *Taq*I and is located in intron 2, while the other is located at the 5' flanking region of human APRT gene and detected by the *Sph*I cut (See genetic map in reference 6). These two RFLP sites (hTaq and Sph) are located about 1.1kb and 3.8kb, respectively, upstream of the *APRT*J* mutation site. Using the two RFLP sites, most of the normal APRT alleles of Japanese were classified almost equally into 4 haplotypes (6). The distribution of *APRT*J* alleles into the 4 haplotypes was quite uneven. Thus, 42 of the 48 *APRT*J* alleles had only one of the 4 haplotypes, and the remaining 6 alleles had another (6).
Cloning of DAP-resistant T-cells: When samples were from normal individuals, no DAP-resistant T-cell clones were obtained (5). When the samples were from 17 putative heterozygotes, however, the resistant clones were obtained at rather high frequencies (5). The frequencies were on the order of 10^{-4} (5). We analyzed DNA from each resistant clone and found that about 80% of all the resistant clones were allele-loss type mutants (5). Figure 1 illustrates a typical result in which mutant clones were obtained from an individual with a genotype of *APRT*1/APRT*J*. In this case, the germinally normal and mutated alleles could be clearly

differentiated by the Sourthern blot analysis. Thus, *APRT*J* allele showed a 2.8kb *Taq*I band while the other normal allele showed a 2.1kb band. In 42 DAP-resistant clones, loss of the 2.1kb band was observed in 33 (79%), while the remaining 9 clones, the normal allele was apparently intact (Figure 1) (5).

It is noteworthy that this type (allele-loss type) of somatic mutations are frequently seen in human neoplasms at such tumor-suppressing loci as RB (8). This type of somatic mutation is rare at another selectable locus hypoxanthine phosphoribosyltransferase (HPRT) probably because it is on the X-chromosome (9). Since all the tumor-suppressing gene loci are on autosomes, results of the in vivo somatic mutation at APRT locus but not those at HPRT locus are relevant to the tumor-causing mutation.

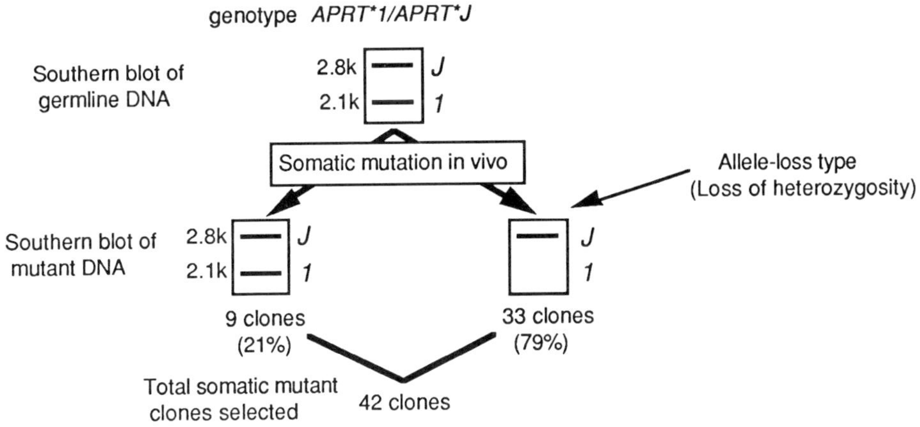

Figure 1. Mechanism of somatic mutation leading to homozygous APRT deficiency in a heterozygous individual with a genotype of *APRT*1/APRT*J* (data from reference 5).

<u>APRT deficient mutant cells in heterozygotes and normal individuals</u>: The above-mentioned results suggest that heterozygotes can be diagnosed by detecting somatic mutant cells resistant to DAP. But results will be ambiguous if some normal individuals also have high frequencies of DAP-resistant mutant cells. To exclude the possibility, we cloned DAP-resistant cells from 432 normal individuals. A total of 3.6×10^9 cells were used, but only 2 DAP-resistant clones were obtained (5). Calculated frequency of DAP-resistant T-cells in normal individuals was 8.4×10^{-9} (5), the value more than 10000 fold lower than the average frequency of DAP-resistant T-cells in heterozygotes (1.4×10^{-4}) (5). Therefore, misdiagnosis is unlikely.

Using this diagnostic method, we examined 13 family members of APRT deficient patients, and 8 of them were heterozygotes (5). Among 425 individuals tested, 2 had high frequencies of DAP-resistant T-cells. Later studies have shown that both of them were real heterozygotes, one having a *APRT*J* allele and the other having a *APRT*Q0* allele (5). We have diagnosed many family members of APRT deficiency patients using a few other methods like measuring enzyme activities in the hemolysates and detecting mutations using the PCR-ASO hybridization method. However, according to our experience, the present method detecting in vivo somatic mutational DAP-resistant T-cells is by far more reliable than other methods. Even if individuals are diagnosed as heterozygotes for APRT deficiency by the enzyme assay or PCR-ASO method, we confirm the diagnosis using the somatic mutation method.

<u>Why do different alleles of Japanese APRT deficient patients have a single mutation?</u>: There are three possible mechanisms for the same mutation in different alleles. Thus, (a) a hot spot (b) the same origin and (c) a limited number of mutations causing a disease may explain this. Our results that all *APRT*J* alleles had only two haplotypes indicate that all of them were probably derived from a

single origin. The presence of the less frequent haplotype in the *APRT*J* alleles is explained by crossovers between the *APRT*J* mutation site and Sph site. Although such events are considered to be quite infrequent, they will occur if they are maintained in a population for a long time. By assuming that the rate of crossovers is about the average, we have calculated the age of the *APRT*J* allele to be 130,000 years (6).

CONCLUSION

About 70% of all the germline APRT deficient alleles among Japanese derived from a single ancestor gene whose mutation occurred many years ago. Using heterozygotes, we identified somatic mutations at the same locus. About 80% of somatic mutant cells in vivo had lost the germinally normal allele. This method could be used for the accurate diagnosis of heterozygotes for APRT deficiency. APRT is the only autosomal locus at which both germline and in vivo somatic mutations can be studied, and the data obtained from our system will provide information as to how germline and somatic mutations occur in the human body.

REFERENCE

(1) H. A. Simmonds, A. S. Sahota and K. J. Van Acker, Adenine phosphoribosyltransferase deficiency and 2,8-dihydroxyadenine lithiasis., in "Metabolic Basis of Inherited Disease," 6th ed. C. R. Scriver, A. L. Beaudet, W. S. Sly and D. Valle, ed., McGraw-Hill, New York (1989) p. 1029

(2) M. Hakoda, K. Nishioka, and N. Kamatani, Homozygous Deficiency at an autosomal locus aprt in somatic cells in vivo Induced by two different (germinal-somatic and somatic-somatic) mechanisms., Cancer Res. 50:1738-1741 (1990)

(3) W. K. Cavenee, A. Kufos, M. F. Hansen, Recessive mutant genes predisposing to human cancer., Mutation Res. 168: 3 (1986)

(4) N. Kamatani, F. Takeuchi, Y. Nishida, H. Yamanaka, K. Nishioka, K. Tatara, S. Fujimori, K. Kaneko, I. Akaoka, and Y. Tofuku, Severe impairment in adenine metabolism with a partial deficiency of adenine phosphoribosyltransferase., Metabolism 34: 164 (1985)

(5) M. Hakoda, H. Yamanaka, N. Kamatani, and N. Kamatani, Diagnosis of heterozygous states for adenine phosphoribosyltransferase deficiency based on detection of in vivo somatic mutants in blood T cells: Application to screening of heterozygotes., Am. J. Hum. Genet. 48:522 (1991)

(6) N. Kamatani, S. Kuroshima, M. Hakoda, T. D. Palella, and Y. Hidaka, Crossovers within a short DNA sequence indicate a long evolutionary history of APRT*J mutation., Hum. Gent. 85:600 (1990)

(7) Y. Hidaka, S. A. Tarle, S. Fujimori, N. Kamatani, W. N. Kelley, and T. D. Palella, Human adenine phosphoribosyltransferase deficiency: Demonstration of a single mutant allele common to the Japanese., J. Clin. Invest. 81:945 (1988)

(8) W. F. Benedict, E. S. Srivastan, C. Mark, A. Banerjee, R. S. Sparkes, and L. Murphree, Complete or partial homozygosity of chromosome 13 in primary retinoblastoma., Cancer Res. 47:4189 (1987)

(9) M. Hakoda Y. Hirai, S. Kyoizumi, M. Akiyama, Molecular analyses of in vivo hprt mutant T cells from atomic bomb survivors., Environ. Mol. Mutagen., 13: 25 (1989)

LONG-TERM EVOLUTION OF TYPE 1 ADENINE PHOSPHORIBOSYLTRANSFERASE (APRT) DEFICIENCY

K.J. Van Acker and H.A. Simmonds

Department of Pediatrics, University of Antwerp, Belgium and Purine Research Laboratory, UMDS Guy's Hospital, London, U.K.

There are two types of inherited adenine phosphoribosyltransferase (APRT) deficiency: type I (APRT*Qo) in which enzyme activity is practically undetectable and which is seen predominantly in Caucasians, and type II (APRT*J) in which there is a partial deficiency of the enzyme and which, until now, has been found only in the Japanese. Most studies on clinical, biochemical and therapeutic aspects have been performed in type I APRT deficiency. From these studies it has become apparent that the only abnormality in these patients is the formation of 2,8.dihydroxyadenine (2,8 DHA) crystals in the urinary tract. It is their presence in the urinary tract and their possible accretion to stones which determines the clinical symptomatology and, through damage to the kidney, the prognosis. Treatment, therefore, is aimed at preventing 2,8.DHA stone formation: this may be obtained by measures such as forced diuresis and dietary purine restriction but in some patients treatment with allopurinol (AP) may be necessary.

The conclusions on the clinical, biochemical and prognostic aspects in type I APRT deficiency are based on observations with a relatively short duration. Whether these conclusions are also valid in the long term is uncertain. Our investigations during more than 15 years in two brothers who are homozygous for type I APRT deficiency and one of whom was treated with AP, now provide data on the long-term evolution.

In the youngest patient (patient 1) the diagnosis was made at the age of 2 1/2 years. The main clinical symptoms were abdominal colic and dysuria since birth. He was treated with a low purine diet and AP in a dose of 250 mg per day. This same dose was maintained during 9 years after which it was increased to 300 mg per day because of an increase in 2,8.DHA excretion. After 13 years the AP treatment was decreased to 100 mg per day and finally stopped 9 months later. The low purine diet was maintained. In his older brother (patient 2) type I APRT deficiency was diagnosed at the age of 6 1/2 years during investigation of the family. There were no complaints or symptoms. He was treated only with a low-purine diet. In both children investigations were performed on a yearly basis.

Purine and Pyrimidine Metabolism in Man VII, Part B
Edited by R.A. Harkness *et al.*, Plenum Press, New York, 1991

The clinical evolution in both patients was as follows. In patient 1 colics and dysuria subsided shortly after initiation of treatment and did not reappear after the AP treatment was stopped. Patient 2 never presented with colics or dysuria. Symptoms suggestive of gout were never observed in either patient. Frequent or severe infections were not a prominent feature. Growth and pubertal development were normal: at age 17, height was on percentile 97 in patient 1 and on percentile 90 in patient 2. Although not tested, intellectual development was considered to be within normal limits: both patients were able to attend school normally.

In patient 2, without AP treatment, 2,8.DHA crystals were always easily identified in the urine under polarized light. In patient 1 the crystals disappeared from the urine during the AP treatment but reappeared when this treatment was stopped. Ultrasonography and IV urography always revealed normal kidneys and urolithiasis or distension of the urinary tract was never observed in either patient. Glomerular filtration rate, measured by Cr_{51}EDTA clearance, remained normal after 14 years: it was 115 and 110 ml/min/1.73 m^2 in patient 1 and 2 respectively. Urinary concentrating capacity was always within the normal range. Anemia never developed in both patients. The bone marrow was examined on one occasion: all cell lines were normal and no 2,8.DHA crystals were seen.

In both patients plasma uric acid values increased after puberty but remained within normal limits (Fig. 1).

In patient 2, without AP treatment, the urinary excretion of total oxypurines (Fig. 2) increased with age, mainly as a consequence of increased uric acid excretion. In patient 1 the same

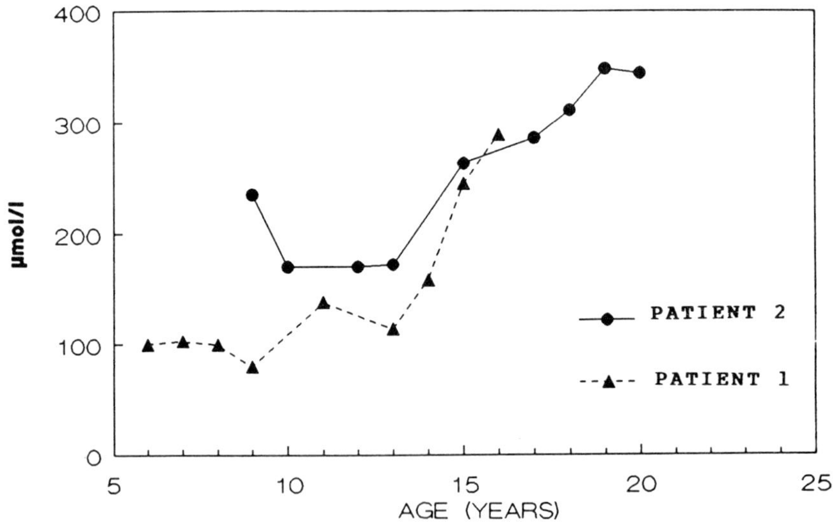

Fig. 1. Evolution of plasma uric acid in both patients.

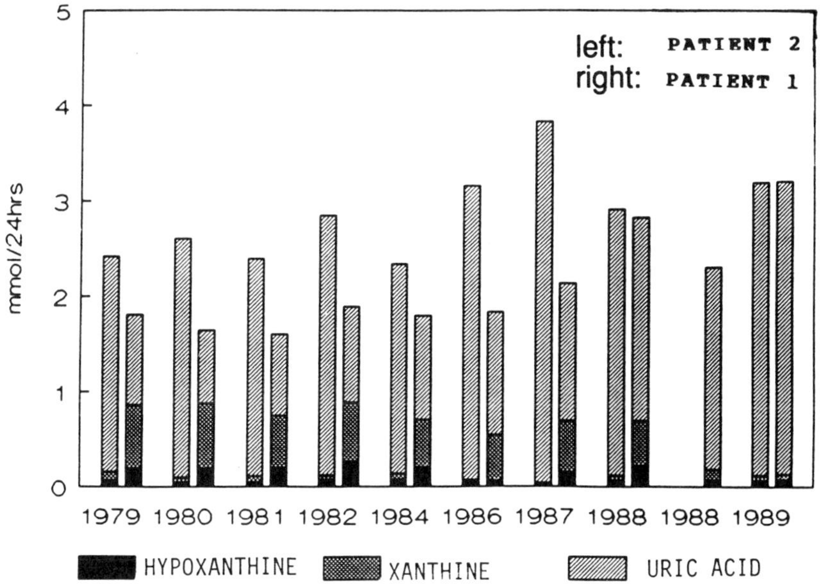

Fig. 2: Urinary excretion of total oxypurines in both patients

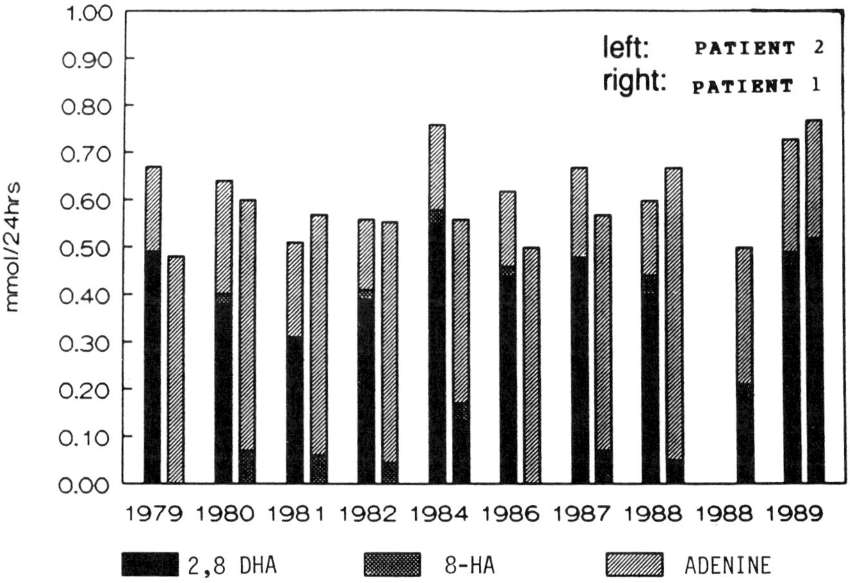

Fig. 3 . Urinary excretion of adenine metabolites in both patients

trend was observed from puberty on; the relative amounts of oxypurines differed, however, from those in patient 2 and reflected the influence of AP. After withdrawal of AP, oxypurines were excreted in the same amount and proportion as in his brother. During the early years the urinary excretion of adenine metabolites showed a similar trend as the excretion of oxypurines, especially in patient 1, but leveled off later (Fig 3). As expected, the relative amounts of the adenine metabolites differed in both patients: in patient 1 they reflected the AP treatment. After withdrawal of AP, the adenine metabolites were excreted in the same amount and proportion as in patient 2.

We conclude from our observations that 1° also in the long-term, type I APRT deficiency is a benign disorder if deposition of 2,8 DHA stones in the urinary tract can be avoided. Somatic and intellectual development are not impaired, the immune and hematopoietic system and the renal function remain normal and there is probably no 2,8 DHA deposition outside the urinary tract 2° while in some patients stone formation can be prevented by forced diuresis and low-purine diet, treatment with AP is necessary in others. Definitive withdrawal of AP treatment seems, however, possible after puberty 3° long-term treatment with AP and a low-purine diet does not seem to have adverse effects in children with normal renal function.

REFERENCE

1. Simmonds H.A., Sahota A.S. and Van Acker K.J., 1989, Adenine phosphoribosyltransferase deficiency and 2,8-dihydroxyadenine lithiasis, in: "Metabolic basis of inherited disease," C.R. Scriver, A.L. Beaudet, W.S. Sly et al., 6th ed., McGraw-Hill, New York, 1029-1044.

A STRATEGY FOR THE CREATION OF MUTATIONS IN HUMAN HPRT-cDNA AND THE EXPRESSION OF RECOMBINANT PROTEINS IN *E.COLI*

R. B. Gordon[1], M. L. Free[2], C. L. Gee[2], D. T. Keough[1], B. T. Emmerson[1] and J. de Jersey[2]

[1]Department of Medicine and [2]Department of Biochemistry, University of Queensland, Queensland, Australia

INTRODUCTION

The Lesch-Nyhan syndrome and gouty arthritis (due to the overproduction of urate), are manifestations of deficiencies of the purine salvage enzyme, HPRT. The molecular lesions responsible for the deficiency of this enzyme activity have been well characterised for numerous human patients. The observed mutations occur over the entire coding region of the HPRT gene, giving rise to proteins with altered amino acid sequence. However, knowledge of the location of these mutations has contributed little information about the active site of the enzyme. We are currently using chemical modification, together with site-directed mutagenesis, to study the relationship between structure and function of the human HPRT enzyme. To this end we have expressed the recombinant protein from normal human HPRT-cDNA sequence in *E.coli* (Free *et al.*, 1990). The recombinant protein was found to have similar properties to human HPRT isolated from erythrocyte and lymphoblast cells. A strategy, using PCR amplification (Saiki *et al.*,1988) and splicing by overlap extension (Horton *et al.*,1989), has been used to reproduce the natural mutations which occur in two patients with partial deficiency of HPRT activity. Expression of these mutant proteins was undertaken to (a) compare the properties of the recombinant and naturally occurring enzymes and (b) produce larger amounts of recombinant enzyme for more detailed studies.

METHODS

The expression system utilised was developed initially by Tabor and Richardson (1985). It consists of two compatible plasmids (Figure 1) transformed into *E.coli* Sϕ606 (a HPRT-negative strain; Jochimsen *et al.*1975). Plasmid pGP1-2 contains the gene for T7 RNA polymerase under the control of the lambda P_L promoter, which is repressed by the product of the cI-857 gene. The other plasmid, T7-7, contains the RNA polymerase promoter ϕ10, and human HPRT-cDNA sequence, derived from pHPT31, has been inserted downstream of a ribosome binding site. Expression of a fusion-less HPRT protein is induced by a temperature shift from 30°C to 42°C, which inactivates the heat sensitive repressor product from the cI-857 gene. Expression was measured by a continuous spectrophotometric assay (Keough *et al.*,1987).

Small fragments of DNA containing the desired mutation within the HPRT coding region are amplified by PCR and inserted in cassette form into the pT7-7/hHPRT plasmid. Exchange of these cassettes is facilitated by unique restriction endonuclease (RE) sites in the middle and at either end of the HPRT-cDNA sequence (XbaI, XhoI, HindIII, MscI). The strategy is depicted in Figure 2. The template used in the PCR was the parent plasmid, pT7-7/hHPRT. The PCR primers 2 and 3 are dictated by the position of the desired mutation. These two primers contain overlapping sequence which includes the mutation. Primers 1 and 4 were chosen to yield a short fragment of amplified DNA and encompass the two RE sites used for the cassette.

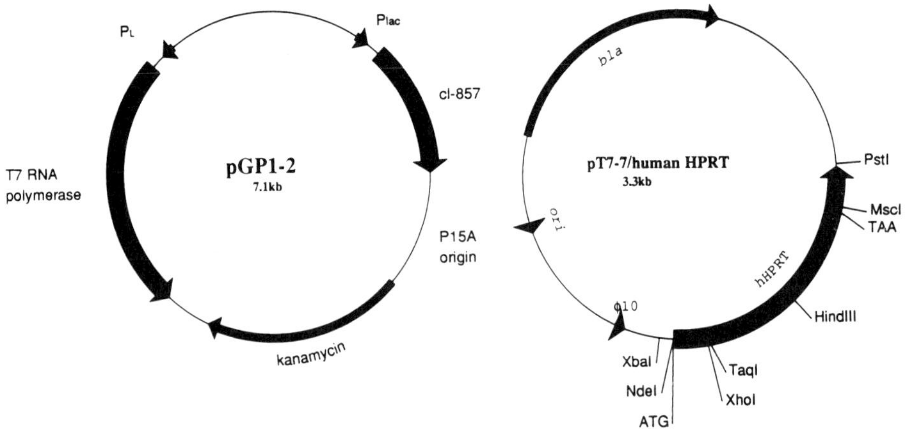

Figure 1. Details of the temperature inducible T7 RNA polymerase/promoter expression system (Tabor and Richardson 1985). Human HPRT-cDNA has been inserted between the NdeI and PstI sites of pT7-7 (Free *et al.*, 1990).

RESULTS

The above strategy was used to create the cDNA sequences coding for $HPRT_{LONDON}$ and $HPRT_{BRISBANE}$. $HPRT_{LONDON}$ has a C > T transition at nucleotide 428 (Davidson *et al.*, 1988), which predicts a serine > leucine substitution at amino acid 110 (confirmed by protein sequencing; Wilson *et al.*, 1983). $HPRT_{BRISBANE}$ has a C > T transition at nucleotide 602 and predicts an amino acid substitution of isoleucine for threonine at position 168 (Gordon *et al.*,1990). Nucleotide numbering is based on nt.100 being the A of the ATG codon for methionine (Jolly *et al.*, 1982).

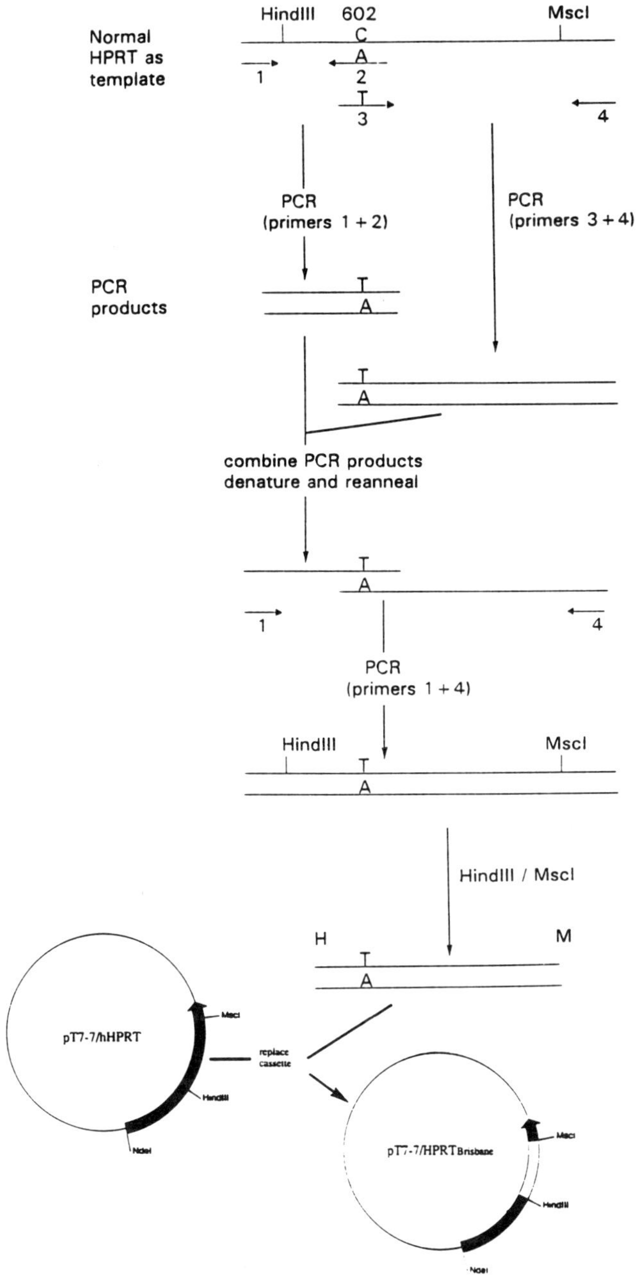

Figure 2. Strategy for creating mutations within the HPRT coding region, using PCR and splicing by overlap extension. The scheme depicts the creation of the C > T transition at nt. 602 of $HPRT_{BRISBANE}$.

Both recombinant proteins have been expressed in the *E.coli* Sϕ606 system. These recombinant enzymes have similar K_m values (Table 1) for the substrates as the naturally occurring mutant enzymes from lymphoblast cells of the HPRT-deficient patients.

Table 1. Comparison between the K_m values of the naturally occurring and recombinant proteins for normal and two mutant forms of human HPRT.

Name	Source[a]	K_m(Guanine) (μM)	K_m(*PRib-PP*) (μM)
Normal	L	2.4 ± 0.4	70 ± 7.0
	R	2.0 ± 0.5	65 ± 20
$HPRT_{LONDON}$	L	20[b]	10[b]
	R	4.4 ± 0.75	2 ± 0.1
$HPRT_{BRISBANE}$	L	2.5[c]	1209 ± 43
	R	2.0 ± 0.6	613 ± 100

[a] L = lymphoblast cells; R = recombinant protein
[b] K_m determined using hypoxanthine as the purine base; Wilson *et al.*,1983
[c] Determined on a human erythrocyte lysate

The above strategy for creating mutations within the human HPRT-cDNA sequence and the *E.coli* expression system will be used to study the relationship between structure and function of HPRT.

ACKNOWLEDGEMENTS

The authors thank the National Health and Medical Research Council of Australia and the Australian Research Council for grants supporting this research. We thank Dr. C.T. Caskey of Baylor College of Medicine, Houston USA for the gift of pHPT31 and Dr. Ifor Beacham of the Division of Science and Technology, Griffith University, Australia for a donation of plasmid pT7-7 and Sϕ606 cells.

REFERENCES

Davidson, B. L., Chin, S-J., Wilson, J. M., Kelley, W. N., and Palella, T. D., 1988, Hypoxanthine-guanine phosphoribosyltransferase: genetic evidence for identical mutations in two partially deficient subjects, *J. Clin. Invest.*, 82: 2164-2167.

Free, M. L., Gordon, R. B., Keough, D. T., Beacham, I. R., Emmerson, B. T., and de Jersey, J., 1990, Expression of active human hypoxanthine-guanine phosphoribosyltransferase in *Escherichia coli* and characterisation of the recombinant enzyme, *Biochim. Biophys. Acta*, 1087: 205-211.

Gordon, R. B., Sculley, D. G., Dawson, P. A., Beacham, I. R., and Emmerson, B. T., 1990, Identification of a single nucleotide substitution in the coding sequence of *in vitro* amplified cDNA from a patient with partial HPRT deficiency, *J. Inher. Met. Dis.*, 13; 692-700.

Horton, R. M., Hunt, H. D., Ho, S. N., Pullen, J. K., and Pease, L. R., 1989, Engineering hybrid genes without the use of restriction enzymes: gene splicing by overlap extension, *Gene*, 77: 61-68.

Jochimsen, B., Nygaard, P., and Vestergaard, T., 1975, Location on the chromosome of *Escherichia coli* of genes governing purine metabolism, *Mol. Gen. Genet.*, 143: 85-91.

Jolly, D. J., Okayama, H., Berg, P., Esty, A. C., Filpula, D., Bohlen, P., Johnson, G. G., Shively, J. E., Hunkapillar, T., and Friedmann, T., 1982, Isolation and characterization of a full-length expressible cDNA for human hypoxanthine phosphoribosyltransferase, *Proc. Natl. Acad. Sci. USA.*, 80: 477-481.

Keough, D. T., McConachie, L. A., Gordon, R. B., de Jersey, J., and Emmerson, B. T., 1987, Human hypoxanthine-guanine phosphoribosyltransferase. Development of a spectrophotometric assay and its use in detection and characterization of mutant forms, *Clin. Chim. Acta.*, 163: 301-308.

Saiki, R. K., Gelfand, D. H., Stoffel, S., Scharf, S. J., Higuchi, R., Horn, G. T., Mullis, K. B., and Erlich, H. A., 1988, Primer-directed enzymatic amplification of DNA with a thermostable DNA polymerase, *Science*, 239: 487-491.

Tabor, S., and Richardson, C. C., 1985, A bacterial T7 RNA polymerase/promoter system for controlled exclusive expression of specific genes, *Proc. Natl. Acad. Sci. USA.*, 82: 1074-1078.

Wilson, J. M., Tarr, G. E., and Kelley, W. N., 1983, Hypoxanthine-guanine phosphoribosyltransferase: an amino acid substitution in a mutant form of the enzyme isolated from a patient with gout, *Proc. Natl. Acad. Sci. USA.*, 80: 870-873.

MOLECULAR ANALYSIS OF HYPOXANTHINE-GUANINE PHOSPHORIBOSYLTRANSFERASE DEFICIENCY IN JAPANESE PATIENTS

Shin Fujimori, Tetsuo Tagaya, Noriko Yamaoka, [1] Naoyuki Kamatani and Ieo Akaoka

Second Department of Internal Medicine, University of Teikyo and [1] Institute of Rheumatology, Tokyo Women's Medical College Tokyo, Japan

Introduction

Hypoxanthine-guanine phosphoribosyltransferase (HPRT) is a purine salvage enzyme that catalyzes the conversion of hypoxanthine and guanine to IMP and GMP, respectively. Complete deficiency of HPRT causes the Lesch-Nyhan syndrome, which is characterized by hyperuricemia, mental retardation, choreoathetosis, and compulsive self-mutilation(1). Partial deficiency of HPRT leads to severe form of gout and nephrolithiasis(2). Several molecular analyses of HPRT deficiency in Caucasian patients have elucidated that the molecular abnormalities in HPRT deficiency are strikingly heterogeneous(3). Although an epidemiological survey has found more than 50 cases with Lesch-Nyhan syndrome in Japan(4), sequence-level analyses have not been performed well(5). We have identified two previously undescribed single point mutations and two unique deletions in the amplified HPRT cDNA derived from four unrelated Japanese patients with Lesch-Nyhan syndrome.

Materials and Methods

Subjects and Cell Lines

Clinical manifestations and HPRT enzyme activities in four patients were summarized in Table 1. Patient T.H., Y.Y. and T.S. presented with typical Lesch-Nyhan syndrome. Patient N.T. presented with mental retardation, hyperrefrexia and choreoathetosis but did not attempt to injure himself. Although HPRT activity was not completely absent in erythrocyte when determined by the routine enzyme assay, the viable T-lymphocyte derived from N.T. was resistant to 6-thioguanine, indicating that N.T. was a patient with complete HPRT deficiecy. Lymphoblastoid cell lines were established from these four patients using Epstein-Barr virus.

Synthesis and Sequencing of Mutant cDNA

Total cellular RNA was isolated from lymphoblastoid cells by the guanidium isothiocynate procedure. The two PCR primers (5'-CCGGtCGaCTCCGTTATG-3', 5'-AACTCAACTTGAAtTCTCATCTTA-3') were synthesized using a DNA synthesizer. The base substitutions (indicated by small letters) were introduced into the synthetic DNA as indicated to create artificial recognition sites for Sal I and

EcoR I, respectively. The DNA fragment amplified by using the set of primers should be 693 bp long which spanned the entire HPRT coding region. First strand of cDNA was synthesized from approximately 10 μg of total cellular RNA, using oligo d(T)15 as a primer and reverse transcriptase. The resulting sample was then treated with 0.2 N NaOH and passed through a Sephadex G-50 spun column. Using the single strand cDNA, thus yielded as template, HPRT double strand cDNA was synthesized by polymerase chain reaction(PCR), under standard conditions, as described(6). Thirty cycles of PCR amplification were performed with each cycle consisting of 94°C for 1 min, 55°C for 2 min, and 72°C for 3 min. The PCR amplified products were digested with Sal I and EcoR I and subcloned into M13 mp19. The recombinant clones were sequenced by the dideoxy chain termination method.

Table 1. Summary of clinical manifestations and HPRT activities in four patients

patients	mental retardation	self-mutilation (onset age)	serum uric acid (mg/dl)	HPRT activity in RBC (% control)
T.H.	+	+ (4 y.o.)	13.7	<0.7
Y.Y.	+	+ (2 y.o.)	8.0	<0.7
N.T.	+	±	17.0	5.0
T.S.	+	+ (2 y.o.)	8.5	< 0.7

Results

693 bp fragments of HPRT cDNAs were sufficiently amplified in patient T.H.and Y.Y.(WR 194). However only a 402 bp fragment was amplified in patient N.T. In patient T.S.(WR 170), two different size of fragments, 693 bp and 516 bp, were produced in the same degree (Fig 1). Sequencing of several recombinant clones showed a C to T substitution at base position 151 in patient T.H. (designated HPRT Fujimi) and a G to A change at position 419 in WR194 (designated HPRT Tokyo). The elimination of a Taq I site (TCGA to TTGA) in the HPRT coding sequence in HPRT Fujimi could prove the existence of the same base change in genomic DNA as well. A 291 bases corresponding to the entire second and third HPRT exon was deleted in patient N.T. Sequencing of 516 bp cDNA revealed a 77 bases deletion corresponding to the entire eight exon, however no nucleotide abnormalities were detected in the 693 bp cDNA derived from WR 170.

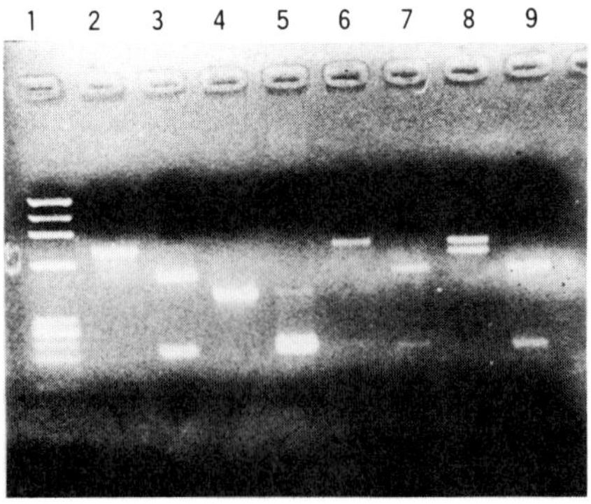

Figure 1. Ethidium bromide stained PCR products electrophoresed in 1.5% agarose gel. **1:** size marker, Hae III digests of Ψ × 174 DNA; **2, 3:** WR160 (normal B cell line); **4, 5:** N.T.; **6, 7:** T.H.; **8, 9:** WR170; **3, 5, 7, 9:** Hind III digests of PCR products.

Discussion

A C to T substitution generates a nonsense codon at amino acid position 51 in HPRT Fujimi. The position of the nucleotide substitution is exactly the same as a previously reported mutation HPRT Toronto(7). In the case of HPRT Toronto, the mutation caused a partial deficiency, whereas in HPRT Fujimi, it caused a complete deficiency. The calculation of the probability of finding substitution mutations at the same base position in the coding region of HPRT indicated that there was no evidence for the presence of a hot spot for substitution mutations in the human HPRT germ line(8). A G to A transition in HPRT Tokyo predicts a Gly to Asp change at the amino acid position 140, located within the putative 5-phosphoribosyl-1-pyrophosphate (PRPP) binding region(9). The increased hydrophilicity around the region of position 135-140 may explain the absence of enzyme activity in HPRT Tokyo. Missense mutations in human HPRT deficient patients thus far reported tend to accumulate in this functionally active region(10,11,12). The deletions in two HPRT variants from patient N.T. and WR 170 predict shortened translation products as well. In patient N.T., the deletion of exon 2 and 3 results in loss of 97 amino acids. All of exon 8 in WR 170 is deleted, leading to a change in the reading frame and the occurrence of a stop codon 15 nucleotides downstream the exon 7 and exon 9 junction. The similar mutation was detected in HPRT Connersville and RJK 888, which was identified in a Caucasian patient with Lesch-Nyhan syndrome(11,13). The deletions in HPRT cDNA that involve exon can arise from errors in RNA splicing. The further additional studies, such as Northern blot analysis and sequencing analysis of the intron in genomic DNA are required to clarify the primary genetic defects in these two variants. These different molecular defects in four Japanese HPRT deficient patients are considered to be the direct evidence of the genetic heterogeneity of HPRT deficiency among Japanese as well as Caucasian.

Table 2. Mutations in HPRT determined by PCR

Patients	Nucleotide changes	Putative amino acid changes
T.H.	C151→T	arg51→stop codon
Y.Y.	G419→A	gly140 →asp
N.T.	Deletion 28→347(exon 2,3)	Loss of 91 amino acids
T.S. (WR 170)	Deletion 532→ 609(exon 8)	Loss of phe178 to asn203

References

1 J. E. Seegmiller, F. M. Rosenbloom and W. N. Kelley, Enzyme defect associated with a sex-linked human neurological disorder and excessive purine synthesis. Science 155:1682 (1967).

2 W. N. Kelley, F. M. Rosenbloom, J. F. Henderson and J. E. Seegmiller, A specific enzyme defect in gout associated with overproduction of uric acid. Proc. Natl. Acad. Sci. USA 57:1735 (1967).

3 J. T. Stout and C. T. Caskey, Hypoxanthine phosporibosyltransferase: the Lesch-Nyhan syndrome and gouty arthritis, in The metabolic basis of inherited disease 6 th, ed. C. R. Scriver, A. L. Beaudet, W. S. Sly, and D. Valle, ed. , McGraw-Hill, New York, pp 1007 (1989).

4 S. Miwa, H. Fujii, K, Tani, T. Miyamoto and Y, Nishida, Lesch-Nyhan syndrome in Japan. Jpn. J. Exp. Med. 56:293 (1986).

5 T. Igarashi, M. Minami and Y. Nishida, Molecular analysis of hypoxanthine-guanine phosphoribosyltransferase mutations in five unrelated Japanese patients. Acta. Paediatr. Jpn. 31:303 (1989).
6 R. K. Saiki, D. H. Gelfand, S. Stoffel, S. J. Sharf, R. Higuchi, G. T. Horn, K. B. Mullis and H. A. Erlich, Primer-directed enzymatic amplification of DNA with a thermostable DNA polymerase. Science 239:487 (1988).
7 J. M. Wilson, P. Frossard, R. L. Nussbaum, C. T. Caskey and W. N. Kelly, Human hypoxanthine guanine phosphoribosyltransferase: detection of a.mutant allele by restriction endonuclease analysis. J. Clin. Invest.72:767 (1983).
8 S. Fujimori, N. Kamatani, Y. Nishida, N. Ogasawara and I. Akaoka, Hypoxanthine guanine phosphoribosyltransferase deficiency:nucleotide substitution causing Lesch-Nyhan syndrome identified for the first time among Japanese. Hum. Genet.84:483 (1990).
9 H. V. Hershey and M. W. Taylor, Nucleotide sequence and deduced amino acid sequence of Esherichia coli adenine phosphoribosyltransferase and comparison with other analogous enzymes. Gene 43:287 (1986).
10 B. L. Davidson, T. D. Palella and W. N. Kelley, Human hypoxanthine guanine phosphoribosyltransferase: a single nucleotide substitution in cDNA clones isolated from a patient with Lesch-Nyhan syndrome (HPRT Midland). Gene 68:85 (1988).
11 R. A. Gibbs, P. N. Nguyen, L. J. McBride, S. M. Koepf and C. T. Caskey, Identification of mutations leading to Lesch-Nyhan syndrome by automated direct DNA sequencing in vitro amplified cDNA. Proc. Natl. Acad. Sci. USA 86:1919 (1989).
12 S. Fujimori, Y. Hidaka, B. L. Davidson, T.D. Palella and W. N. Kelley, Identification of a single nucleotide change in a mutant gene for hypoxanthine-guanine phosphoribosyltransferase (HPRT Ann Arbor). Hum. Genet. 79:39 (1988).
13 B. L. Davidson, S. A. Tarle, T. D. Palella and W. N. Kelley, Molecular basis of hypoxanthine-guanine phosphoribosyltransferase deficiency in ten subjects determined by direct sequencing of amplified transcripts. J. Clin. Invest. 84:342 (1989).

EXPRESSION OF NORMAL AND VARIANT HUMAN HYPOXANTHINE-GUANINE PHOSPHORIBOSYLTRANSFERASE IN *E. COLI*

Beverly L. Davidson, Blake J. Roessler, and Thomas D. Palella

Department of Internal Medicine and the Rackham Arthritis Research Center, University of Michigan Medical School, Ann Arbor, Michigan 48109-0680 USA

INTRODUCTION

Hypoxanthine-guanine phosphoribosyltransferase (HPRT) is a purine salvage enzyme which catalyzes the conversion of hypoxanthine and guanine to their respective mononucleotide forms, inosine 5' monophosphate and guanosine 5' monophosphate. A deficiency of HPRT in humans has two distinct clinical consequences. Complete deficiency is associated with the Lesch-Nyhan syndrome, a disease characterized by hyperuricemia, hyperuricaciduria, spasticity, choreathetosis, and a bizarre tendency to self-mutilate. Partial HPRT deficiency leads to severe precocious gout .

Subjects deficient in HPRT are strikingly heterogeneous with regard to levels of enzyme activity, levels of protein expressed, and kinetic properties of the mutant enzymes. The three dimensional structure of both the normal and naturally occurring mutants would provide us with natural probes to analyze the mechanisms of substrate binding and catalysis by human HPRT. As a preliminary step towards these determinations, we have expressed normal and mutant human HPRT in *E. coli*, and purified the normal human protein to approximately ninety five per cent homogeneity.

RESULTS AND DISCUSSION

Synthesis of normal and mutant HPRT expression cassettes by the polymerase chain reaction and cloning into pSP72

We used the polymerase chain reaction to 'clamp on' sequences necessary for high expression levels in *E.coli*. These sequences include the bacteriophage gene 10 leader sequence (Olin et al., 1986), the bacterial Shine-Delgarno sequence, an AT rich spacer and an appropriate restriction enzyme site for cassette insertion into the expression vector.

First strand synthesis was done using total cellular RNA from B-lymphoblasts derived from a normal individual. The reaction mixture contained 1 μg total cellular RNA, 3 μg oligo $d(T)_{12}$ or an HPRT specific primer, 0.5 mM dNTP's, and 10 Units AMV reverse transcriptase. Reverse transcription was carried out at 41°C for 1 hour.

The primer containing expression cassette sequences is shown in Figure 1, and was used in combination with a 3' primer also containing a BAm HI cloning site. The expression cassette PCR consisted of 0.25 μg each primer, 0.5 mM dNTP's, and 10 μl first strand in 50 μl total volume. Thermocyler conditions were 5 cycles of 94°C 1 min, 22°C 1 min, 72°C 1 min, and 30 cycles of 94°C 1 min, 72°C 1 min. The PCR fragment was purified on low melting point agarose, cut with Bam HI, and further

Purine and Pyrimidine Metabolism in Man VII, Part B
Edited by R.A. Harkness *et al.*, Plenum Press, New York, 1991

purified on a Qiagen™ column. The resulting DNA was then cloned into Bam HI cut pSP72 plasmid in the proper orientation such that transcription off the T7 promoter results in human *hprt* transcript.

AGA GGA TCC AAT AAT TTT GTT TAA CTT TAA GAA GGA GAT ATA TCC *ATG GCG ACC*

Bam HI — Epsilon — Shine-Delgarno — Human HPRT Sequence

CGC AGC CCT GGC GTC

Figure 1. Expression cassette primer for HPRT. The significant sequences are underlined.

Expression of normal and mutant HPRT in *E. coli* strain BL21(DE3)

E. coli BL21(DE3) has a prophage containing the T7 polymerase gene under control of the lacUV5 promoter. Induction of transcription of T7 polymerase with IPTG will result in transcription of sequences under the control of the T7 promoter, in this case normal and mutant HPRT expression cassettes. Translation of these cassettes will result in the production of HPRT.

The plasmids containing normal HPRT (pSPHP3), $HPRT_{London}$ (pSPDB1), $HPRT_{Ashville}$ (pSPPC24), and $HPRT_{Ann\ Arbor}$ (pSPKC4) were transformed into BL21(DE3), and analyzed for induction and synthesis of human HPRT by SDS polyacrylamide gel electrophoresis (PAGE), western blotting, and native PAGE and enzyme assay. As seen in Figure 2, induction of pSPHP3 results in the production of significant quantities of HPRT in as little as 30 minutes and these levels increase up to 8 hours. Longer induction times did not significantly increase the amount of human HPRT produced (data not shown).

Activity gel assays and western blotting were done to check that the expressed protein was enzymatically active, and to confirm that pSPHP3 was producing authentic human HPRT. HPRT was active and exhibited normal electrophoretic migration when compared to HPRT from human B-lymphoblasts. Immunoreactivity was also normal (data not shown).

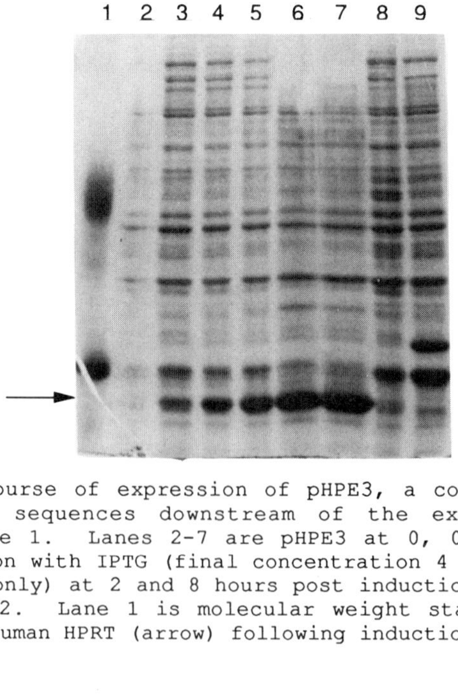

Figure 2. Time course of expression of pHPE3, a construct containing normal HPRT cDNA sequences downstream of the expression cassette described in Figure 1. Lanes 2-7 are pHPE3 at 0, 0.5, 1, 2,4, and 8 hours post induction with IPTG (final concentration 4 mM). Lane 8 and 9 are pSP72 (vector only) at 2 and 8 hours post induction. Lane 13 and 14 are uninduced pSP72. Lane 1 is molecular weight standards. Note the rapid increase of human HPRT (arrow) following induction with IPTG.

In order to use this technique for producing large amounts of protein for examining the structural and/or functional differences between native and mutant HPRT proteins, we had to insure ourselves that the mutant proteins would be physically indistinguishable from their lymphoblast counterparts. We chose three kinetic variants, to test this expression system, $HPRT_{London}$ (Davidson et al., 1988), $HPRT_{Ashville}$ (Davidson et al., 1989), and $HPRT_{Ann\ Arbor}$ (Fujimori et al., 1988). Variant HPRT proteins were expressed from plasmids pSPDB1,pSPPC24, and pSPKC4 and analyzed by SDS or native PAGE. HPRT from pSPDB1 migrates faster than HPRT from pSPHPE3 in SDS PAGE gels, as does $HPRT_{London}$ from B-lymphoblasts. $HPRT_{Ashville}$ demonstrates more anodal migration in native gels, as does HPRT from pSPPC24 (Figure 3). $HPRT_{Ann\ Arbor}$ has normal electrophoretic migration.

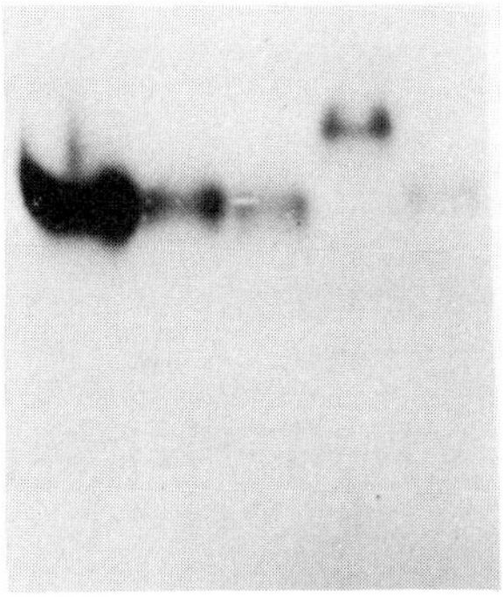

Figure 3. Native PAGE followed by direct enzyme assay of normal and mutant human HPRT expressed in *E. coli* . Aliquots of pHPE3, pHPPC21, pHPDB2 , and pHPKC4 were analyzed 4 hours post IPTG induction for human HPRT activity. Note the cathodal migration of pHPPC21 (lane 4) compared to pHPE3 (lane 5) or HPRT from a normal lymphoblast cell line (lane 1). Lanes 2 and 3 are HPRT from pHPDB1 and pHPKC4, respectively.

Purification of the expressed proteins

Following induction of pHPE3 by IPTG and incubation for 12 hours, bacterial cells were pelleted, washed, and resuspended in buffer A (100 mM Tris pH 7.4, 10 mM $MgCl_2$, 1 mM DTT). The cells were lysed by French press, and cellular debris removed by centrifugation. The supernatant was then loaded onto a Sephacryl S-300 molecular exclusion column. Fractions containing the peak of human HPRT activity was pooled, concentrated, and loaded onto a GMP agarose affinity column equilibrated in buffer A. After column washing, human HPRT was eluted with buffer A containing 5 mM GMP. Fractions containing HPRT activity were pooled and concentrated. As seen in Figure 4, this two-step purification results in greater than 95% pure human HPRT.

SUMMARY

1. ECPCR is a rapid and effective means for generating recombinant human HPRT.
2. The Bl21(DE3) T7 polymerase/T7 promoter system provides high level expression of human HPRT constructs after induction of the T7 polymerase gene with IPTG.
3. Human HPRT constructs expressed in *E.coli* mimic the variant properties originally demonstrated in lymphoblast extracts from affected individuals.
4. Human HPRT expressed in *E.coli* can be rapidly purified to near homogeneity by a two step purification scheme.

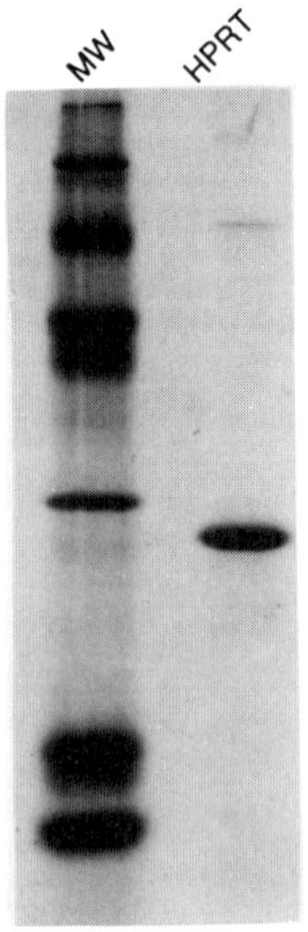

Figure 4. Purified human HPRT. A five microliter aliquot of a concentrated solution of recombinant human HPRT eluted from GMP agarose was analyzed by SDS-PAGE and stained with Coomasie blue. MW, molecular weight markers; HPRT, partially purified recombinant human HRPT.

REFERENCES

Olins, P. O., Devine, C. S., Rangwala, S. H., and Kavka, K.S., 1988, The T7 phage gene *10* leader RNA, a ribosome-binding site that dramatically enhances the expression of foreign genes in *Escherichia coli*, Gene, 73:227.

Davidson, B. L., Chen, S-J., Wilson, J. M., Kelley, W.N., and Palella, T. D., 1988, Hypoxanthine-guanine phosphoribosyltransferase. Genetic evidence for identical mutations in two partially deficient subjects, J Clin Invest, 82:2164.

Davidson, B. L., Pashmforoush, M., Kelley, W. N., and Palella, T. D., 1989, Human hypoxanthine guanine phosphoribosyltransferase deficiency: The molecular defect in a patient with gout ($HPRT_{Ashville}$), J Biol Chem, 264:520.

Fujimori, S., Hidaka, Y., Davidson, B.L., Palella, T.D., and Kelley, W.N., 1988, Identification of a single nucleotide change in a mutant HPRT gene ($HPRT_{Ann\ Arbor}$), Human Genet, 79:39.

HPRT GENE MUTATIONS IN A FEMALE LESCH-NYHAN PATIENT

Nobuaki Ogasawara, Yasukazu Yamada and Haruko Goto

Department of Genetics, Institute for Developmental Research, Aichi Prefectural Colony, Kasugai, Aichi 480-03, Japan

INTRODUCTION

Lesch-Nyhan disease is an X-linked recessive disorder characterized by hyperuricemia, physical and mental retardation, choreoathetosis, and compulsive self-mutilation. The disease is associated with absence of activity of an enzyme involved in purine metabolism, namely hypoxanthine guanine phosphoribosyl transferase (HPRT). HPRT is X-linked and because of inability of reproduction in Lesch-Nyhan patient, this syndrome occurs only in males. We have, however, an unusual case of a girl with Lesch-Nyhan syndrome.[1,2]

BACKGROUND OF THE STUDY

The patient was born in 1975. At the age of 5 years, she was diagnosed as Lesch-Nyhan syndrome because of the absence of HPRT activity in erythrocytes, T-lymphocytes and fibroblasts. The patient's karyotype was 46, XX and there was no abnormality such as X-autosome translocation. Very importantly, patient's mother is not heterozygous for a deficiency of HPRT.

Northern analysis showed no mRNA in the patient's B-lymphoblast. Although the patient had two X chromosomes, only one copy of HPRT gene was detectable by Southern analysis. The analyses of restriction fragment length polymorphism of DXS10, which was anonymous human X chromosome specific gene, can identify the maternal or paternal origin of X chromosome. To find out on which of the chromosome the HPRT gene is totally deleted, the patient's fibroblast was fused with TK^-mouse cell, and cells were selected in HAT medium. Southern analyses of DXS10 and HPRT gene of the hybrid clones showed that the clone having the maternal X chromosome alone deleted totally human HPRT gene, but the clone having the paternal X chromosome had human HPRT gene.[3]

There must be a mutation or a modification on the paternal X chromosome which is not detectable in Southern analysis yet results in non-functional gene, since Northern analysis showed no mRNA in the patient's lymphoblast. One clone having the maternal X chromosome alone expressed human glucose-6-phosphate dehydrogenase (G6PDH). Six independent clones having the paternal X chromosome alone did not expressed human G6PDH. These results indicate that the paternal X chromosome is always inactive and the maternal X chromosome always active. Thus, the most likely genetic and molecular mechanism of this rare female Lesch-Nyhan patient is the total HPRT gene deletion on the maternal X chromosome and the rare specific inactivation of the paternal X chromosome.[3]

RESULTS AND DISCUSSION

Previous studies showed that the most likely mechanism of the female Lesch-Nyhan patient is the HPRT gene deletion on the maternal X chromosome and the specific inactivation of the paternal X chromosome. However, the specific inactivation of X chromosome is only observed on the normal X chromosome in the female cells having balanced X-autosome translocation, or on the abnormal X chromosome in the cells having the X chromosome with the deletion and the normal X chromosome.

Very fortunately, the HPRT active gene can be distinguished from the inactive gene through changes in DNA methylation on the first intron.[4,5] Southern analysis after Bam HI digestion showed 10 kb band when the first intron probe was used. After Bam HI plus Hha I digestion, the male showed consistently 9.7 kb band, while the female 9.7 kb band and additional two smaller bands about 2 kb derived from the inactive HPRT gene. The female Lesch-Nyhan patient showed 9.7 kb band and two bands about 2 kb, indicating the presence of the active and inactive forms.

Bam HI plus Hpa II digestion showed 8.4 kb band in male, while 8.4 kb band 5.8 kb band in female, indicating that 8.4 kb band was derived from the active copy, and 5.8 kb band from the inactive copy. Female Lesch-Nyhan patient showed two band, 5.8 kb band and 10 kb band instead of active copy specific 8.4 kb band (Fig. 1). The presence of two distinct methylation patterns of the paternal gene indicates that the female patient has apparently two subpopulation of cells and the Lyonization occurs normally.

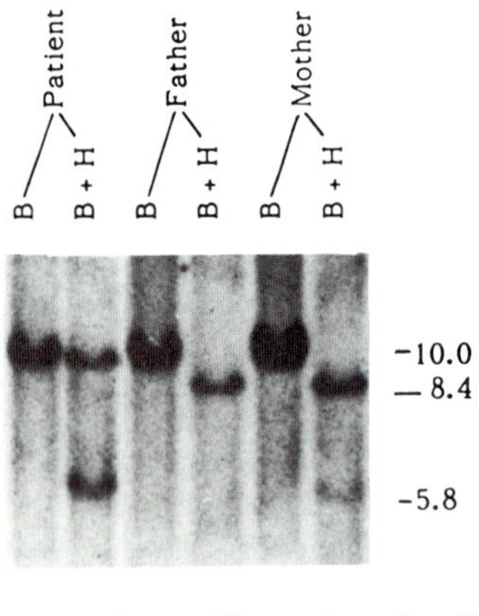

Fig. 1. Southern analyses after digestion by Bam HI (B) and Bam HI plus Hpa II (H), Bam HI - Pst I fragment at 5' region of 1st intron was used as the probe.

The loss of Hpa II digestion site specific for the active HPRT gene in this rare female patient can be derived from the methylation of Hpa II site or the mutation which can make Hpa II insensitive. The sequence analyses showed no mutation on the first intron of the patient, indicated the unusual methylation on specific Hpa II site on patient's first intron.

Hpa II sites located at 66 bp (H1), 704 bp (H2) 768 bp (H3) from the Bam HI site of the first intron. H2 and H3 are too close to be distinguished by Southern analysis. Therefore, it is not clear whether the active copy was digested by Hpa II at H2 or H3, or at both H2 and H3. As shown in Fig 2, PCR primers were designed and synthesized. Using a Hpa II digested male DNA as template, PCR product was obtained only by combination of A1-B2, but not by A2-B1. These results clearly indicate that H2 was methylated and H3 was not methylated in the active HPRT gene, but in patient both H2 and H3 sites were methylated.

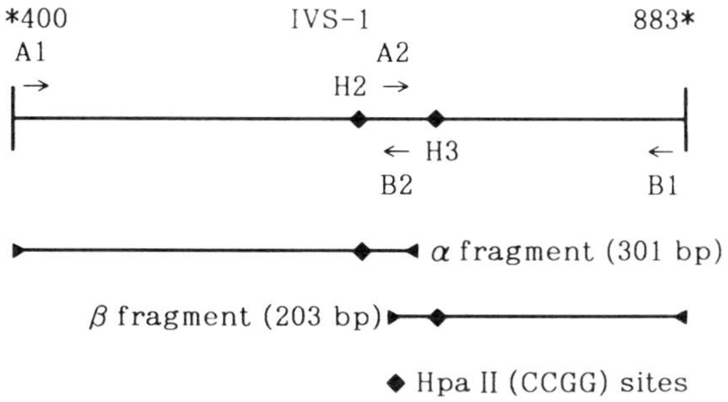

PCR primers

A1. 5'-GCT CAT GGC CTC ATT GAA GC-3' (400-419)
A2. 5'-GCA TGA TCA GAA CGG TTG AG-3' (681-700)
B1. 5'-CGA GGA CCT CTT ACA AGC CA-3' (883-864)
B2. 5'-CTC AAC CGT TCT GAT CAT GA-3' (700-681)

Fig. 2. Location of Hpa II sites (H2, H3), and position and sequence of synthesized PCR primers. If H2 is methylated, PCR product α can be obtained by A1-B2 combination. If H3 is methylated, product β can be produced by A2-B1.

Thus, the molecular mechanisms of this rare female Lesch-Nyhan patient are 1) a total maternal HPRT gene deletion and 2) a gene inactivation due to the methylation at the active copy special Hpa II site on the first intron of the paternal HPRT gene.

ACKNOWLEDGMENTS

This work was supported by a Gout Research Foundation grant and an Intractable Disease grant from Ministry of Health and Welfare of Japan.

REFERENCES

1. K. Hara, S. Kashiwamata, N. Ogasawara, H. Ohishi, R. Natsume, T. Yamanaka. S. Hakamada, S. Miyazaki, and K. Watanabe, A female case of the Lesch-Nyhan syndrome, *Tohoku J. Exp. Med.* 137:275 (1982).
2. N. Ogasawara, S. Kashiwamata, H. Oishi, K. Hara, K. Watanabe, S. Miyazaki, T. Kumagai, and S. Hakamada, Hypoxanthine-guanine phosphoribosyl transferase (HGPRT) deficiency in a girl, *Adv. Exp. Med.* 165A:13 (1984).
3. N. Ogasawara, J.T. Stout, H. Goto, S. Sonta, A. Matsumoto, and C.T. Caskey, Molecular analysis of a female Lesch-Nyhan patient, *J. Clin. Invest.* 84: 1024 (1989).
4. P.H. Yen, P. Patel, A.C. Chinault, T. Mohandas and L.J. Shapiro, Differential methylation of hypoxanthine phosphoribosyltransferase gene on active and inactive human X chromosomes, *Proc. Natl. Acad. Sci. USA,* 81:1759 (1984).
5. S.F. Wolf, D.J. Jolly, K.D. Lunnen, and T. Friedmann, Methylation of the hypoxanthine phosphoribosyltransferase locus on the human X chromosome: Implications for X-chromosome inactivation, *Proc. Natl. Acad. Sci. USA,* 81:2806 (1984).

MOLECULAR ANALYSIS OF HUMAN *hprt* GENE DELETIONS AND DUPLICATIONS

Raymond J. Monnat, Jr., Alden F.M. Hackmann, Teresa A. Chiaverotti, and Grace A. Maresh*

Department of Pathology SM-30, University of Washington, Seattle, WA 98195 and *Bristol Myers-Squibb Research Institute, Seattle, WA 98125

Deletion and duplication mutations play important roles in heritable human disease, in tumorigenesis and tumor progression, and in genome evolution. The mutation pathways that generate these rearrangements are not as yet well understood, in part due to the molecular scale on which deletions and duplications can occur (from 1bp to >1Mb), which often falls inconveniently between facile molecular and cytogenetic analytical techniques. We have been defining the nucleotide sequence substrates for and products of deletion and duplication mutagenesis in the human *hprt* gene as a first step to understanding how these mutation pathways operate in human cells.

We used a common approach to locate deletion and duplication breakpoints in the human *hprt* gene and recover rearrangement junctions for DNA sequence analysis. Fine-structure Southern and, in the case of duplications, partial digest-end label (Smith-Birnstiel) blot hybridization mapping were used to locate rearrangement breakpoints in the *hprt* gene. Oligonucleotide primer pairs that flanked the location of predicted rearrangement junctions were then used to amplify deletion and duplication novel junctions from mutant DNAs for sequence analysis.

DELETIONS

The structure of the *hprt* gene in 10 independent intragenic human *hprt* deletion mutants was determined by this strategy. These mutants were isolated as spontaneous thioguanine(TG)-resistant forward mutations in an SV40-transformed Werner syndrome cell line (PSV-811)[1] or in the HL-60 human myeloid leukemia cell line[2]. We have previously demonstrated a spontaneous deletion mutator phenotype in cell lines derived from patients with Werner syndrome (WS), a rare autosomal recessive disorder with features of premature aging[1]. These 10 deletions span a wide size range (57bp to 19.3kb) and involve a majority of the *hprt* locus (44kb of DNA and 6 of the 9 human hprt exons; Figure 1). The nucleotide sequence analysis of these mutations revealed three types of deletion junction: (1) five simple deletion junctions, which had 1-5bp of nucleotide sequence identity between donor sequences at the deletion junction, though little sequence identity either 5′ or 3′ to the junction; (2) two compound (deletion-

Purine and Pyrimidine Metabolism in Man VII, Part B
Edited by R.A. Harkness *et al.*, Plenum Press, New York, 1991

insertion) deletion junctions, which contained the insertion of 4 contiguous nucleotides not found in either donor duplex at the junction. These additional "orphan" or "filler" nucleotides were AT-rich (6 Ts, 1 A and 1 G), and created exact, direct 6 or 9bp direct repeats with sequences found 4 and 1bp downstream (3´), respectively. These direct repeats may have arisen by slipped mispairing at a replication fork of similar, closely spaced DNA sequences; and (3) complex deletion junctions consisting of tandem duplications of from 57-1991bp that were associated with no, 1 or 2 DNA inversions of 20-716bp.

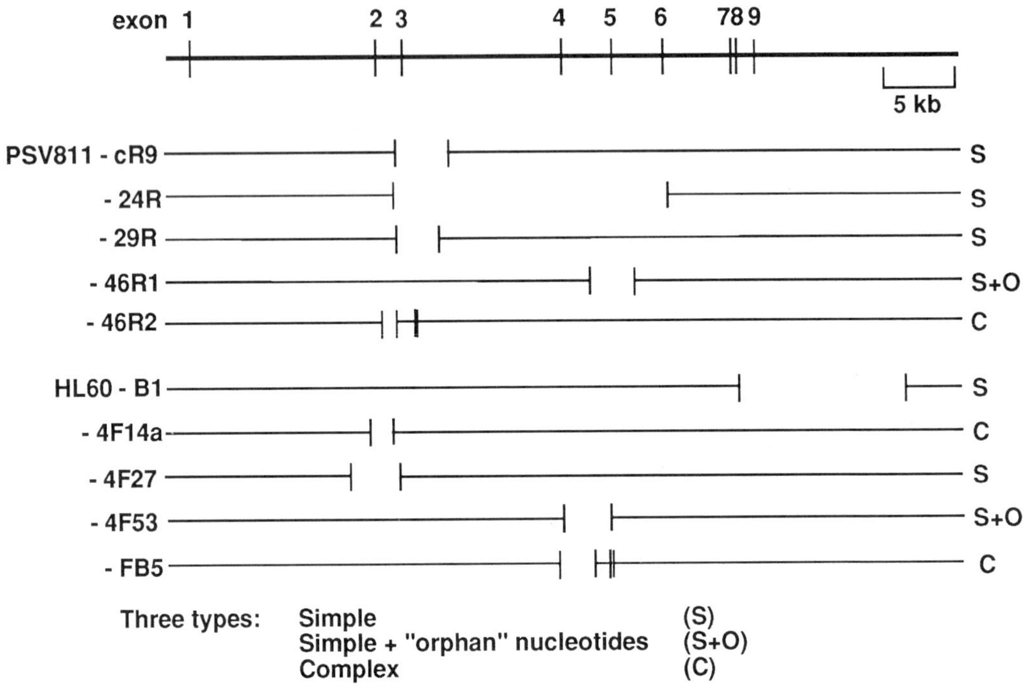

Figure 1. Sequenced human hprt deletions. The structure of the human *hprt* gene is shown at top with the positions of exons marked by vertical lines(┼). Deleted segments of the *hprt* gene in each mutant are indicated by blank spaces bounded by verticals(|). Individual mutant designations are given to the left, and deletion junction types to the right, of each mutant *hprt* gene.

All but one of the deletion junctions we have examined appear to have arisen by illegitimate recombination between *hprt* sequences having little nucleotide sequence identity. The single exception is the shortest deletion, a curious 57bp intra-*Alu* deletion in mutant HL60-4F14a that contains a 42bp region of perfect nucleotide sequence identity between donor duplexes immediately 5´ to the junction, a 6bp junction containing 3 inserted "orphan" nucleotides, and a 3´ flanking region containing little sequence identity between donor duplexes. The only consistent sequence motif we identified in the 200bp of deletion donor and junction DNAs centered on the junction was a 2-4-fold elevation in observed vs. expected frequencies of polypurine/polypyrimidine and alternating purine-pyrimidine tracts >5bp in length. There was no consistent association of these elements with breakpoints or junctions, however.

DUPLICATIONS

The three independent duplications we have studied were derived from HL-60 cells (3F22 and HG4;[2]) and from a patient with Lesch-Nyhan syndrome (GM6804;[3]). We determined the genetic stability of each of these duplications by quantifying the HAT resistance reversion rate using a modification of the Luria-Delbrück fluctuation test. All three mutants were genetically unstable, having HAT reversion rates of $4x10^{-6}$ (HG4) to $9.6x10^{-5}$(GM6804)/cell/generation. These reversion rates, which were corrected for the colony forming efficiency of each mutant, are approximately 100-fold higher than estimates of the forward mutation rate for duplication formation at the human *hprt* locus[2].

The molecular structures of these 3 duplications were determined using a strategy similar to that used to map and sequence deletion breakpoints and junctions (Figure 2). We also used a modification of the partial digest-end label mapping method of Smith and Birnstiel[4] to determine *hprt* gene structure in each mutant by restriction mapping from flanking, unduplicated DNA into duplicated regions. This mapping method can be performed with unfractionated mutant DNA, and thus should aid the analysis of other X- or Y-chromosomal gene rearrangements.

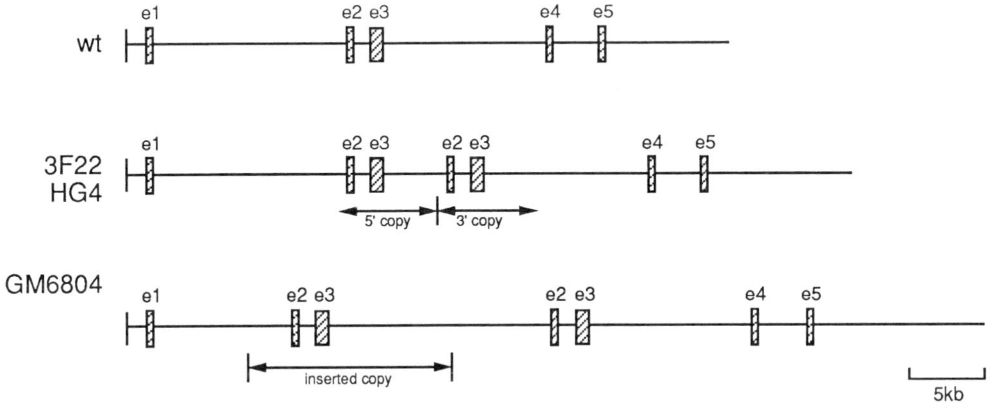

Figure 2. Structures of sequenced human *hprt* gene duplications. The structure of the 5′ half of *hprt* genes (horizontal solid lines) contained in control (wt) and duplication mutants was deduced from a combination of blot hybridization and cDNA sequence analyses. The positions of exons in each gene are indicated by hatched boxes. 3F22 and HG4 contain virtually identical 6.8kb tandem duplications (↔) of the *hprt* exon 2-3 region with single intron 3 - intron 1 novel junctions (|), while GM6804 contains a 13.7kb duplication of the *hprt* exon 2-3 region (↔) that has been inserted into intron 1 with the creation of 5' intron 1 - intron 1 and 3' intron 3 - intron 1 novel junctions (|).

DNA sequence analysis of the duplication junctions contained in 3F22 and HG4 revealed that both junctions had been generated by unequal homologous recombination between a single pair of *hprt Alu* repeats. Crossover regions were separated by 110bp within this *Alu* repeat pair, and demonstrate substantial nucleotide sequence identity and the potential for direct donor duplex pairing both at and 5′ and 3′ to the junction regions. The DNA sequence analysis of duplication junctions in GM6804, in contrast, revealed illegitimate recombination junctions having no and 4bp of sequence identity at, though little donor nucleotide sequence identity flanking the junctions. A 10bp deletion of intron 1 DNA was also observed in association with this insertion/illegitimate recombination event.

The model we find most attractive to explain the generation of independent duplications of different lengths of the same human *Alu* region involves the generation of a partially replicated *hprt* gene segment from an endogenous *hprt* exon 2-3 region DNA replication origin, followed by either homologous (3F22, HG4) or illegitimate recombination of these DNAs into the chromosomal *hprt* locus. It should be possible to explore this model further by determining whether an endogenous replication origin is present in the human *hprt* exon 2-3 region, and by characterizing the structure of additional exon 2-3 region duplications generated by the targeted recombination of different length *hprt* exon 2-3 fragments into the human or mouse *hprt* loci.

SPECULATION

Why are so few human somatic *hprt* gene rearrangements generated by homologous recombination between *Alu* repeats? The molecular analysis of additional germinal *hprt* gene deletions and duplications should reveal whether homologous human *hprt* gene rearrangements are indeed uncommon. If this proves to be the case, could nucleotide sequence divergence among the 49 human *hprt Alu* repeats be suppressing, rather than stimulating, homologous recombination? The advantage of approaching mechanistic questions such as these using mammalian *hprt* loci are that both germinal and somatic mutations can be readily examined, and *hprt* somatic cell, molecular and reverse genetics are sufficiently well-developed so that mechanistic hypotheses can be tested as well as devised.

ACKNOWLEDGEMENTS

We thank Al Edwards and Tom Caskey for human *hprt* DNA sequence data, Doug Jolly and Pragna Patel for probes, Stephanie Davis for help with DNA sequencing and Kris Carrol and Mary Bohidar for help with computer graphics. This work was supported by Public Health Service grants R29 CA48022 to R.J.M., Jr., and PO1 AG01751 to George M. Martin.

REFERENCES

1. K-i Fukuchi, G.M. Martin and R.J. Monnat, Jr., Mutator phenotype of Werner syndrome is characterized by extensive deletions, Proc. Natl. Acad. Sci. USA 86:5893 (1989).
2. R.J. Monnat, Jr., Molecular analysis of spontaneous hypoxanthine phosphoribosyltransferase mutations in thioguanine-resistant HL-60 human leukemia cells, Cancer Res. 49:81 (1989).
3. R.P. Gottlieb, M.M. Koppel, W.L. Nyhan, B. Bakay, E. Nissinen, M. Borden, and T. Page, Hyperuricaemia and choreoathetosis in a child without mental retardation or self-mutilation - a new HPRT variant. J. Inherit. Metab. Dis. 5:183 (1982).
4. H.O. Smith and M.L. Birnstiel, A simple method for DNA restriction site mapping, Nuc. Acids Res. 3:2387 (1976).

RAT HYPOXANTHINE PHOSPHORIBOSYLTRANSFERASE cDNA CLONING AND SEQUENCE ANALYSIS

Teresa A. Chiaverotti, Narayana Battula* and Raymond J. Monnat, Jr.†

Department of Pathology SM-30, University of Washington, Seattle, WA 98195 and *FDA Division of Antiviral Drug Products HFD-530, Kensington, MD 20895

We have determined the nucleotide sequence of the rat *hprt* mRNA coding region and of adjacent 5′ and 3′ untranslated sequence, and have designed an oligonucleotide primer pair for efficient PCR amplification of the complete rat *hprt* coding region as a cDNA. These sequence data and primer pair will aide workers interested in coupling well-developed rat toxicologic and carcinogenicity bioassays with quantitative and molecular analyses of somatic mutation at the *hprt* locus in rat cells in vivo and in vitro.

METHODOLOGY

Our strategy for determining the rat *hprt* cDNA sequence consisted of direct PCR amplification and sequencing of *hprt* cDNA sequences from two different rat cDNA libraries. Rat *hprt* cDNA sequence data derived from these cDNA libraries was used to synthesize a primer pair that was specific for the rat *hprt* cDNA coding region. This rat *hprt*-specific cDNA primer pair (97(-) and 763 (+), Figure 1.) was then used to amplify a third, independent rat *hprt* coding region from cDNA prepared from NRLM rat liver epithelial cell RNA[1].

The cDNA libraries were a λgt11 rat liver cDNA library (Clontech #RL1023b) prepared from adult male Sprague-Dawley rat liver, and a λZAPII neonatal rat smooth muscle cDNA library prepared from cultured aortic medial smooth muscle cells from 12-day old Wistar-Kyoto rats (library kindly supplied by Drs. Ceci Giachelli and Steve Schwartz, Department of Pathology, University of Washington, Seattle, WA). Initial library screens were performed using pairs of vector-specific and mammalian *hprt* consensus primers (441(-): 5′-CAG TCA ACA GGG GAC ATA-3′ and 461(+): 5′-TCC AGT TAA AGT TGA GAG AT-3′). Amplified rat *hprt* library and cDNA fragments were sequenced by the dideoxy chain termination method using modified T7 DNA polymerase (Sequenase, USB) and end-labelled primers (11 total, including the 4 described above).

Purine and Pyrimidine Metabolism in Man VII, Part B
Edited by R.A. Harkness *et al.*, Plenum Press, New York, 1991

```
                                                   10        20        30        40
                                          GGCACGAGGGACTTACCTCACTGCTTTTCCTAGCGGTACGACCTCCT

50        60        70        80        90       100         110         120
CCGCCAGCTTCCTCCTCAGACCGCTTTTCCCGCGAGCCGACCGGTTCTGTC ATG TCG ACC CTC AGT CCC AGC
                            |___________________|  Met SER Thr LEU Ser Pro Ser
                                     97-

       130         140         150         160         170         180
GTC GTG ATT AGT GAT GAT GAA CCA GGT TAT GAC CTA GAT TTA TTT TGC ATA CCT AAT CAT
Val Val Ile Ser Asp Asp Glu Pro Gly Tyr Asp Leu Asp Leu Phe Cys Ile Pro Asn His

       190         200         210         220         230         240
TAT GCT GAA GAT TTG GAA AAG GTG TTT ATT CCT CAT GGA CTG ATT ATG GAC AGG ACT GAA
Tyr ALA Glu Asp Leu Glu Lys Val Phe Ile Pro His Gly LEU Ile Met Asp Arg Thr Glu

       250         260         270         280         290         300
AGA CTT GCT CGA GAT GTC ATG AAG GAG ATG GGA GGC CAT CAC ATT GTG GCC CTC TGT GTG
Arg Leu Ala Arg Asp Val Met Lys Glu Met Gly Gly His His Ile Val Ala Leu Cys Val

       310         320         330         340         350         360
CTG AAG GGG GGC TAT AAG TTC TTT GCT GAC CTG CTG GAT TAC ATT AAA GCG CTG AAT AGA
Leu Lys Gly Gly Tyr Lys Phe Phe Ala Asp Leu Leu Asp Tyr Ile Lys Ala Leu Asn Arg

       370         380         390         400         410         420
AAT AGT GAT AGG TCC ATT CCT ATG ACT GTA GAT TTT ATC AGA CTG AAG AGC TAC TGT AAT
Asn Ser Asp Arg Ser Ile Pro Met Thr Val Asp Phe Ile Arg Leu Lys Ser Tyr Cys Asn

       430         440         450         460         470         480
GAC CAG TCA ACG GGG GAC ATA AAA GTT ATT GGT GGA GAT GAT CTC TCA ACT TTA ACT GGA
Asp Gln Ser Thr Gly Asp Ile Lys Val Ile Gly Gly Asp Asp Leu Ser Thr Leu Thr Gly

       490         500         510         520         530         540
AAG AAC GTC TTG ATT GTT GAA GAT ATA ATT GAC ACT GGT AAA ACA ATG CAG ACT TTG CTT
Lys Asn Val Leu Ile Val Glu Asp Ile Ile Asp Thr Gly Lys Thr Met Gln Thr Leu Leu

       550         560         570         580         590         600
TCC TTG GTC AAG CAG TAC AGC CCC AAA ATG GTT AAG GTT GCA AGC TTG CTG GTG AAA AGG
Ser Leu Val Lys GLN Tyr SER PRO Lys Met Val Lys Val Ala Ser Leu Leu Val Lys Arg

       610         620         630         640         650         660
ACC TCT CGA AGT GTT GGA TAC AGG CCA GAC TTT GTT GGA TTT GAA ATT CCA GAC AAG TTT
Thr Ser Arg Ser Val Gly Tyr Arg Pro Asp Phe Val Gly Phe Glu Ile Pro Asp Lys Phe

       670         680         690         700         710         720
GTT GTT GGA TAT GCC CTT GAC TAT AAT GAG CAC TTC AGG GAT TTG AAT CAT GTT TGT GTC
Val Val Gly Tyr Ala Leu Asp Tyr Asn Glu HIS Phe Arg Asp Leu Asn His VAL Cys Val

       730         740         750         760         770         780
ATC AGC GAA AGT GGA AAA GCC AAG TAC AAA GCC TAA AAGACAGCGGCAAGTTGAATCTACAAGAGTC
Ile Ser Glu SER Gly Lys Ala Lys Tyr Lys Ala End     |____________________|
                                                             763+

790       800       810       820       830       840       850
CTGTTGATGTGGCCAGTAAAGAACTAGCAGACGTTCTAGTCCTGTGGCCATCTACTTAGTAAAGCTT
```

Figure 1. Nucleotide and deduced amino acid sequence of rat HPRT. The 853bp sequence includes a 657bp coding region and 98bp each of 5′ and 3′ untranslated sequence. The nucleotide numbering follows that of the human *hprt* cDNA sequence of Jolly et al.[2]. **UPPER CASE BOLD** amino acid residues differ between rat and hamster HPRTs. Single-underline residues differ between rat and human HPRT only, while double-underline residues differ between rat and mouse and rat and human HPRT. The binding sites for rat-specific primers 97(-) and 763(+) are indicated by brackets in untranslated sequence adjacent to the HPRT coding region.

RESULTS AND DISCUSSION

We determined the nucleotide sequence of the 657bp rat *hprt* cDNA coding region and of 98bp each of 5′ and 3′ untranslated sequence. The coding region sequence was determined from all 3 sources, while the flanking untranslated DNA sequence data was derived from both libraries. These sequence data appear in the EMBL, GenBank, and DDBJ DNA Databases under Accession No. M63983.

A comparison of the nucleotide and deduced amino acid sequences of rat *hprt* cDNA with other mammalian *hprt* cDNAs revealed coding region nucleotide sequence identities of 93.3% (human), 94.4% (hamster) and 96.7% (mouse)(Figure 1;)[2,3], and fifteen different amino acid substitutions between rat HPRT and the human, mouse and hamster HPRT amino acid sequences. Ten substitutions were observed between rat and human or hamster HPRTs, and 4 substitutions between the rat and mouse HPRTs. These amino acid substitutions are located in the N- and C-terminal portions of HPRT that flank a 109 residue region (amino acids 42-150) containing evolutionarily conserved HPRT regions that are believed to be hypoxanthine and α-D-5-phosphoribosyl-1-pyrophosphate (PRPP) binding sites[4,5]. Of the 15 residues that vary among mammalian HPRTs only two, ser_7 (gly_7 in human HPRT) and leu_{41} (val_{41} in hamster) have been identified as sites of mutation in Lesch-Nyhan or gout patients (gly_7 to asp_7 in $HPRT_{Gravesend}$, and leu_{41} to pro_{41} in $HPRT_{Detroit}$[5,6].

Hypoxanthine phosphoribosyltransferase (HPRT; EC 2.4.2.8.) plays an important role in purine nucleotide salvage in mammalian cells, and is the best-characterized of a larger group of phosphoribosyltransferases that utilize PRPP in the synthesis of purines, pyrimidines, histidine and tryptophan[7,8]. The genes encoding several different HPRTs have also been intensively studied. The mammalian *hprt* genes in particular have been widely used as targets for quantitative and molecular mutation analyses in somatic cells. The particular utility of the mammalian *hprt* locus for mutation analyses stems from its X-linkage and functional hemizygosity, simple purine analogue selections for isolating HPRT-deficient mutants, and a wealth of restriction maps, probes, and nucleotide sequence data from several mammalian *hprt* genes[2,3,8].

The sequence data and primer pair reported here should be useful to workers interested in *hprt* somatic mutation analysis in rat, and in the enzymology and structural biology of HPRT and related phosphoribosyltransferases. These small, soluble proteins can be efficiently expressed and assayed in mammalian and bacterial cells[2,3,9], and are thus attractive enzymes for analyzing the relationship of protein structure to function. Preliminary structural models for these enzymes have been developed[7,8], and there is a large and rapidly growing collection of well-characterized mutant alleles that could be used to test the effects of amino acid substitution on HPRT stability, activity and substrate specificity (see, e.g.,[5,6,8]). A better understanding of how phosphoribosyltransferase structure determines function would be intrinsically interesting, and may have practical utility given the central role these enzymes play in purine, pyrimidine and amino acid biosynthesis, in inherited diseases and in protozoal infection[7,8,10].

ACKNOWLEDGEMENTS

We thank Stephanie Davis for DNA sequencing, Alden F.M. Hackmann for help with sequence assembly and analysis, and Dr. Beverly Davidson for communicating results prior to publication. This research was supported by NIH grant R29 CA48022 to R.J.M., Jr.

REFERENCES

1. J.B. McMahon, W.L. Richards, A.A. del Campo, M.-K.H. Song and S.S. Thorgeirsson, Differential effects of transforming growth factor-β on proliferation of normal and malignant rat liver epithelial cells in culture, Cancer Res. 46:4665 (1986).
2. D.J. Jolly, H. Okayama, P. Berg, A.C. Esty, D. Filpula, P. Bohlen, G.G. Johnson, J.E. Shively, T. Hunkapillar, and T. Friedman, Isolation and characterization of a full-length expressible cDNA for human hypoxanthine phosphoribosyltransferase, Proc. Natl. Acad. Sci. USA 80:477 (1983).
3. D.S. Konecki, J. Brennand, J.C. Fuscoe, C.T. Caskey, and A.C. Chinault, Hypoxanthine-guanine phosphoribosyltransferase genes of mouse and Chinese hamster: construction and sequence analysis of cDNA recombinants, Nuc. Acids Res. 10:6763 (1982).
4. H.V. Hershey and M.W. Taylor, Nucleotide sequence and deduced amino acid sequence of *Escherichia coli* adenine phosphoribosyltransferase and comparison with other analogous enzymes, Gene 43:287 (1986).
5. B.L. Davidson, S.A. Tarlé, T.D. Pallela, and W.N. Kelly, Molecular basis of hypoxanthine-guanine phosphoribosyltransferase deficiency in ten subjects determined by direct sequencing of amplified transcripts, J. Clin. Invest. 84:342 (1989).
6. B.L. Davidson, S.A. Tarlé, M.V. Antwerp, D.A. Gibbs, R.W.E. Watts, W.N. Kelly, and T.D. Pallela, Identification of 17 independent mutations responsible for human hypoxanthine-guanine phosphoribosyltransferase (HPRT) deficiency, Am. J. Hum. Genet. 48:951 (1991).
7. W.D.L. Musick, Structural features of the phosphoribosyltransferases and their relationship to the human deficiency disorders of purine and pyrimidine metabolism, CRC Crit. Rev. Biochem. 11:1 (1981).
8. J.T. Stout and C.T. Caskey, Hypoxanthine phosphoribosyltransferase deficiency: The Lesch-Nyhan syndrome and gouty arthritis. in "The Metabolic Basis of Inherited Disease", 6th Ed., C.R. Scriver, A.L. Beaudet, W.S. Sly, and D. Valle, eds, McGraw-Hill, New York (1989).
9. M.L. Free, R.B. Gordon, D.T. Keough, I.R. Beacham, B.T. Emmerson, and J. de Jersey, Expression of active human hypoxanthine-guanine phosphoribosyltransferase in *Escherichia coli* and characterization of the recombinant enzyme. Biochim. Biophys. Acta 1087:205 (1990).
10. C.C. Wang, Parasite enzymes as potential targets for antiparasitic chemotherapy, J. Med. Chem. 27:1 (1984).

IDENTIFICATION OF TWO INDEPENDENT JAPANESE MUTANT HPRT GENES USING THE PCR TECHNIQUE

Yasukazu Yamada, Haruko Goto and Nobuaki Ogasawara

Department of Genetics, Institute for Developmental
Research, Aichi Prefectural Colony
Kasugai, Aichi 480-03, Japan

INTRODUCTION

Deficiency of a purine salvage enzyme, hypoxanthine guanine phosphoribosyltransferase (HPRT), is resulted in two distinct clinical disorders, which are inherited as an X linked recessive trait. Complete deficiency of HPRT leads to Lesch-Nyhan syndrome,[1] whereas partial deficiency causes severe form of gout.[2] The gene for HPRT located on the long arm of X chromosome consists of nine exons and eight introns, and recently, the sequence of 57 kb of the entire human HPRT gene locus was determined.[3] The molecular analysis of mutations at the human HPRT locus has been greatly facilitated by the recent development of polymerase chain reaction (PCR), coupled with direct-sequencing. The technique of PCR amplification of reverse-transcribed mRNA has recently been used to identify the molecular basis of HPRT deficiency in number of subjects.[4-8] Further, the elucidation of the complete DNA sequence of the human HPRT gene locus[3] has been enabled the analysis by the amplification of all nine HPRT exons from genomic DNA.[9] In this studies, we report two independent mutant HPRT genes from a patient of HPRT partial deficiency associated with gout and from a patient of Lesch-Nyhan syndrome, identified by the analyses of PCR amplified product of cDNA synthesized from mRNA and also the product of each exon from genomic DNA.

MATERIAL AND METHODS

Case 1 subject is a patient of HPRT partial deficiency associated with hyperuricemia (12.8 mg/dl), severe gout and attenuated mental retardation. HPRT activity in erythrocytes was 0.09 nmol/min/mg hemoglobin of 5 % of control value (1.76 ± 0.06), and APRT activity (0.80 nmol/min/mg hemoglobin) was increased to about 2-fold compared with control value (0.43 ± 0.06). Case 2 subject is diagnosed as Lesch-Nyhan syndrome from the symptoms of hyperuricemia, hyperuricaciduria, severe neurologic dysfunction including self-mutilation, and a much lower level of HPRT activity (less than 1 % of control). Lymphoblastoid cell lines were established from these two patients using Epstein-Barr (EB) virus.

Total cellular RNA was isolated from the EBV-transformed B-lymphoblasts (1×10^8 cells) derived from the patients, and mRNA was separated by oligo(dT)-cellulose chromatography. The first strand cDNA

Purine and Pyrimidine Metabolism in Man VII, Part B
Edited by R.A. Harkness *et al.*, Plenum Press, New York, 1991

was synthesized using oligo-(dT)15 primer and reversetranscriptase. Then the entire coding region of HPRT was amplified from the synthesized cDNA by PCR using *Taq* DNA polymerase and the HPRT specific oligonucleotide primers (Table 1). The nucleotide sequence of the amplified HPRT cDNA fragment was defined directly using the specific oligonucleotides as sequencing primers, and the sequence of pUC recombinant containing the cDNA fragment was also analyzed.

Table 1. Specific primers for the HPRT cDNA amplification and sequencing.

	Nucleotide sequence	Location
HCA1	5'-GaaTTCCTCCTCCTGAGCAGTCAG-3' *Eco* RI	-47 - -24, F
HCA2	5'-TGCTCGAGATGTGATGAAGG-3' *Xho* I	146 - 165, F
HCB2	5'-CCTTCATCACATCTCGAGCA-3' *Xho* I	165 - 146, R
HCA3	5'-AAGGTCGCAAGCTTGCTGGT-3' *Hind* III	475 - 494, F
HCB3	5'-ACCAGCAAGCTTGCGACCTT-3' *Hind* III	494 - 475, R
HCB4	5'-AAAGCTtTACTAAGCAGATGGCCAC-3' *Hind* III	760 - 736, R

The amplification using HCA1 and HCB4 resulted in the 807 bp fragment including the entire coding region of HPRT cDNA. The base position numbers were counted by the A of the initiation codeine ATG as base 1. F, sence primer; R, antisence primer.

The genomic DNA primers for the amplification of each exon from the genomic DNA prepared from lymphoid cells of the patients, were designed from the nucleotide sequence of human HPRT gene locus.[3,9] The DNA fragment (334 bp) including exon 4 of HPRT gene was amplified using a sence primer (E4A, 5'-TAGCTAGCTAACTTCTCAAATCTTCTAG-3') at the third intron (IVS 3) and an antisence primer (E4B, 5'-ATTAACCTAGACTGCTTCCAA GGGT-3') at IVS 4, and the amplification using a sence primer (E78A, 5'-CCATGGTACACTCAGCACGGATGA-3') at IVS 6 and an antisence primer (E78B, 5'-TATGAGGTGCTGGAAGGAGAAAAC-3') at IVS 8 gives a fragment (499 bp) including exon 7 and 8.

RESULTS AND DISCUSSION

The amplification of the entire HPRT cDNA from Case 1 patient of HPRT partial deficiency, resulted in the same size of DNA fragment as a normal control. By the direct-sequencing of the fragment, a single nucleotide substitution of T to C at base position 563 was found. Further, all 10 recombinant pUC clones sequenced have the same mutation (T to C at the base 563). Since there is the mutation in exon 8 of HPRT gene locus, the DNA fragment (499 bp) including exon 7 and 8 from genomic DNA of Case 1 patient was amplified using the specific oligonucleotides primers E78A and E78B, and sequenced directly using the same oligonucleotides as sequence primers. The substitution of T to C was also detected in the genomic DNA of the patient. In this case, it was suggested that a single amino acid substitution of valine (GTT) to alanine (GCT) at the codon 188 caused decreased HPRT activity (Fig. 1). This amino acid substitution might be exert comparative large effect for protein structural change, because two valine at the codon 188 and 189 in normal HPRT combined strongly. The point mutation of the Case 1 subject is not identical with any single nucleotide substitution reported previously, whereas some mutants having the mutation near the codon 188 in exon 8 have been found.[4-9]

```
                                         *563
Mutant  547.ATT.CCA.GAC.AAG.TTT.GCT.GTA.GGA.TAT.GCC.CTT.579
            Ile-Pro-Asp-Lys-Phe-Ala-Val-Gly-Tyr-Ala-Leu
            *** *** *** *** ***     *** *** *** *** ***
Normal      ATT.CCA.GAC.AAG.TTT.GTT.GTA.GGA.TAT.GCC.CTT
        183.Ile-Pro-Asp-Lys-Phe-Val-Val-Gly-Tyr-Ala-Leu.193
                                *188
```

Fig. 1. Comparison of deduced amino acid sequences of the Case 1 mutant and normal HPRT gene.

```
(A) genomic DNA
                                 (IVS 3)*(exon 4)
Normal 5'-tagtttttttttttttaactag.AATGACCAGTCAACAGGGGACATAA-3'
          *********************     *********************
Mutant    tagtttttttttttttaactag.----ACCAGTCAACAGGGGACATAA

(B) cDNA
                (exon 3)*(exon 4)
Normal  AAG.AGC.TAT.TGT.AAT.GAC.CAG.TCA.ACA.GGG.GAC.ATA.AAA
        Lys-Ser-Tyr-Cys-Asn-Asp-Gln-Ser-Thr-Gly-Asp-Ile-Lys

Type A  AAG.AGC.TAT.TGT.----.ACC.AGT.CAA.CAG.GGG.ACA.TAA.AA
        Lys-Ser-Tyr-Cys-     -Thr-Ser-Gln-Gln-Gly-Thr.***

Type B  AAG.AGC.TAT.TGT.---.---.---.TCA.ACA.GGG.GAC.ATA.AAA
        Lys-Ser-Tyr-Cys-   -   -   -Ser-Thr-Gly-Asp-Ile-Lys

Type C  AAG.AGC.TAT.TGT.--(del exon 4, 66bp)---.AAT.GTC.TTG
        Lys-Ser-Tyr-Cys-  (del 22 amino acid)  -Asn-Val-Leu
                (exon 3)                       (exon 5)
```

Fig. 2. The alteration of the genomic DNA (A) and cDNA (B) including HPRT gene in the Case 2 subject.

In the case of Lesch-Nyhan syndrome (Case 2), the HPRT mutation seems to be 4 bases deletion at the 5'-end of exon 4. Amplifying the entire HPRT cDNA from Case 2 patient, one more miner DNA fragment shorter than the main fragment having normal size, was also observed. The direct-sequencing of the amplified main and the short cDNA fragment could not clarify its alteration, because the ghost bands appeared around the 5'-end of exon 4. In the analysis of pUC recombinant, three types of HPRT cDNA with the different sequence were detected. The most of clones (9/12, type A) were deleted 4 bases at the 5'-end of exon 4. The clones deleted 9 bases at the 5'-end of exon 4 (2/12, type B) and that deleted 66 bp of whole the exon 4 (1/12, type C) were also found, but no clone having the normal HPRT coding region was discovered. The short fragment observed in the amplification of cDNA might be from the mutant mRNA lacked whole the exon 4 (type C). In this case, the mutation must be occurred around the 5'-end of exon 4 of HPRT gene locus. Therefore, the DNA fragment (334 bp) including the 3'-end of IVS 3, the exon 4 and the 5'-end of IVS 4 was amplified from genomic DNA of Case 2 patient using the specific oligonucleotides primers E4A and E4B, and the nucleotide sequence was determined by direct-sequencing using the same oligonucleotides as sequence primers. From the analysis of genomic DNA, it is demonstrated that the HPRT gene of the Case 2 patient deletes 4 bp AATG at the 5'-end of exon 4 (Fig. 2A). Thus, in Case 2, the splicing of RNA might be missing because of the deletion at the 5'-end of exon 4. Type B mRNA can make HPRT enzyme protein deleted three amino acids, Asn-Asp-Gln, at the codon 107 to 109 (Fig. 2B). Type C results in the protein deleted 22 amino acids of the exon 4, and in type A, the stop codon (TAA) appears at the 7th codon because of frame-shift of the codon. The further studies will make clear the mechanisms of the RNA splicing of this firstly discovered interesting mutation, why the 4 bp deletion at the 5'-end of exon 4 results in three different types of abnormal mRNA.

ACKNOWLEDGMENTS

This work was supported by a Gout Research Foundation grant and an Intractable Disease grant from Ministry of Health and Welfare of Japan.

REFERENCES

1. J.E. Seegmiller, F.M. Rosenbloom, and W.N. Kelley, Enzyme defect associated with a sex-linked human neurological disorder and excessive purine synthesis, *Science* 155:1682 (1967).
2. W.N. Kelley, F.M. Rosenbloom, J.F. Henderson, and J.E. Seegmiller, A specific enzyme defect in gout associated with overproduction of uric acid, *Proc. Natl. Acad. Sci. USA* 57:1735 (1967).
3. A. Edwards, H. Voss, P. Rice, A. Civitello, J. Stegemann, C. Schwager, J. Zimmermann, H. Erfle, C.T. Caskey, and W. Ansorge, Automated DNA sequencing of the human HPRT locus, *Genomics* 6:593 (1990).
4. B.L. Davidson, S.C. Chen, J.M. Wilson, and W.N. Kelley, Hypoxanthine guanine phosphoribosyltransferase: genetic evidence for identical mutations in two partially deficient subjects, *J. Clin. Invest.* 82:2164 (1988).
5. B.L. Davidson, S.A. Tarle, T.D. Palella, and W.N. Kelley, Molecular basis of hypoxanthine-guanine phosphoribosyltransferase deficiency in ten subject determined by direct sequencing of amplified transcripts, *J. Clin. Invest.* 71:1331 (1989).
6. R.A. Gibbs, P.-N. Nguyen, L.J. McBride, S.M. Koepf, and C.T. Caskey, Identification of mutations leading to the Lesch-Nyhan syndrome by automated direct DNA sequencing of in vitro amplified cDNA, *Proc. Natl. Acad. Sci. USA* 86:1735 (1989).
7. S. Fujimori, N. Kamatani, Y. Nishida, N. Ogasawara, and I. Akaoka, Hypoxanthine guanine phosphoribosyltransferase deficiency: nucleotide substitution causing Lesch-Nyhan syndrome identified for the first time among Japanese, *Hum. Genet.* 84:483 (1990).
8. B.L. Davidson, S.A. Tarle, M. Van Antwerp, D.A Gibbs, R.W.E. Watts, W.N. Kelley, and T.D. Palella, Identification of 17 independent mutations responsible for human hypoxanthine-guanine phosphoribosyltransferase (HPRT) deficiency, *Am. J. Hum. Genet.* 48:951 (1991).
9. R.A. Gibbs, P.-N. Nguyen, A. Edwards, A.B. Civetello, and C.T.Caskey, Multiplex DNA deletion detection and exon sequencing of the hypoxanthine phosphoribosyltransferase gene in Lesch-Nyhan families, *Genomics* 7:234 (1990).

IDENTIFICATION OF DISTINCT PRS1 MUTATIONS IN TWO PATIENTS WITH X-LINKED PHOSPHORIBOSYLPYROPHOSPHATE SYNTHETASE SUPERACTIVITY

Blake J. Roessler*, Nimrod Golovoy*, Thomas D. Palella*, Steven Heidler# and Michael A Becker#

*Department of Internal Medicine, University of Michigan Ann Arbor Michigan 49104
#Department of Medicine, University of Chicago, Chicago Illinois 60637

INTRODUCTION

Phosphoribosylpyrophosphate synthetase (PRS) (EC 2.7.6.1) catalyzes the reaction: ribose-5-phosphate + ATPMg2+ → 5-phosphoribosyl 1-pyrophosphate (PRPP) + AMP (Fox et al., 1971). Superactivity of PRS is an X chromosome linked inherited disorder in which excessive enzyme activity is associated with uric acid overproduction and gout (Yen et al., 1978; Becker et al., 1979). Uric acid overproduction in individuals with PRS superactivity appears to result from increased production of PRPP and subsequent acceleration of de novo and salvage purine nucleotide synthesis (Zoref et al., 1975; Becker et al., 1973).

To date, multiple affected kindred have been identified (Becker et al., 1986, 1988). Among these are at least two patients (N.B. & S.M.) in whom PRS superactivity and purine nucleotide feedback resistance have been associated with neurodevelopmental abnormalities in addition to hyperuricemia and gout (Becker et al., 1980; Losman et al., 1985). Previous analysis of partially purified erythrocyte PRS from patient S.M. showed evidence for a mutant enzyme with a combination of loss of sensitivity to feedback inhibition and increased maximal reaction velocity (Becker et al., 1980). Analysis of partially purified erythrocyte PRS from patient N.B. showed evidence for a mutant enzyme with diminished purine nucleotide feedback response and increased affinity for inorganic phosphate (Losman et al., 1985). These studies of the partially purified superactive enzymes suggest that structural alteration of the protein underlies superactivity. Thus gene mutation as the basis for enzyme superactivity is a likely possibility.

Two distinct X-linked loci for PRS (PRS 1 & PRS2) have been identified and the respective PRS cDNAs have been cloned (Becker et al., 1990; Roessler et al., 1989; Taira et al., 1989b). In an attempt to determine the possible genetic basis for these functional PRS abnormalities we have sequenced the coding region of PRS1 and PRS2 cDNA of lymphoblast and fibroblast cell lines derived from patients N.B. and S.M. Using the technique of direct sequencing of amplified transcripts, we have identified two distinct point mutations occurring in the PRS1 cDNA of N.B. and S.M. associated with the phenotypic expression of PRS superactivity.

MATERIALS AND METHODS

Maintenance of Cell Lines - Fibroblasts and lymphoblasts were derived from skin biopsies and whole blood respectively from normals, N.B. and S.M. as previously described (Losman et al., 1985). All cell lines were propogated in RPMI supplemented with 2mM glutamine, 5% fetal calf serum at 37°C in a humidified environment containing 5% CO_2. Lymphoblasts were harvested by centrifugation. Fibroblasts were harvested by trypsinization followed by centrifugation.

RNA isolation and PCR amplification of PRS cDNAs - Total cellular RNA was isolated from approximately 1 gram (wet weight) of lymphoblasts or fibroblasts using described methods (Davidson et al., 1991). 1 ug of whole cellular RNA was primed with either a PRS1 or PRS2 specific antisense primer and reverse transcribed using AMVRT (Sekai) at 42°C containing 50 mM KCl, 20mM Tris-Cl (pH 8.4) 2.5mM $MgCl_2$, 1mM deoxynucleotide triphosphates and 0.1mg/ml nuclease free bovine serum albumin.

A portion of this reaction mixture was diluted 1:5 in the above buffer (50 ul total volume), containing 1 ug of each PRS1 or PRS2 sense and antisense primer and 1-2 units of Taq DNA polymerase (Perkin-Elmer Cetus) were added to each sample. The samples were then overlaid with mineral oil and DNA amplification was performed for 30 cycles using a thermocycler (Perkin-Elmer Cetus). Each cycle consisted of 94°C, 1.25 min; 55°C, 1 min; 72°C, 3min. Each cell line was reverse transcribed and amplified a minimum of three times in order to insure authenticity of mutations found upon sequencing of amplified transcripts.

Sequencing - Sequencing of the amplified PRS1 and PRS2 transcripts was done directly without cloning using a modification of the dideoxynucleotide chain termination method (Sanger et al., 1977). All of the coding sequence of both strands of PRS1 and PRS2 were sequenced using 10 PRS consensus primers that hybridize to both the PRS1 and PRS2 coding sequences internal to the PCR amplification primers. The primers were phosphorylated with [γ^{32}P]dATP (6000Ci/mmol)(Amersham), and T4 polynucleotide kinase direct sequencing was performed using previously reported methods (Davidson et al., 1991). An aliquot of each reaction mixture (4ul) was electrophoresed at constant wattage (60W) for either 3 or 1.5 hrs through 4% polyacrylamide gels containing 8M urea and autoradiographed at -70°C using intensifying screens.

RNAse mapping - A 651 bp BalI fragment of normal PRS1 cDNA spanning the putative mutation sites in both N.B. and S.M. was cloned into pGEM3Z (Promega Biosystems) in an orientation such that transcription using the SP6 promoter results in the generation of an antisense probe. The ^{32}P labeled RNA probe was hybridized with 100 ug of total RNA isolated from normal, N.B. or S.M., transferred to message affinity (polyU) paper, and digested with RNAse A and RNAse T1 using described conditions (Gibbs et al., 1987). The digestion products were electrophoresed on a 6% polyacrylamide gel containing 8M urea, and autoradiographed at -70°C using intensifying screens.

RESULTS

Sequencing of PRS1 and PRS2 transcripts -Direct dideoxynucleotide sequencing of the PCR amplified PRS2 transcripts showed completely normal sequence in both N.B. and S.M. However direct sequencing of the PCR amplified PRS1 transcripts revealed two distinct mutations present in N.B. and S.M. For both N.B. and S.M. the same mutation was identified in at least threeindependent amplifications and was the sole departure from the normal PRS1 sequence. N.B. PRS1 differed from normal PRS1 by a single nucleotide substitution: an A to G transition at nucleotide position 341 (Table 1). This mutation alters a AAT codon to AGT, predicting the substitution of serine for asparagine. S.M. PRSI differed from normal PRS1 by a single nucleotide substitution: a G to C transition at nucleotide position 547 (Table 1). This mutation alters a GAC codon to CAC, predicting the substitution of histidine for aspartic acid. No other differences among N.B. PRS1, S.M. PRS1 and normal PRS1 were detected.

Table 1

Patient	Nucleotide Mutation	Amino Acid Alteration
N.B.	A341 → G	Asn113 → Ser
S.M.	G547 → C	Asp182 → His

RNase mapping analysis - In the case of S.M. the 651 bp antisense PRS1 probe was cleaved into two fragments of 99 and 552 bp upon digestion with RNase A and RNase T1 when hybridized to S.M. RNA but not when hybridized to normal (data not shown). Similarly the same PRS1 antisense probe was cleaved into fragments of 305 and 346 bp upon digestion with RNase A and RNase T1 when hybridized to N.B. RNA but not when hybridized to normal (data not shown). In the case of S.M. this predicts a mutation at position bp 547 while in the case of N.B. this predicts a mutation at position bp 341.

DISCUSSION

The supernormal activities of the mutant forms of PRS previously studied appear to arise from structural alterations of the protein itself. These probably arise from point mutations within one of the PRS structural genes. Most inherited metabolic disorders which have been characterized at the molecular level are caused by diverse mutations that result in a relative deficiency of enzymatic activity. For example, individuals with partial or complete deficiency of hypoxanthine-guanine phosphoribosyltransferase are very heterogeneous with regard to enzymatic properties and levels of enzyme produced (Wilson et al., 1986; Davidson et al., 1991). While the class of disorders characterized by supernormal enzyme activity is probably much smaller, the above precedents suggest that mutations of the PRS1 gene may be expressed as enhanced enzymatic activity.

We have identified two separate mutations in the PRS1 cDNA associated with erythrocyte PRS superactivity in two individuals with a clinical syndrome of precocious gout, hyperuricemia and neurodevelopmental abnormalities. Both mutations were found within the PRS1 gene transcripts while transcripts from the PRS2 gene were completely normal. In the individual N.B. an asparagine to serine substitution at codon 113 is the predicted result of a A to G transition at nucleotide 341. In the individual S.M. an aspartic acid to histidine substitution at codon 182 is the predicted result of a G to C transition at nucleotide 547. These mutations were identified by direct sequencing of amplified transcripts and confirmed by RNase mapping analysis.

PRS has been purified from human erythrocytes and retains substantial catalytic activity during the final purification step (Fox et al., 1974; Becker M. A., unpublished observations). Amino acid sequencing of purified PRS from human erythrocytes has shown that PRS1 is the major isoform present. Despite analysis of over 75% of the purified PRS protein isolated from human erythrocytes, we have failed to detect evidence for the presence of PRS2 protein (Roessler et al., 1990; Becker et al., 1991). It has been shown that PRS2 cDNA is a distinct 2.7 kb transcript which is expressed in a number of tissues (Taira et al., 1989a). Although the putative protein product has a predicted amino acid sequence which is >97% homologous to PRS1, to date, there is no information regarding the functional properties of this protein.

Despite the identification of distinct point mutations occurring in the PRPS1 gene from individuals with PRS superactivity and gout, the exact functional relationship between PRS1 and PRS2 isoforms as the basis for the phenotypic expression of enzyme superactivity remains to be determined. Recent cloning of PRS1 cDNA into a bacterial expression system has revealed that recombinant PRS1 protein retains catalytic activity after purification (Nosal, J., et al., unpublished observations). Analysis of these two mutant PRS1 proteins synthesized in this bacterial expression system will permit a more precise evaluation of these cDNA changes as the basis for the distinctive purine nucleotide feedback resistant phenotypes found in these affected individuals.

REFERENCES

Becker, M. A., Heidler, S. A., Bell, G. I., Seino, S., LeBeau, M. M., Westbrook, C. A., Neuman, W., Shapiro, L. J., Mohandas, T. K., Roessler, B. J., and Palella, T. D., 1990, Cloning of cDNAs for human phosphoribosylpyrophosphate synthetases 1 and 2 and X chromosome localization of PRPS1 and PRPS2 genes, Genomics, 8:555.

Becker, M. A., Losman, M. L., Rosenberg, A. L., Mehlman, I., Levinson, D. J., and Holmes, E. W., 1986, Phosphoribosylpyrophosphate synthetase superactivity: A study of five patients with catalytic defects in the enzyme, Arthritis Rheum., 29:880.

Becker, M. A., Puig, J. G., Mateos, F. A., Jimenez, M. L., Kim, M., and Simmonds, H. A., 1988, Inherited superactivity of phosphoribosylpyrophosphate synthetase: Association of uric acid overproduction and sensorineural deafness, Am J Med., 85:383.

Becker, M. A., Raivio, K. O., and Seegmiller, J. E., 1979, Synthesis of phosphoribosylpyrophosphate in mammalian cells, Adv Enzymol Relat Areas Mol Biol., 49:281.

Becker, M. A., Yen, R. C. K., Itkin, P., Goss, S. J., Seegmiller, J. E., and Bakay, B., 1979, Regional localization of the gene for human phosphoribosylpyro-phosphate synthetase on the X chromosome, Science, 203:1016.

Davidson, B. L., Tarle, S. A., Van Antwerp, M., Gibbs, D.A., Watts, R. W. E., Kelley, W. N., and Palella, T.D., 1991, Identification of 17 independent mutations responsible for human hypoxanthine-guanine phosphoribosyl-transferase (HPRT) deficiency, Am J Hum Genet., 48:951.

Fox, I. H., and Kelley, W. N., 1971, Human phosphoribosylpyrophosphate synthetase: distribution, purification and properties, J Biol Chem., 246:5739.

Gibbs, R. A., and Caskey, C. T., 1987, Identification and localization of mutations at the Lesch-Nyhan locus by ribonuclease A cleavage. Science, 236:303.

Roessler, B. J., Bell, G., Heidler, S., Seino, S., Becker, M. A., and Palella, T. D., 1990, Cloning of two distinct copies of human phosphoribosylpyrophosphate synthetase, Nucl Acid Res., 18:193.

Sanger, F., Nicklin, S., and Coulson, A. R., 1977, DNA sequencing with chain-terminating inhibitors, Proc Natl Acad Sci USA., 74:5463.

Taira, M., Ishijima, S., Kita, K., Yamada, K., Iizasa, T., and Tatibana, M., 1989, Tissue differential expression of two distinct genes for phosphoribosylpyrophosphate synthetase and existence of a testis specific transcript, Biochim Biophys Acta., 1007: 203-208.

Taira, M., Kudoh, J., Minoshima, S., Iizasa, T., Shimada, H., Shimizu, Y., Tatibana, M., and Shimizu, N., 1989, Localization of human phosphoribosylpyrophosphate synthetase subunit I and II genes (PRPS1 and PRPS2) to different regions of the X chromosome and assignment of two PRPS-related genes to autosomes, Somat Cell Mol Genet., 15:29.

Wilson, J. M., Stout, J. T., Palella, T. D., Davidson, B. L., Kelley, W. N., and Caskey, C. T., 1986, A molecular survey of hypoxanthime-guanine phosphoribosyltransferase deficiency in man., J CLin Invest., 77:188.

Yen, R. C. K., Adams, W. B., Lazar, C., and Becker, M. A., 1978, Evidence of X-linkage of human phosphoribosylpyrophosphate synthetase, Proc Natl Acad Sci. USA., 75:482.

Zoref, E., deVries, A., and Sperling, O.,1975, Mutant feedback resistant phosphoribosylpyrophosphate synthetase associated with purine overproduction and gout, J Clin Invest., 56:1093.

HUMAN PHOSPHORIBOSYLPYROPHOSPHATE SYNTHETASE (PRS) 2: AN INDEPENDENT ACTIVE, X CHROMOSONE-LINKED PRS ISOFORM

M.A. Becker, S.A. Heidler, J.M. Nosal, R.L. Switzer, M.M. LeBeau, L.J. Shapiro, T.D. Palella, and B.J. Roessler

University of Chicago, Chicago, IL; University of Illinois, Urbana, IL; Harbor-UCLA Medical Center, Torrance, CA; University of Michigan, Ann Arbor, MI

Cloning of rodent and human (h)phosphoribosylpyrophosphate synthetase (PRS) cDNAs has resulted in identification of multiple PRS transcripts,[1-3] at least two of which are encoded by separate X chromosome-linked genes (PRPS1 and PRPS2) in man.[4,5] The normal hPRPS1 gene product, hPRS1, has been studied in detail in enzyme preparations purified from erythrocytes.[6-9] In two unrelated patients in whom PRS superactivity is accompanied by purine nucleotide feedback-resistance, gout with uric acid overproduction, and neurodevelopmental impairment, single base substitutions in hPRS1 cDNA which predict changes in hPRS1 primary structure have recently been identified.[10] Although PRS2 (as well as PRS1) transcripts are present in RNA preparations from all organs thus far tested,[2] little is known about the role or regulation of hPRS2 activity. We have applied molecular genetic, cytogenetic, and biochemical methods to study the hPRPS2 gene and the corresponding hPRS2 cDNA, RNA transcript, and protein.

Localization of the hPRPS2 gene, previously assigned to the interval Xpter-q21,[4] was undertaken.[5] Confirmation of X chromosome linkage of both hPRPS2 and hPRPS1 loci was provided by the results of Southern blot analysis of enzyme-restricted human genomic DNA with hPRS cDNAs as probes (figure 1). For both BamHI- and HindIII-restricted DNA preparations, some restriction fragments showed distinctly greater intensity of hybridization in lanes containing DNA from a 48,XXXX cell line than in corresponding lanes containing equal amounts of 46,XY DNA. The presence of restriction fragments showing equal intensity of hybridization suggests the existence of autosomal hPRPS genes (or pseudogenes), such as hPRPS3.[11] The patterns of restriction fragments identified with labeled hPRS1 and hPRS2 cDNA probes were distinct (figure 1), providing further evidence for separate PRPS1 and PRPS2 genes.

In situ chromosomal hybridization and human x rodent hybrid cell line analysis utilizing appropriately labeled hPRS2 cDNA probes permitted the hPRPS2 gene to be localized to the short arm of X in the interval Xp22.2-p22.3.[5] Study

Purine and Pyrimidine Metabolism in Man VII, Part B
Edited by R.A. Harkness *et al.*, Plenum Press, New York, 1991

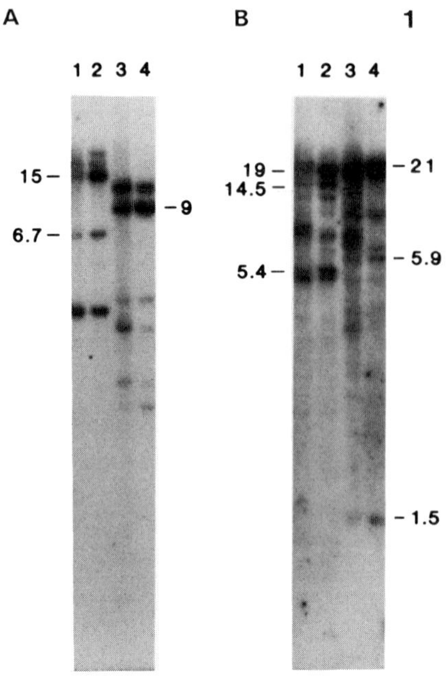

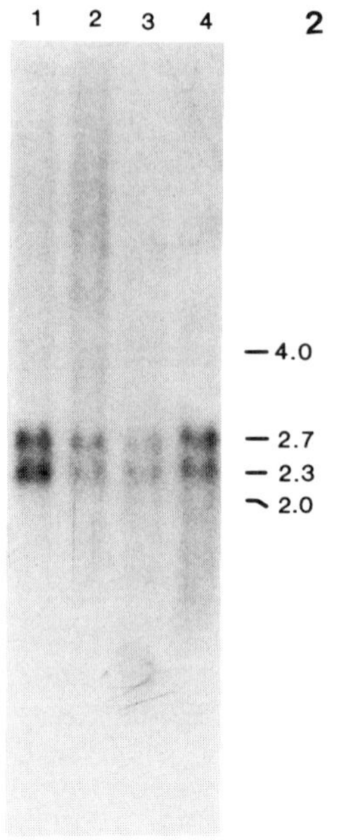

Figure 1. (left) Southern blot analysis of human genomic DNA extracted from lymphoblast lines CO (46,XY; lanes 1 and 3) and GM01416B (48,XXXX; lanes 2 and 4) and digested with BamHI (lanes 1 and 2) and HindIII (lanes 3 and 4). Enzyme-restricted DNA (10ug per lane) was electrophoresed in 0.8% agarose gel, transferred to a Gene Screen filter, hybridized overnight at 65°C with oligo-labeled hPRS1 (panel A) or 2 (panel B) cDNA, and autoradiographed. Note that hPRS1 and hPRS2 cDNAs identify different restriction fragments in both BamHI and HindIII digests. The location and sizes (in kb) of bands hybridizing with greater intensity in lanes containing DNA from the 48,XXXX cell line than in lanes containing an equal quantity of 46,XY DNA are indicated.

Figure 2. (right) Northern blot analysis of human lymphoblast total RNAs probed sequentially with oligo-primer labeled hPRS1 cDNA and hPRS2 cDNA. Size in kb of hybridizing bands and ribosomal RNA are indicated. Lymphoblast RNAs were applied to a 1% agarose-formaldehyde gel which was electrophoresed at 30v for 19h prior to transfer to a nitrocellulose filter. The filter was hybridized overnight at 45°C (in the presence of 50% formamide) with labeled hPRS1 cDNA and was then washed prior to autoradiographic exposure at -70°C for 72h. At this point, only bands corresponding to 2.3kb were observed. The filter was then hybridized with labeled hPRS2 cDNA, washed, and exposed under identical conditions. Lane 1, 10ug normal lymphoblast RNA; lane 2, 4ug HGPRT-deficient lymphoblast RNA; lanes 3 and 4, lymphoblast RNA from patients SM (3ug) and NB (6ug), respectively. These patients have feedback-resistant, superactive PRPP synthetases.[14,15] RNA samples, identically prepared and hybridized with labeled hPRS2 cDNA alone, showed bands at 2.7kb only.

of somatic hybrid cell lines with human chromosomal translocations indicate that within this interval PRPS2 is located distal to the POLA and ZFX loci but proximal to the markers AMEL and STS. In contrast to ZFX and STS, genes which bracket the PRPS2 locus, PRPS2 expression does not escape random X chromosome inactivation in female cells.[12]

Although the 2.7kb hPRS2 RNA transcript is identifiable in all organs thus far examined,[2] the abundance of this mRNA relative to 2.3kb hPRS1 transcript varies widely. In peripheral blood lymphocytes and reticulocytes, for example, hPRS2 transcripts are undetectable on Northern blot analysis and can only be identified after reverse transcription of the respective RNA and PCR amplication of the cDNA with hPRS2-specific primers. However, both hPRS1 and 2 transcripts are readily identifiable in B lymphoblasts derived from peripheral blood lymphocytes (figure 2), and in lymphocytes stimulated with concanavalin-A. These findings raise the possibility of regulation of PRPS2 expression during mitogenic stimulation and/or transformation. hPRS1 and 2 transcript size and abundance are comparable in total RNA preparations from normal lymphoblasts and fibroblasts and from the corresponding cells of patients with inherited purine nucleotide feedback resistant forms of PRS (figure 2).

Sequence homology between hPRS1 and hPRS2 cDNAs is 80% in the 954bp coding regions of the cDNAs, but the 5' and 3'-flanking sequences are divergent.[3] The predicted amino acid sequences of hPRS1 and 2 are 97% identical. Normal hPRS purified from erythrocytes is all or virtually all hPRS1.[3] In the course of gas phase protein sequencing of erythrocyte PRS, we have found no evidence for the presence of hPRS2 at amino acid residues where hPRS1 and hPRS2 sequences diverge. Moreover, these preparations retain PRS activity through the final purification step, suggesting independent activity of at least the hPRS1 isoform. The results of efforts to express hPRS1 and 2 in a bacterial host confirm this suggestion.

In one such study, an hPRS2 cDNA containing the entire coding sequence was engineered by PCR methodology into the plasmid pKK223-3 expression vector to yield a construct (pBC105) which was in turn used to transfect E.coli HO-700, a multiply auxotrophic strain lacking bacterial PRS by virtue of insertion of a kanamycin-resistance marker into the PRS gene.[15] Introduction of pBC105 into HO-700 restored the capacity of this strain to grow in minimal medium and was associated with expression of PRS at a specific activity nearly 300-fold that of normal human erythrocyte PRS. SDS-PAGE analysis confirmed the appearance of a protein band corresponding to the predicted hPRS2 subunit molecular weight in strains transfected with pBC105 but not with a construct containing hPRS2 cDNA cloned into the vector in reverse orientation. Kinetic characteristics of hPRS2 expressed in strain HO-700 were strikingly similar to those of human erythrocyte PRS, representative of hPRS1.

Acknowledgments - This research was supported by Grants DK-28554, AR-20557, and DK-38932 from the National Institutes of Health and by a grant from the Arthritis Foundation, Illinois Chapter. M.M.L. is a Scholar of the Leukemia Society of America. The authors acknowledge the excellent manuscript preparation by Ms. Danette Shine.

REFERENCES

1. M. Taira, S. Ishijima, K. Kita, K. Yamada, T. Iizasa, and M. Tatibana, J. Biol. Chem. 262:14,867-14,870 (1987).
2. M. Taira, T. Iizasa, K. Yamada, H. Shimada, and M. Tatibana, Biochim. Biophys. Acta 1007:203-208 (1989).
3. B.J. Roessler, G. Bell, S. Heidler, S. Seino, M. Becker, and T.D. Palella, Nucl. Acids Res. 18:193 (1990).
4. M. Taira, J. Kudoh, S. Minoshima, T. Iizasa, H. Shimada, Y. Shimuzu, M. Tatibana, and N. Shimuzu, Somat. Cell Mol. Genet 15:29-37 (1989).
5. M.A. Becker, S.A. Heidler, G.I. Bell, S. Seino, M.M. LeBeau, C.A. Westbrook, W. Neuman, L.J. Shapiro, T.K. Mohandas, B.J. Roessler, and T.D. Palella, Genomics 8:555-561 (1990).
6. I.H. Fox, and W.N. Kelley, J. Biol. Chem. 246:5739-5748 (1971).
7. I.H. Fox, and W.N. Kelley, J. Biol. Chem. 247:2126-2131 (1972).
8. M.A. Becker, P.J. Kostel, and L.J. Meyer, J. Biol. Chem. 250:6822-6830 (1975).
9. M.A. Becker, L.J. Meyer, W.H. Huisman, C. Lazar, and W.B. Adams, J. Biol. Chem. 252:3911-3918 (1977).
10. B.J. Roessler, T.D. Palella, S.A. Heidler, and M.A. Becker, In Press, these proceedings.
11. M. Taira, T. Iizasa, H. Shimada, J. Kudoh, N. Shimuzu, and M. Tatibana, J. Biol. Chem. 265:16,491-16,497 (1990).
12. T.K. Mohandas, R.S. Sparkes, B. Hellkuhl, K.H. Grzeschik, and L.J. Shapiro, Proc. Natl. Acad. Sci. U.S.A. 77:6759-6763 (1980).
13. M.A. Becker, K.P. Raivio, B. Bakay, W.B. Adams, and W.L. Nyhan, J. Clin. Invest. 65:109-120 (1980).
14. M.A. Becker, M.J. Losman, J. Wilson, and H.A. Simmonds, Biochim. Biophys. Acta 882:168-176 (1986).
15. B. Hove-Jensen, Mol. Microbiol. 3:1487-1492 (1989).

RESCUE OF A LETHAL PURINE NUCLEOSIDE PHOSPHORYLASE MUTATION IN THE MOUSE VIA A SECOND LOCUS INTERACTION

F. F. Snyder and E. R. Mably

Departments of Paediatrics and Medical Biochemistry
University of Calgary
Calgary, Alberta, Canada

INTRODUCTION

We have recovered five independent mutations at the purine nucleoside phosphorylase (NP) locus in the mouse among the offspring of mutagenized male mice[1,2]. The biochemical and metabolic features of two of these mutations, the NP-1E and NP-1F alleles, have been described[2]. One of the five mutations, NP-1G, is lethal on the C57BL/6J background, but viable homozygous PNP deficient mice have been recovered in crosses with a second inbred strain, DBA/2J.

MATERIALS AND METHODS

Mice have been typed for purine nucleoside phosphorylase activity by enzymatic assay of erythrocyte lysates[3]. Test mating results have been compared to predicted genetic models using the goodness of fit analysis[4].

RESULTS AND DISCUSSION

We have demonstrated that the NP-1G allele is lethal on the C57BL/6J (B6) background. Erythrocyte PNP activity is 48.3 nmole/min/mg protein (U) for B6 ($Np\text{-}1^{d/d}$) and 26.6 U for $Np\text{-}1^{d/g}$ carriers. No homozygous mutants, $Np\text{-}1^{g/g}$, were recovered in matings between carriers: observed, 0/68; expected, 17/68. The compound mutants, $Np\text{-}1^{e/g}$ and $Np\text{-}1^{f/g}$, were recovered in the expected frequency (0.5) from matings of $Np\text{-}1^{d/g}$, crossed with either of the homozygous mutants $Np\text{-}1^{e}$ or $Np\text{-}1^{f}$. Thus a single copy of the G allele in combination with either the E or F alleles is not lethal.

We then asked the question as to whether the G allele was lethal on another inbred background, DBA/2J (D2), having 15.2 U of PNP activity. By crossing carriers $Np\text{-}1^{d/g}$ (B6) with $Np\text{-}1^{a}$ (D2), we obtained the desired carrier progeny, $Np\text{-}1^{a/g}$ (B6.D2)f1, in the expected frequency (0.5) having 7.4 U of PNP activity. The cross between the f1 x f1 ($Np\text{-}1^{a/g}$) carriers did yield the homozygous mutant $Np\text{-}1^{g/g}$ (B6.D2)f2, 0.94 U PNP activity, but again at a lower than expected frequency: observed 3/54; expected 13/54.

There are at least two models which accommodate these observations. One is that there is a second locus which determines viability of the homozygous G mutation. We postulate and consider here that there is a suppressor locus, Npsu, which controls manifestation of viability for the G allele and B6 is $Npsu^{-/-}$ (lethal) and D2 is $Npsu^{+/+}$ (viable). This model predicts that no homozygous mutant mice would be recovered from the B6 background as observed. The frequency of homozygous mutant mice recovered from the B6.D2 carrier cross would be less than 0.25 and would vary depending upon the Npsu locus being dominant or recessive.

The mating, $Np\text{-}1^{d/g}$, (presumed $Npsu^{-/-}$) (B6) x $Np\text{-}1^{g/g}$, (presumed $Npsu^{-/+}$ or $Npsu^{+/+}$) (B6.D2)f2, gave both the expected progeny, $Np\text{-}1^{d/g}$ (0.71) and $Np\text{-}1^{g/g}$ (0.29) consistent with a dominant single copy suppressor locus model. If two copies of the $Npsu^{+}$ gene were required for viability, no homozygous mutants would be recovered from this mating.

Two locus models have previously been described in the mouse. In the dactylaplasia model both dominant and recessive alleles were found at the modifying locus[5]. In as much as complex patterns of inheritance have impeded investigations of human disease, the understanding of a two locus enzyme-effector gene interaction at the biochemical and molecular level in this model would be of considerable value.

Another possible interpretation of our findings is that there has been a second mutational event, coincident yet distinct from the production of the G mutation at the PNP locus. The observations are compatible with a model in which a second locus lethal mutation has occurred. Our data is consistent with the second mutational site being linked to Np-

1G on mouse chromosome 14. Present studies are designed to distinguish between the second locus suppressor and the second locus mutation models.

We also wished to investigate the timing and cause of the G allele associated embryolethality. As all animals typed in this study were examined by enzyme assay following weaning at day 21, we considered the possibility that homozygous G mutants might be viable prior to birth on the B6 background. In a cross between $Np\text{-}1^{d/g}$ x $Np\text{-}1^{d/g}$ on the B6 background, fetal liver was used for typing at day 18 of gestation. We obtained 3/9 NP-1D, 6/9 NP-1DG and 0/9 NP-1G embryos. One embryo had undergone resorption. These results are consistent with the homozygous null being lethal prior to birth. A second mating between $Np\text{-}1^{d/g}$ (B6) female by $Np\text{-}1^{g/g}$ (B6.D2)f2 male, was examined at day 13 of gestation. In this mating 7/11 implantations had undergone resorption with an estimate of arrest at day 10. The four viable embryos typed as 2/4 NP-1DG, and 2/4 NP-1G. These limited analyses are consistent with the G allele associated effect being lethal on the B6 background at approximately day 10 of embryonic development.

Additional test mating studies will clearly establish the genetic interaction of the NP-1G allele and the rescuing of the associated embryolethality. The recent cloning and sequencing of mouse purine nucleoside phosphorylase cDNA[6] will also enable the molecular characterization of the mutant G allele.

ACKNOWLEDGEMENTS

This work was supported by the Medical Research Council of Canada grant MT-6376.

REFERENCES

1. Chapman, VM, Miller, DR, Armstrong, D, and Caskey, CT. Recovery of induced mutations for X chromosome-linked muscular dystrophy in mice. Proc Natl Acad Sci USA 86:1292-1296 (1989).
2. Mably, ER, Fung, E, Snyder, FF. Genetic deficiency of purine nucleoside phosphorylase in the mouse. Characterization of partially and severely enzyme deficient mutants. Genome 32:1026-1032 (1989).

3. Snyder, FF, Mendelson, J, and Seegmiller, JE. Adenosine metabolism in phytohemagglutinin-stimulated human lymphocytes. J Clin Invest 58:654-666 (1976).
4. Sokal, RR, and Rohlf, FJ. Biometry. The principles and practise of statistics in Biological Research, WH Freeman, San Francisco (1969).
5. Chai, CK. Dactylaplasia in mice. A two-locus model for developmental anomalies. J Heredity 72:234-237 (1981).
6. Jenuth, JP and Snyder, FF. Nucleotide sequence of murine purine nucleoside phosphorylase cDNA. Nucl Acids Res 19:1708 (1991).

GENETIC MODELS OF PURINE NUCLEOSIDE PHOSPHORYLASE DEFICIENCY IN THE MOUSE

Floyd F. Snyder

Department of Paediatrics and Medical Biochemistry, University of Calgary, Calgary, Alberta, Canada, T2N 4N1

Purine nucleoside phosphorylase deficiency is associated with severe T cell immunodeficiency disease[1] and more recently behavioral and neurological features have also been described[2,3]. With respect to presentation, this disease appears intermediary in phenotype to the severe combined immunodeficiency disease of adenosine deaminase deficiency and the severe neurological impairment associated with complete HPRT deficiency.

We have recovered five independent mutations at the PNP locus in the mouse. These were identified in the progeny of ethylnitrosourea mutagenized male mice mated to untreated females[4]. Blood from approximately 2500 mice was screened by quantitative enzyme assay or isoelectric focusing and we detected the five PNP variants in the carrier state. As the mutants were generated from the cross between C57BL/6J (B6) and C3H/HeHa mice[5], we have been in the process of constructing congenic mutant stocks on the B6 background. The production of congenic mutant strains serves two purposes. In the first instance, the occurrence of extraneous mutations is eliminated by this process and, secondly, having the mutations on a homogenous genetic background ensures reproducible expression of the mutant phenotype from animal to animal.

Four of the PNP mutants are viable in the homozygous state and one is lethal (Table 1). Of the four viable mutants, we have examined the enzymology, metabolism and pathology of two, the NP-1E and NP-1F mutants. The NP-1F mutant is more severely deficient in PNP activity than NP-1E as shown by tissue survey, but both have retained normal affinity for nucleoside and phosphate substrates[4]. In the NP-1F mutant, PNP activity expressed as a percentage of control is for: spleen leucocytes, 3.6; thymocytes, 1.4; kidney, 0.2 and liver, 0.1. Thus, lymphoid cells have retained significantly greater levels of activity than other cells or organs.

The mutants accumulate nucleoside substrates in proportion to the severity of the enzyme deficiency[4]. Metabolic analysis has shown that the NP-1E mutants excrete inosine and guanosine at greater than 10-fold normal levels but they do not excrete deoxyribonucleosides. The NP-1F mutants excrete inosine, guanosine, deoxyinosine and deoxyguanosine, together at greater than 100-fold the level of control mice. Examination of cellular nucleotide pools has shown that the mutant mice have normal levels of NAD and ATP, marginally increased levels of GTP and no detectable dGTP[6]. In addition, neither spleen leucocytes nor thymocytes showed evidence for dGTP accumulation by HPLC analysis. Thus by comparison to human PNP deficiency, the mouse model differs in that NAD is not increased, GTP is not reduced and if there are increases in dGTP, they are rather subtle and were not detectable.

Table 1. Purine nucleoside phosphorylase deficient mutants on the C57BL/6J (B6) inbred background.

Strain	Genotype	Erythrocyte PNP Activity (nmole/min/mg protein)
C57BL/6J	$Np\text{-}1^d/Np\text{-}1^d$	48.1
B6-NPE	$Np\text{-}1^e/Np\text{-}1^e$	2.62
B6-NPF	$Np\text{-}1^f/Np\text{-}1^f$	0.96
B6-NPG	$Np\text{-}1^g/Np\text{-}1^g$	lethal
B6-NPH	$Np\text{-}1^h/Np\text{-}1^h$	0.61
B6-NPI	$Np\text{-}1^i/Np\text{-}1^i$	1.14

As dGTP accumulation in PNP deficiency is presumed to be causally related to the T cell lymphopenia, we began to examine deoxyguanosine metabolism in the mouse. Our initial studies revealed a marked decrease in deoxyguanosine kinase in PNP deficient erythrocytes: 8.1% for NP-1E; and 3.5% for NP-1F[6]. This secondary enzyme deficiency was apparent to varying degrees in other cells and tissues including thymocytes. Metabolic modelling[7] of deoxyguanosine metabolism via phosphorylation or phosphorolysis revealed that in the PNP deficient mutant phosphorylation would be the principal route of metabolism if deoxyguanosine kinase activity were unchanged. As a consequence of the decrease in deoxyguanosine kinase, phosphorolysis is the predominant route of metabolism at all deoxyguanosine concentrations. Thus the secondary deficiency of deoxyguanosine kinase appears to be a protective mechanism which prevents the marked accumulation of dGTP in PNP deficient mouse cells.

There were no changes in hematological parameters for mutant mice as compared to controls. In addition pathological examination did not reveal any gross changes. Examination of lymphocyte populations by fluorescence activated cytometry revealed a partial reduction in Thy-1 positive thymocytes in the more severe NP-1F mutation. Further, the thymocyte population in general showed significant perturbation with a 10-fold increase to 40% of the total cells being double negative $CD4^-CD8^-$ prothymocytes as compared to controls[6]. Thus there appears to be an accumulation of precursor cells in the thymus of PNP deficient mice.

The NP-1G mutation is lethal in the homozygous state on the B6 background as no mutant mice have been recovered. Our studies also suggest that G allele associated lethality occurs at approximately day 10 of embryonic development[6]. In crosses of the NP-1G mutation with a second inbred strain, DBA/2J (D2), we have recovered the homozygous mutant (Table 2). These observations suggest there may be a second locus interaction for which B6 and D2 differ that permits the homozygous mutant to be rescued on the alternate background. Another possibility that has not been conclusively eliminated is that there has been a second recessive lethal mutation which is tightly linked to the Np-1 locus on mouse chromosome 14. We have shown by cloning and sequencing of mouse PNP cDNA that the $Np\text{-}1^a$ and $Np\text{-}1^d$ alleles differ by six base changes, only one of which confers a codon and amino acid change[8]. Further analysis will define the molecular basis of the mutant $NP\text{-}1^g$ allele.

Table 2. Crosses of the lethal NP-1G mutation.

Strain	PNP Genotype	Erythrocyte PNP activity (nmole/min/mg protein)
DBA/2J	$Np\text{-}1^a/Np\text{-}1^a$	15.2
(B6.D2)f1	$Np\text{-}1^a/Np\text{-}1^g$	7.43
(B6.D2)f2	$Np\text{-}1^g/Np\text{-}1^g$	0.94

The mouse model presents us with a system in which a variety of metabolic and therapeutic strategies can be examined. The endpoints against which treatment strategies can be assessed are: purine nucleoside excretion, reduced levels of deoxyguanosine kinase, altered thymocyte profiles and the NP-1G associated embryolethality. The severity of expression of these mutations may also be further manipulated by the production of compound mutants among the five variants. The metabolic or pharmacologic administration of PNP substrates or inhibitors may also exacerbate the observed phenotypes.

Three modes of therapy are readily testable against the previously described endpoints. Metabolite therapy may be directed towards manipulation of nucleotide pools by provision of nucleosides or bases to study their consequence in the relevant cells and tissues. In addition the potential depletion of available ribose-phosphates in PNP deficiency may be assessed by provision of metabolites which spare PP-ribose-P utilization. The second mode of treatment which can be explored is that of enzyme therapy which has already proved beneficial in human adenosine deaminase deficiency[9]. Finally, the efficacy of bone marrow transplantation has been well established in immunodeficiency disease but the availability of matched donors remains a limiting constraint. Gene therapy is an alternate to stem cell transplantation and the PNP deficient mouse is a valuable model for the introduction of PNP expression retroviruses into somatic cells.

ACKNOWLEDGEMENTS

We thank Drs. V. Chapman for provision of mutant progeny, L. Bryan for pathological analyses and F.G. Biddle for practical genetic advice. This work was supported by the Medical Research Council of Canada, Grant MT-6376.

REFERENCES

1. Giblett ER, Ammann AJ, Sandman R, Wara DW, and Diamond LK. Nucleoside phosphorylase deficiency in a child with severely deficient T-cell immunity and normal B-cell immunity. Lancet, 3:1010-1013 (1975).
2. Rijksen G, Kuis W, Wadman SK, Spaapen LJM, Duran M, Voorbrood BS, Staal GEJ, Stoop JW, and Zegers BJM. A new case of purine nucleoside phosphorylase deficiency: Enzymologic, clinical, and immunologic characteristics. Pediat Res 21:137-141 (1987).
3. Simmonds HA, Fairbanks LD, Morris GS, Morgan G, Watson AR, Timms P, and Singh B. Central nervous system dysfunction and erythrocyte guanosine triphosphate depletion in purine nucleoside phosphorylase deficiency. Arch Dis Child 62:385-392 (1987).
4. Mably ER, Fung E, and Snyder FF. Genetic deficiency of purine nucleoside phosphorylase in the mouse. Characterization of partially and severely enzyme deficient mutants. Genome 32:1026-1032 (1989).
5. Chapman VM, Miller DR, Armstrong D, and Caskey CT. Recovery of induced mutations for X chromosome-linked muscular dystrophy in mice. Proc Natl Acad Sci USA 86:1292-1296 (1989).
6. Jenuth JP, Dilay JE, Fung E, Mably ER, and Snyder FF. (in preparation).
7. Snyder FF and Lukey T. Kinetic considerations for the regulation of adenosine and deoxyadenosine metabolism in mouse and human tissues based on a thymocyte model. Biochim Biophys Acta 696:299-307 (1982).
8. Jenuth JP and Snyder FF. Nucleotide sequence of murine purine nucleoside phosphorylase cDNA. Nucleic Acids Res 19:1708 (1991).
9. Hershfield MS, Buckley RH, Greenberg ML, Melton AL, Schiff R, Hatem C, Kurtzberg J, Markert ML, Kobayashi RH, Kobayashi AL, and Abuchowski A. Treatment of adenosine deaminase deficiency with polyethylene glycol-modified adenosine deaminase. N Eng J Med 316:589-596 (1987).

A GENETIC DEFECT IN MUSCLE PHOSPHOFRUCTOKINASE DEFICIENCY, A TYPICAL CLINICAL ENTITY PRESENTING MYOGENIC HYPERURICEMIA

Hiromu Nakajima[1], Norio Kono[1], Tomoyuki Yamasaki[1], Kikuko Hotta[1], Masanori Kawachi[1], Tomoya Hamaguchi[1], Takamichi Nishimura[1], Ikuo Mineo[1], Masamichi Kuwajima[1], Tamio Noguchi[2], Takehiko Tanaka[2] and Seiichiro Tarui[1]

Osaka University Medical School, the Second Department of Internal Medicine[1] and the Department of Nutrition and Physiological Chemistry[2], Osaka, JAPAN

INTRODUCTION

Myogenic hyperuricemia is a common pathophysiologic feature of muscle glycogenoses[1]. Type VII glycogenosis is the one which lacks catalytic activity of phosphofructokinase, a key enzyme of glycolysis, in muscle (PFK-M)[2]. Energy crisis in the exercising muscles of the patient results in the abnormal purine degradation leading to a typical manifestation of myogenic hyperuricemia[1,3]. As a step to clarify the molecular basis of this abnormal purine metabolism, we have reported the cloning of full-length human PFK-M cDNA[4]. The sequence data enabled us to reconfirm the gene duplication mechanism as hypothesized in rabbit PFK-M gene[5,6,7]. We have also elucidated the existence of the alternative splicing and the alternative promoter system of human PFK-M gene[8,9]. In addition, we have recently reported the complete structure of human PFK-M gene[10]. In this paper, we report on the genetic defect in a patient with PFK-M deficiency as the typical clinical entity presenting myogenic hyperuricemia.

EXPERIMENTAL PROCEDURES

Muscle biopsy was performed on a male patient with typical PFK-M deficiency. The clinical features and definite diagnosis of the enzyme defect were described[2]. He was also the first case of metabolic myopathy presenting myogenic hyperuricemia[3]. Northern analysis and cDNA cloning of PFK-M of this patient have been reported recently[11]. Polymerase chain reaction (PCR) was performed using 1 μg of genomic DNA and 2.5 units of *Taq* DNA polymerase. The reaction condition was 25 cycles of denaturation, annealing and polymerization at 94°C, 60°C and 72°C, respectively. Amplified DNA fragments were subcloned into pUC119 and then they were sequenced.

Allele specific oligonucleotide (ASO) dot hybridization analysis was performed on the PCR-amplified genomic DNA. PCR product was dot-blotted and was hybridized with the ASO-probes.

Abnormal PFK-M structure of the patient was predicted on the basis of the crystallographic analysis of *Bacillus stearothermophilus* PFK[12]. The secondary structure was also predicted by the method of Chou and Fasman[13].

RESULTS AND DISCUSSION

Northern analysis revealed the apparently normal expression of PFK-M mRNA in the patient muscle. From the muscle cDNA library, PFK-M cDNAs including a full-length one were cloned and were sequenced. In addition to a non-pathological polymorphic base transition at nt. 517 [ACT (Thr) to ACC (Thr)], a 75-base in-frame deletion was identified in the coding region corresponding to a part of exon 15 of PFK-M gene.

The genomic DNA region of exon 15, intron 15 and exon 16 was amplified by PCR. From sequence analysis, a point mutation at the splice donor site of intron 15 was identified (Fig. 1). In the consensus sequence of the 5' splice site [(C/A)AG:GT(A/G)AGT], the first two bases are reported to be highly restricted to GT[14]. For example, a nucleotide substitution of A for the first G completely abolished the splicing at this site in a case of β_0-thalassemia[15]. On the other hand, cryptic splice sites are known to be activated by such substitutions[16]. In our case, consensus GT dinucleotide was broken by the single base transversion of G to T, resulting in the inactivation of the splice donor site. The mechanism of the abnormal pre-mRNA splicing at the cryptic donor located 75-base upstream was elucidated. ASO analysis indicated that this mutation existed on both alleles of the patient PFK-M gene (Fig. 2).

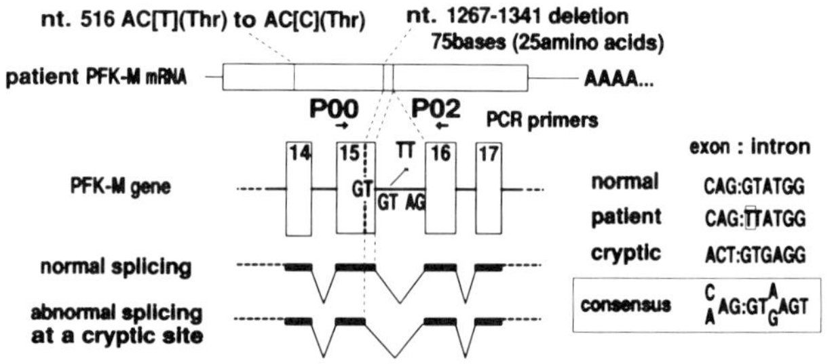

P00:5'-AGTGGTTCGCACACAGTGGCTGTGATGAAC-3' P02:5'-CAAGTTTAGAGCCACCTTGGCCAGTCCAGC-3'

Fig. 1. *Mutation and the mechanism of the abnormal splicing* : A transition from T to C at nt. 516 does not change the codon for Thr. At nt. 1267 to 1341, 75 bases were deleted in-frame. Amplification of exon 15, intron 15 and exon 16 by PCR using primers P00 and P02 revealed a point mutation from G to T at the splice donor of intron 15. This changed the splice site of CAG:GTATGG to CAG:TTATGG. A cryptic splice site of ACT:GTGAGG at 75 bases upstream was alternatively activated.

Our preliminary observation indicates that immunologically active, biologically inactive PFK-M is present (Tarui, *et al.*, unpublished). The abnormal gene product, which possesses in-frame deletion, would be translated. By the mutation, 25 amino acid residues around ADP/AMP activation site[6,12] would be lost in the patient PFK-M. A drastic configurational alteration of the enzyme (Fig. 3) would have led to the complete loss of catalytic activity of PFK-M in the patient muscle resulting in the energy crisis to cause myogenic hyperuricemia.

```
                            exon 15 intron 15
PFK-M gene     5'-...CTGGCCAAGGGGCAG GTATGGGGACTATCT...-3'
               3'-...GACCGGTTCCCCGTC CATACCCCTGATAGA...-5'
ASO probes
  P1N (Tm=64°C)           3'-TTCCCCGTC CATACCCCTGA-5'
  P1P (Tm=62°C)           3'-********* A**********-5'
```

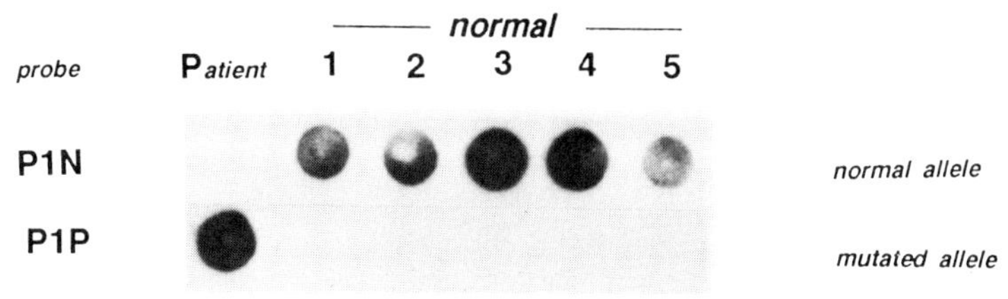

Fig. 2. *ASO dot hybridization analysis:* PCR product by P00 and P02 was hybridized with oligonucleotides P1N and P1P at low stringency. Washing was performed at high stringency in 6×SSC. Control DNA (lanes 1 - 5) hybridized only to P1N (normal sequence). Patient DNA hybridized only to P1P (G to T transversion).

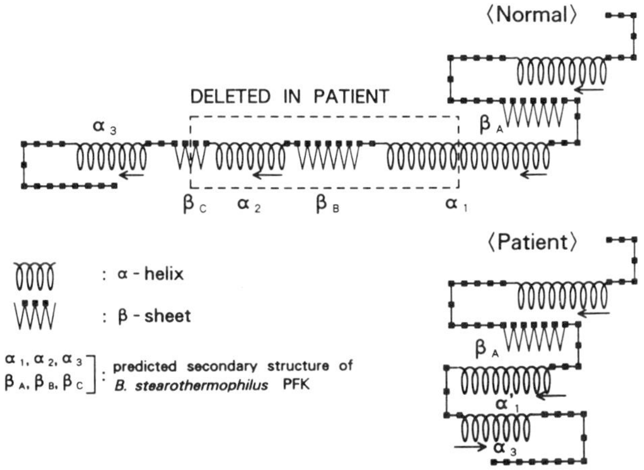

Fig. 3. *The secondary structure of the abnormal gene product* : The secondary structure of amino acid residues coded by exons 14, 15 and 16 was predicted by the method of Chou and Fasman[13].

CONCLUSION

The genetic defect of PFK-M deficiency was identified at both DNA and mRNA levels for the first time in all types of glycogenoses. This is the case as well, among all the clinical entities known to present myogenic hyperuricemia. The results would contribute to the future studies in clarifying the precise molecular mechanism of myogenic hyperuricemia.

ACKNOWLEDGEMENTS

The authors are grateful to Drs. K. Moriwaki, T. Hanafusa, A. Ono, K. Imamura, M. Takenaka and K. Yamada for helpful discussion. This work was supported by the grants from the Uehara Memorial Foundation, Yamanouchi Foundation for Research on Metabolic Disorders and the Gout Research Foundation of Japan. It was also supported by a Grant-in-Aid for Scientific Research, and a Grant-in-Aid for Research on Priority Areas from the Ministry of Education, Science and Culture, Japan.

REFERENCES

1. I. Mineo, N. Kono, N. Hara, T. Shimizu, Y. Yamada, M. Kawachi, H. Kiyokawa, Y. L. Wang, and S. Tarui, Myogenic hyperuricemia: A common pathophysiologic feature of glycogenosis types III, V and VII, *N. Engl. J. Med. 317*:75 (1987).
2. S. Tarui, G. Okuno, Y. Ikura, T. Tanaka, M. Suda, and M. Nishikawa, Phosphofructokinase deficiency in skeletal muscle: A new type of glycogenosis, *Biochem. Biophys. Res. Commun. 19*:517 (1965).
3. N. Kono, I. Mineo, T. Shimizu, N. Hara, Y. Yamada, K. Nonaka, and S. Tarui, Increased plasma uric acid after exercise in muscle phosphofructokinase deficiency, *Neurology 36*:106 (1986).
4. H. Nakajima, T. Noguchi, T. Yamasaki, N. Kono, T. Tanaka, and S. Tarui, Cloning of human muscle phosphofructokinase cDNA, *FEBS Lett. 223*:113 (1987).
5. H. Nakajima, T. Noguchi, I. Mineo, T. Yamasaki, N. Kono, T. Tanaka, and S. Tarui, Molecular aspect of myogenic hyperuricemia: Cloning of human muscle phosphofructokinase cDNA, *Adv. Exp. Med. Biol. 253A*:485 (1989).
6. R. A. Poorman, A. Randolph, R. G. Kemp, and R. L. Heinrikson, Evolution of phosphofructokinase - gene duplication and creation of new effector sites, *Nature 309*:467 (1984).
7. C. -P. Lee, M. -C. Kao, B. A. French, S. D. Putney, and S. H. Chang, The rabbit muscle phosphofructokinase gene: Implication for protein structure, function and tissue specificity, *J. Biol. Chem. 262*:4195 (1987).
8. H. Nakajima, T. Yamasaki, T. Noguchi, T. Tanaka, N. Kono, and S. Tarui, Evidence for alternative RNA splicing and possible alternative promoters in the human muscle phosphofructokinase gene at the 5' untranslated region, *Biochem. Biophys. Res. Commun. 166*:637 (1990).
9. H. Nakajima, N. Kono, T. Yamasaki, T. Hamaguchi, K. Hotta, M. Kuwajima, T. Noguchi, T. Tanaka, and S. Tarui, Tissue specificity in expression and alternative RNA splicing of human phosphofructokinase-M and -L genes, *Biochem. Biophys. Res. Commun. 173*:1317 (1990).
10. T. Yamasaki, H. Nakajima, N. Kono, K. Hotta, K. Yamada, E. Imai, M. Kuwajima, T. Noguchi, T. Tanaka, and S. Tarui, Structure of the entire human muscle phosphofructokinase-encoding gene: A two promoter system, *Gene* in press.
11. H. Nakajima, N. Kono, T. Yamasaki, K. Hotta, M. Kawachi, M. Kuwajima, T. Noguchi, T. Tanaka, and S. Tarui, Genetic defect in muscle phosphofructokinase deficiency: Abnormal splicing of the muscle phosphofructokinase gene due to a point mutation at the 5'-splice site, *J. Biol. Chem. 265*:9392 (1990).
12. P. R. Evans, G. W. Farrants, and P. J. Hudson, Phosphofructokinase: Structure and control, *Phil. Trans. R. Soc. Lond. B 293*:53 (1981).
13. P. Y. Chou and G. D. Fasman, Prediction of the secondary structure of proteins from the amino acid sequence, *Adv. Enzymol. 47*:45 (1978).
14. S. M. Mount, A catalogue of splice junction sequences, *Nucl. Acids Res. 10*:459 (1982).
15. R. Treisman, N. J. Proudfoot, M. Shander, T. Maniatis, A single-base change at a splice site in a β_0-thalassemic gene causes abnormal RNA splicing, *Cell 29*:903 (1982).
16. M. R. Green, Pre-mRNA splicing, *Ann. Rev. Genet. 20*:671 (1986).

5′-NUCLEOTIDASE - AN OVERVIEW OF THE LAST THREE YEARS

Linda F. Thompson

Immunobiology and Cancer
Oklahoma Medical Research Foundation
Oklahoma City, OK 73104

INTRODUCTION

There have been many exciting developments regarding the structure, function, and regulation of 5′-nucleotidase (5′-NT) during the three years since the 6th International Symposium on Purine and Pyrimidine Metabolism in Man. First, the gene for 5′-NT was cloned and sequenced from both rat liver and human placenta by Dr. Yukio Ikehara and colleagues[1,2]. The sequencing of the gene, along with other structural information[3,4], has led to the confirmation that 5′-NT is indeed anchored into the plasma membrane via covalent linkage to glycosyl phosphatidylinositol (GPI). Second, 5′-NT was assigned to cluster of differentiation #73 (CD73) by the Fourth International Workshop on Human Leukocyte Differentiation Antigens[5]. The assignment of a CD number gives new stature to 5′-NT as a lymphocyte differentiation antigen in the eyes of the immunological community. Furthermore, the availability of uniform reagents will greatly facilitate studies of the expression and function of this interesting molecule on lymphocytes, as well as in other tissues. New insight into the function of 5′-NT comes from the observation that purified enzyme from chicken gizzard can serve as a receptor for the extracellular matrix proteins fibronectin and laminin[6,7] and from the finding that anti-5′-NT antibodies cause activation of human lymphocytes[8,9]. I will use the terms 5′-NT and CD73 interchangably.

5′-NUCLEOTIDASE STRUCTURE

During the last three years, several papers appeared comparing the properities of various membrane bound and soluble nucleotidases[10-13]. For the purpose of this review, however, I will limit my discussion to the enzyme commonly referred to "ecto-5′-nucleotidase" or low K_M 5′-nucleotidase (E.C. 3.1.3.5) which catalyzes the dephosphorylation of purine and pyrimidine ribo- and deoxyribonucleoside monophosphates[14]. The enzyme has a broad substrate specificity in that it utilizes a variety of purine or pyrimidine bases and either ribose or deoxyribose as the sugar. However, only 5′-monophosphates are hydrolyzed. The K_Ms for most of the substrates are in the range of 5-50 μM and the enzyme has a high turnover number. 5′-NT activity can be distinguished from other nucleotidases and phosphatases in crude extracts, by the fact that its activity is inhibited by the ADP analogue, α,β-methylene adenosine 5′-diphosphate, or AOPCP. 5′-NT has been purified from several tissues including placenta, which we also chose because of the high level of enzyme activity and the ready availability of this tissue.

We found 5'-NT activity in both soluble and membrane bound fractions of human placenta. The purification of the membrane bound form was greatly facilitated by its release from plasma membranes by a highly purified phosphatidylinositol specific phospholipase C (PI-PLC) from Bacillus thuringiensis (gift of Dr. Martin Low)[15]. This treatment yielded completely soluble enzyme which could then be further purified by affinity chromatography without the use of detergent-containing buffers. This material was used to produce a polyclonal anti-5'-NT antiserum in a goat[16] and also for the production of murine monoclonal antibodies[17]. There was some residual membrane-bound 5'-NT activity, even after two PI-PLC treatments, and this material was also purified after solubilization by 1% Triton X-100 + 2% sodium deoxycholate. All three forms of placental 5'-NT (i.e., soluble, membrane bound and PI-PLC releasable, and membrane bound and PI-PLC resistant) showed a similar molecular weight when subjected to SDS-PAGE, both before and after N-glycanase treatment to remove N-linked sugars (73 kDa vs. 56 kDa). We will hear more about the biosynthesis of 5'-NT and the carbohydrate side chains from Dr. Burgemeister[18] (this volume). All three forms of placental 5'-NT were recognized by both our polyclonal antiserum and our monoclonal antibodies, suggesting that they were related structurally. Because we were very much interested in cloning the gene for 5'-NT, we purified all three forms to near homogeneity and obtained N-terminal amino acid sequences[4]. The sequences were very similar and agreed well with the sequence which was deduced from the cloned placental cDNA published by Dr. Ikehara and colleagues[2].

In order to determine the relationship among these three forms of 5'-NT, we also determined the inositol content by mass spectrometry in collaboration with Dr. William Sherman[4]. The PI-PLC releasable form of 5'-NT contained nearly 1 (0.74) mol of myo-inositol per mol of purified protein, as would be expected if it has a GPI membrane anchor. The finding of 0.78 mol myo-inositol/mol of soluble 5'-NT strongly suggests that this form of the enzyme is generated from a previously GPI-anchored form. Later on in this session, Dr. Spychala will present the cDNA cloning of a soluble 5'-NT from human placenta and liver, so we may hear the final word as to the molecular origin of this form of 5'-NT. The membrane bound, but PI-PLC-resistant form of 5'-NT contained 0.47 mol myo-inositol/mol of protein, again suggesting that at least a portion of it also has a GPI anchor and leaving open the possibility of a fourth form of the enzyme with a conventional transmembrane domain. Additional biochemical or molecular genetic evidence will be necessary before the existence of this form of 5'-NT can be confirmed. There are many examples of GPI-anchored proteins which are resistant to cleavage by PI-PLC due to modifications on the inositol ring[19]. In the coming years it will be of great interest to learn whether the different forms of 5'-NT have different functions and to understand factors which may regulate the conversion of the GPI-anchored form to the soluble form. There may also be heterogeneity in the molecular forms of 5'-NT expressed on human lymphocyte subsets. For example, most 5'-NT can be released from T cells by treatment with PI-PLC, while B cell 5'-NT is largely resistant[17]. We have also observed a large increase in 5'-NT activity upon release from lymphocytes in some experiments, and different forms of the enzyme may interact with anti-5'-NT antibodies in different ways. These issues will be discussed by Dr. Christensen (this volume).

GENE CLONING

The greatest amount of structural information about 5'-NT was derived from the cDNA cloning achieved by Dr. Yukio Ikehara and colleages[2]. The published cDNA contains 3547 bp and includes an open reading frame encoding a 574-amino acid polypeptide to give a calculated molecular weight of 63,375 Da. The amino acid sequence contains a 26-residue signal peptide at the amino terminus, 4 N-linked glycosylation sites, and a 25-residue hydrophobic sequence at the carboxy terminus, which is missing in the mature protein. A 14 amino acid peptide isolated from the carboxy terminus of the mature protein contains covalently attached ethanolamine, glycosamine, mannose, inositol, palmitic acid, and stearic acid, components of other GPI-anchored proteins. These results suggest that during the biosynthesis of 5'-NT, the terminal 25 amino acids are cleaved as the GPI anchor is attached. We will hear more about this from Dr. Burgemeister. Northern blots of human placental RNA, probed with this 3.5 kb cDNA, reveal only a single band of approximately 4.1 kb. Thus, there is no obvious suggestion of alternate mRNAs encoding additional molecular forms of 5'-NT, but a more detailed analysis is necessary to rule this out. We anticipate that during the next three years before the next Purine Meeting that much will be learned regarding the regulation of 5'-NT expression, now that the molecular tools are available. Cultured rat mesangial cells will undoubtedly be exploited for this purpose as 5'-NT expression in these cells is upregulated by the cytokines IL-1β and TNFα[20].

MONOCLONAL ANTIBODIES

About 3 1/2 years ago, we finally succeeded in making murine monoclonal antibodies (mAbs) against human 5'-NT[17] after trying on and off for over seven years. We were successful thanks to a suggestion of an immunologist colleague, Dr. Alex Lucas, that we try immunizing a mouse strain other than Balb/c, and when we finally adopted the screening strategy that had worked for Wolf Gutensohn, namely, to test culture supernatants for the ability to block enzyme activity[21]. 5'-NT purified from PI-PLC treated placental plasma membranes was used as the immunogen. Although we were long past the official deadline for the 4th International Workshop on Human Leukocyte Differentiation Antigens, we submitted two antibodies, 1E9 and 7G2 and they were accepted into the B cell panel. Dr. Max Cooper also submitted an antibody, AD2, which was made by immunizing with a pre-B leukemic cell line. When the results of the workshop investigations were tabulated, it was concluded that AD2, 1E9, and 7G2 recognized a previously unclustered lymphocyte antigen which was then designated CD73[5]. Over the last several years, several other groups have made anti-5'-NT mAbs[21,22]. These antibodies recognize at least three distinct epitopes. mAbs 1E9, 7G2, and 5N2 (one of Gutensohn's mAbs) appear to recognize the same or closely related epitopes; AD2 recognizes a second epitope, and three additional mAbs from Gutensohn (5N1, 5N4, and 5N5) recognize an epitope different that that of the above three antibodies. The epitopes recognized by the other antibodies are not yet known.

<u>CD73 Expression in Lymphocytes from Normals and Immunodeficient Patients</u>

The availability of anti-CD73 mAbs has allowed an extensive characterization of the subsets of normal human lymphocytes which express this protein utilizing two color immunofluorescence[17,23]. These reagents have also made it possible to analyze CD73 expression on subsets of lymphocytes from immune deficient patients who often have reduced CD73 expression. CD73 is expressed on approximately 20% of normal human ($CD3^+$) T cells, 12% of helper/inducer ($CD4^+$) T cells, 50% of cytotoxic/suppressor ($CD8^+$) cells, and 75% of B ($CD19^+$) cells. In patients with HIV infection, CD73 expression is virtually absent on $CD4^+$ T cells and is also substantially reduced on $CD8^+$ T cells and B cells[23]. In contrast, in patients with infectious mononucleosis, CD73 expression is virtually absent in both $CD4^+$ and $CD8^+$ T cells, but remains normal on B cells. Although there is a strong correlation between reduced numbers of CD73$^+$ T cells in patients with a variety of immunodeficiency diseases[16], purified CD73$^-$ T cells prepared by cell sorting are fully functional in a variety of clinical assays used to evaluate immune function[24]. Thus, the significance of a lack of CD73 expression on subsets of human T cells has proved elusive. Massaia and colleagues[25], however, have demonstrated that precursors of cytotoxic T cells are primarily found in the CD8$^+$,CD73$^+$ subset.

<u>CD73 Expression in Leukemia</u>

CD73 mAbs have also been utilized to study 5'-NT expression in leukemic cells of B cell lineage. This subject has been of interest since two groups have reported that high 5'-NT expression is a poor prognostic sign for children with acute lymphoblastic leukemia (ALL)[26,27]. The ability to identify children at greater risk for relapse will allow them to receive more potent chemotherapy. Dr. Rob Pieters and colleages[28] compared CD73 expression to that of other B cell differentiation antigens on leukemic cells from 100 patients. The leukemic cells were assigned to four developmental stages according to the expression of a panel of differentiation antigens. Stage I cells (pro-B) are CALLA$^-$, cμ^-, Stage II cells (pre-pre B) are CALLA$^+$, cμ^-, Stage III (pre-B) cells are CALLA$^+$, cμ^+, and Stage IV cells (B) are CALLA$^-$, κ/λ^+. CD73 expression was highest in Stages II and III and correlated positively to the level of CALLA (CD10) expression and inversely to the surface expression of κ/λ, a marker of the most mature stage. It will now be important to determine whether this pattern of CD73 expression also holds for normal B cells at the corresponding stages of development, or whether abnormal CD73 expression is functionally important for the malignant process. CD73 expression was also found to be unexpectedly high on the CD4$^+$ cells of patients with cutaneous T cell lymphoma and adult T cell leukemia/lymphoma[22]. It is curious that the gene encoding 5'-NT has been localized to chromosome 6[29] where cytogenetic alterations are frequently associated with childhood ALL in relapse. Dr. Morgan and colleagues are investigating the role of 5'-NT and other purine salvage enzymes in rescueing malignant cells from chemotherapeutic drugs (this volume).

FUNCTION OF CD73 ON LYMPHOCYTES

New insights into possible unique functions for 5'-NT on lymphocytes have been gained by the observation that some CD73 mAbs have profound effects on T cell proliferation. The CD73 mAb, 1E9, in soluble form and in combination with submitogenic concentrations of the phorbol ester, phorbol myristate acetate (PMA), can drive human T cells to express IL-2 receptors, secrete IL-2, and proliferate[8]. This antibody causes an increase in intracellular [Ca^{2+}]. Proliferation is probably a result of this Ca^{2+} signal in combination with an activation of protein kinase C brought about by PMA. It is curious that mAbs against a variety of other human, murine, and rat GPI-anchored proteins can also induce lymphocyte proliferation in combination with submitogenic PMA[30]. Robinson[31] recently reported that a murine Class I gene could be molecularly engineered to encode a signal transmitting molecule by replacing the normal transmembrane and cytoplasmic domain sequences with an oligonucleotide sequence coding for the attachment of the GPI-anchor of Qa-2. When the Class I molecule was GPI-anchored, anti-Class I antibodies which were not normally mitogenic, stimulated T cells to proliferate. These findings raise the possibility that GPI-anchored proteins are somehow uniquely suited for transmitting activation signals in spite of the fact that they lack cytoplasmic domains. Whether there are natural ligands for some, or all, GPI-anchored lymphocyte proteins, including 5'-NT, which are capable of inducing activation signals by a mechanism similar to that seen with specific monoclonal antibodies is unknown. Mannherz and colleagues have demonstrated that 5'-NT purified from chicken gizzard can interact with the extracellular matrix proteins laminin and fibronectin[6,7]. However, whether laminin and/or fibronectin can serve as natural ligands for cell surface 5'-NT on lymphocytes is not yet known.

More recently, Dr. Massaia and colleagues have devised a more "physiologic" system to demonstrate the potent mitogenic effects of certain CD73 mAbs. It is known that activation of T cells through the CD3/T cell receptor (TCR) complex (i.e., through the antigen receptor) can be mimicked in vitro through the use of antibodies which bind to either CD3 or the TCR. Such antibodies have been invaluable in unravelling the biochemical events which take place during the activation process (such as inositol phosphate turnover, Ca^{2+} fluxes, protein phosphorylations, etc.). Thus, Dr. Massaia investigated whether signals generated by CD73 mAbs could synergize with signals emanating through the CD3/TCR complex. In his system, the CD73 mAb 1E9, when immobilized on plastic, synergizes with the immobilized CD3 mAb OKT3 at submitogenic concentrations to cause vigorous T cell proliferation[9]. More details of this system can be found in Dr. Massaia's contribution to this volume. The degree of T cell proliferation is larger than expected given the fact that only 20-30% of T cells are $CD73^+$ at the initiation of the cultures. By 92 hours, >90% of the cells express IL-2 receptors, suggesting that virtually all T cells eventually become activated. To determine whether these findings could be explained by an induction of CD73 expression on all T cells during activation, we followed CD73 expression at various times after the initiation of the cultures. Much to our surprise, CD73 expression was virtually completely lost during the first four hours of culture. Although the mechanism responsible for this loss of CD73 expression has not yet been identified, one possibility is that the enzyme is being cleaved by an endogenous phospholipse C or D.

It is a major goal of our laboratory to determine whether there is a so-called "natural ligand" which interacts with cell surface CD73 to cause lymphocyte activation in much the same way as our mAb 1E9. Such a ligand could be either a soluble molecule or a molecule on the surface of another cell such as an antigen presenting cell. The CD73 ligand could share structural features with 5'-NT substrates or it could interact with a part of the enzyme separate from the active site. In any event, we hypothesize that the CD73 ligand interacting with CD73 would serve to reduce the threshhold for activation from signals delivered through the CD3/TCR complex in cases where antigen density is low or affinity of antigen for the TCR is sub-optimal. We will hear more on this subject from Dr. Massaia. According to our model, the interaction of CD73 with its ligand would not only generate an activation signal, but it would also lead to the release of CD73 from the cell surface, thus rendering the cell anergic to similar signals at least for some period of time. Our model would further predict that lymphocytes from immune deficient patients with low 5'-NT expression or immature lymphocytes which have not yet expressed 5'-NT would be incapable of receiving activation signals in the above situations.

REFERENCES

1. Misumi, Y., S. Ogata, S. Hirose, and Y. Ikehara. 1990. *J. Biol. Chem.* *265*:2178.
2. Misumi, Y., S. Ogata, K. Ohkubo, S. Hirose, and Y. Ikehara. 1990. *Eur. J. Biochem.* *191*:563.
3. Bailyes, E.M., M. A. J. Ferguson, C. A. L. S. Colaco, and J. P. Luzio. 1990. *Biochem. J.* *265*:907.
4. Klemens, M.R., W. R. Sherman, N. J. Holmberg, J. M. Ruedi, M. G. Low, and L. F. Thompson. 1990. *Biochem. Biophys. Res. Comm.* *172*:1371.
5. Dorken, B., P. Moller, A. Pezzutto, R. Schwartz-Albiez, and G. Moldenhauer. 1989. In *Leucocyte Typing IV: white cell differentiation*

antigens. W. Knapp,et al., ed. Oxford University Press, New York, p. 102.

6. Stochaj, U., J. Dieckhoff, J. Mollenhauer, M. Cramer, and H.G. Mannherz. 1989. *Biochim. Biophys. Acta 992*:385.
7. Stochaj, U., H. Richter, and H.G. Mannherz. 1990. *Eur. J. Cell Biol. 51*:335.
8. Thompson, L.F., J. M. Ruedi, A. Glass, M. G. Low, and A. H. Lucas. 1989. *J. Immunol. 143*:1815.
9. Massaia, M., L. Perrin, A. Bianchi, J. Ruedi, C. Attisano, D. Altieri, G. T. Rijkers, and L. F. Thompson. 1990. *J. Immunol. 145*:1664.
10. Hoglund, L. and P. Reichard. 1990. *J. Biol. Chem. 12*:6589.
11. Spychala, J., V. Madrid-Marina, and I.H. Fox. 1988. *J. Biol. Chem. 263*:18759.
12. Itoh, R. and K. Yamada. 1990. *Int. J. Biochem. 22*:231.
13. Skladanowski, A.C. and A.C. Newby. 1990. *Biochem. J. 268*:117.
14. Naito, Y. and J. M. Lowenstein. 1981. *Biochemistry 20*:5188.
15. Thompson, L.F., J. M. Ruedi, and M. G. Low. 1987. *Biochem. Biophys. Res. Comm. 145*:118.
16. Thompson, L.F., J. M. Ruedi, M. G. Low, and L. T. Clement. 1987. *J. Immunol. 139*:4042.
17. Thompson, L.F., J. M. Ruedi, A. Glass, G. Moldenhauer, P. Moller, M. G. Low, M. R. Klemens, M. Massaia, and A. H. Lucas. 1990. *Tissue Antigens 35*:9.
18. Burgemeister, R., I. Danescu, and W. Gutensohn. 1990. *Biol. Chem. Hoppe Seyler 371*:355.
19. Low, M. 1989. *FASEB J. 3*:1600.
20. Savic, V., V. Stefanovic, N. Ardaillou, and R. Ardaillou. 1991. *Immunology*, in press.
21. Kummer, U., J. Mysliwietz, W. Gutensohn, S. Buschette, H. Hahn, D. Neuser, and R. Munker. 1984. *Immunobiology 166*:203.
22. Fukunaga, Y., S. S. Evans, M. Yamamoto, Y. Ueda, K. Tamura, T. Kakakuwa, D. Gebhard, J. Allopenna, S. Demaria, B. Clarkson, L. F. Thompson, B. Safai, and R. L. Evans. 1989. *Blood 74*:2486.
23. Thompson, L.F., J. M. Ruedi, A. Glass, and A. H. Lucas. 1989. In *Leucocyte Typing IV: White cell differentiation antigens*. W. Knapp,et al., ed. Oxford University Press, New York, p. 104.
24. Thompson, L.F. and J. M. Ruedi. 1989. *J. Immunol. 142*:1518.
25. Dianzani, U., M. Massaia, A. Pileri, C. E. Grossi, and L. T. Clement. 1986. *J. Immunol. 137*:484.
26. Veerman, A.J.P., P. H. G. Hogeman, C. H. van Zantwijk, and P. D. Bezemer. 1985. *Leukemia Research 9*:1227.
27. Gutensohn, W. and E. Thiel. 1990. *Cancer 66*:1755.
28. Pieters, R., L. F. Thompson, G. J. Broekema, D. R. Huismans, G. J. Peters, S. T. Pals, E. Horst, K. Hahlen, and A. J. P. Veerman. 1991. Submitted.
29. Boyle, J.M., Y. Hey, K. -H. Grzeschik, L. Thompson, E. Munro, and M. Fox. 1989. *Hum. Genet. 83*:179.
30. Robinson, P.J. 1990. *Immunol. Today 12*:35.
31. Robinson, P.J., M. Millrain, J. Antoniou, E. Simpson, and A. L. Mellor. 1989. *Nature 342*:85.

CYTOSOLIC PURINE 5'-NUCLEOTIDASE FROM CHICKEN HEART : AN ISOZYME OF THE LIVER ENZYME AS EVIDENCED BY ANTIBODIES

Jun Oka, Roichi Itoh, and Hisashi Ozasa

The National Institute of Health and Nutrition (JO, RI) and Kidney Center, Tokyo Women's Medical College (HO), Shinjuku-ku, Tokyo 162, Japan

INTRODUCTION

Purine 5'-nucleotidase (1) is one of soluble nucleotidases which include pyrimidine 5'-nucleotidase (2), deoxyribonucleotidase (3), and AMP-specific 5'-nucleotidase (4). Purine 5'-nucleotidase preferentially hydrolyzes IMP, GMP, and, to a lesser extent, AMP in the presence of magnesium ions. The enzyme is allosterically activated by ATP (5), diadenosine tetraphosphate (6), 2,3-diphosphoglycerate (7) and decavanadate (8), and inhibited by Pi. Purine 5'-nucleotidase is located in the cytoplasmic matrix of the cell (9, 10).

In chicken, cytosolic purine 5'-nucleotidase is highly purified from the liver (11) and heart (12), but a structural relationship between the enzymes from the two organs remains obscure. We previously reported that the antiserum against chicken liver cytosolic purine 5'-nucleotidase did not precipitate the heart enzyme, and concluded that the two enzymes were differentiated immunochemically (12). However, the above conclusion seemed to be immaturely drawn from a limited number of experiments. We shortly thereafter raised another batch of antibody against the liver enzyme, and showed that this antibody precipitated the heart enzyme (13), suggesting their related structures. In this study, immunological analysis was added to resolve apparent discrepancy in our own data (12, 13), and it was revealed that cytosolic purine 5'-nucleotidases from chicken liver and heart are structurally related isozymes.

EXPERIMENTAL PROCEDURES

From chicken liver and heart were purified cytosolic purine 5'-nucleotidases (12), both of which are assumed to be products of proteolysis (10, 14). An antibody was raised against the purified liver enzyme (13, 14). Immunoprecipitation of the enzymes, and Ouchterlony double diffusion analysis with the antibody and enzymes were done as described previously (9).

Nondenaturing gel electrophoresis of the purified enzymes

followed by immunodiffusion was performed as described below; cytosolic purine 5'-nucleotidases purified from chicken liver and heart were electrophoresed on an agarose gel (Universal electrophoresis film, Corning) in a 0.06 M barbital buffer (pH 8.6) at 15 mA for 25 min. After electrophoresis, a solution containing the antibody was poured into a slot between the two lanes to diffuse in the gel for two days at 4 ℃. Then, the gel was rinsed in phosphate-buffered saline overnight at 4 ℃, and stained with Coomassie blue.

The localization of cytosolic purine 5'-nucleotidase in chicken heart was investigated by indirect immunostaining using horseraddish peroxidase-labeled anti-rabbit IgG as the second antibody (13); chicken heart was *in situ* fixed with 4% formaldehyde-1% glutaraldehyde.

RESULTS AND DISCUSSION

From chicken heart was highly purified cytosolic purine 5'-nucleotidase which is of kinetic properties similar to those of the liver enzyme (12). The antibody raised against the purified liver enzyme (13, 14) is shown to precipitate the heart enzyme, suggesting a relationship between the structures (Fig. 1). We also carefully observed that the equivalence points of the immunoprecipitation were different between the liver and heart enzymes (Fig. 1). In Ouchterlony double diffusion, it was demonstrated that a single precipitin line was produced between the antibody, and the liver and heart enzymes, respectively. Two precipitin lines were fused each other with a small spur, indicating some structural difference between the two enzymes (Fig. 2). As shown in Fig. 3, cytosolic purine 5'-nucleotidases purified from chicken liver and heart apparently possessed different electrophoretic properties on a nondenaturing gel. Taken together, these results indicate that cytosolic purine 5'-nucleotidases from chicken liver and heart are distinct but structurally related isozymes as evidenced by the antibodies.

Cytosolic purine 5'-nucleotidase has a wide tissue distribution (7). The enzymes were immunohistochemically detected in various chicken tissues, *e. g.* the liver, heart, kidney, brain and aorta (13), and the enzyme activity in various mammalian tissues was recently immunotitrated (15), lacking an explanation for the possible existence of isozymes of cytosolic purine 5'-nucleotidase. It would be interesting to distinguish the isozymic pattern, *e. g.* liver-, heart- or other type(s), in various tissues.

Adenosine has established a regulatory role of coronary blood flow in the heart (16). Whether adenosine is formed extra- or intracellularly is a major controversy. Ecto-5'-nucleotidadase is responsible only for adenosine formation from extracellular AMP, most probably derived from released ATP. On the other hand, we once argued that cytosolic purine 5'-nucleotidase is intracellularly producing adenosine as a vasodilator during hypoxia in the heart (17, 18). As shown in Fig. 4, however, immunohistochemical study showed that the enzyme is mainly localized in capillary endothelial cells of the heart (13) which are less possible sites of adenosine production (19, 20). In fact, a role of cytosolic purine 5'-nucleotidase as an enzyme producing adenosine (17, 18) was recently

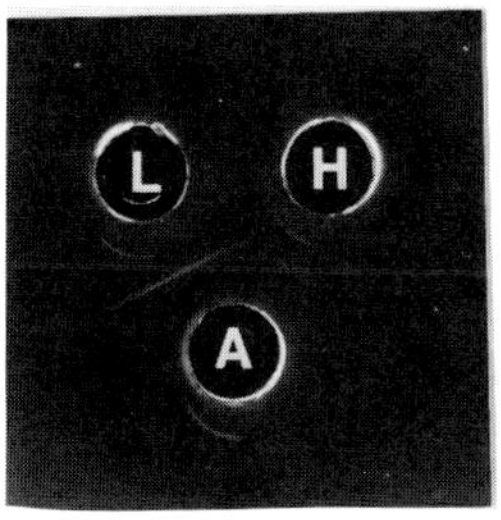

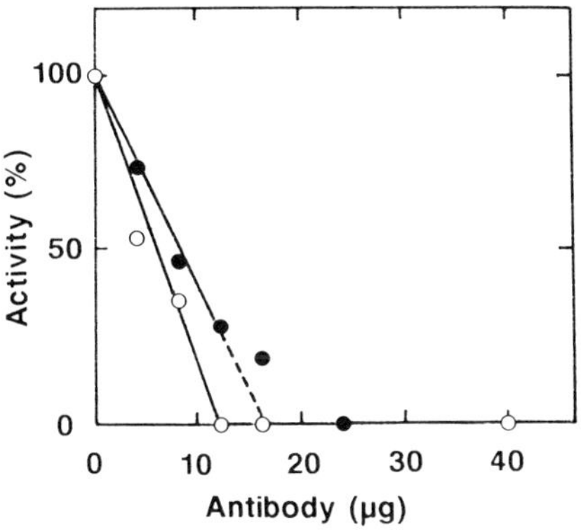

Fig. 1. (left) Immunoprecipitation of cytosolic purine 5'-nucleotidases (0.04 unit each) purified from chicken liver (○) and heart (●) with the antibody against the liver enzyme. Fig. 2. (right) Ouchterlony double diffusion. Well A, the antibody (160 μg); Well L and H, the enzymes purified from chicken liver (0.45 unit) and heart (0.10 unit), respectively.

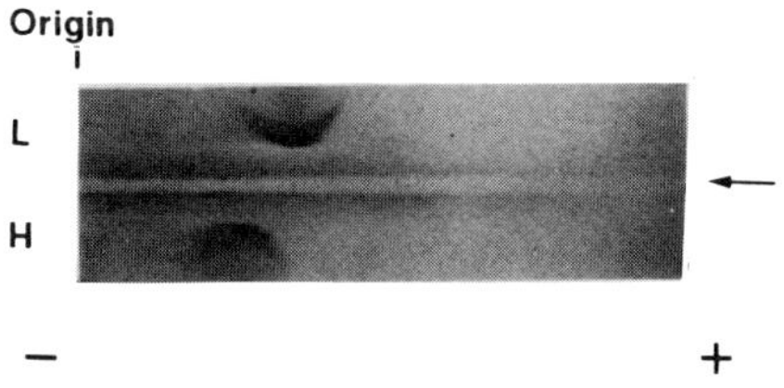

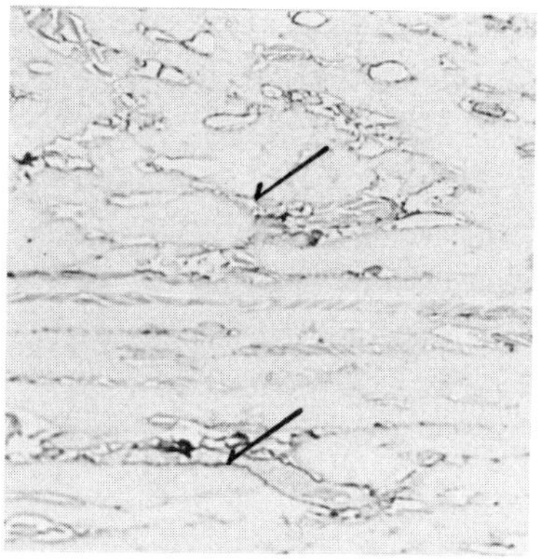

Fig. 3. (left) Immunodiffusion after nondenaturing gel electrophoresis. Lanes L and H, the enzymes purified from chicken liver (0.074 unit) and heart (0.125 unit), respectively. An arrow indicates a slot where a solution containing the antibody (120 μg) was poured. Fig. 4. (right) Chicken heart: immunostaining of capillary endothelial cells (arrows), x 780.

reconsidered through the results in the study using cultured endothelial cells (21). The principal function of cytosolic purine 5'-nucleotidase is assumed to degrade IMP at the first step in the pathway of uric acid formation (1).

AMP-specific 5'-nucleotidase has recently been characterized to be responsible for intracellularly producing adenosine as a signal of net ATP breakdown in the heart (4), although there remains a crucial question whether the enzyme is truely localized in the cytosol of cardiomyocytes (see DISCUSSION in Ref. 10).

ACKNOWLEDGMENTS

This work was supported in part by a Research Grant from Gout Research Foundation of Japan.

REFERENCES

1. Tsushima, K. (1986) Adv. Enzyme Regul. 25, 181-200
2. Paglia, D. E., and Valentine, W. N. (1975) J. Biol. Chem. 250, 7973-7979
3. Höglund, L., and Reichard, P. (1990) J. Biol. Chem. 265, 6589-6595
4. Yamazaki, Y., Truong, V. L., and Lowenstein, J. M. (1991) Biochemistry 30, 1503-1509
5. Van den Berghe, G., Van Pottelsberghe, C., and Hers, H. -G. (1977) Biochem. J. 162, 611-616
6. Pinto, R. M., Canales, J., Sillero, M. A. G., and Sillero, A. (1986) Biochem. Biophys. Res. Commun. 138, 261-267
7. Bontemps, F., Vincent, M. F., Van den Bergh, F., van Waeg, G., and Van den Berghe, G. (1989) Biochim. Biophys. Acta 997, 131-134
8. Le Hir, M. (1991) Biochem. J. 273, 795-798
9. Yokota, S., Oka, J., Ozasa, H., and Itoh, R. (1988) J. Histochem. Cytochem. 36, 983-989
10. Oka, J., Ozasa, H., Itoh, R., and Yokota, S. (1989) Adv. Exp. Med. Biol. 253B, 113-118
11. Naito, Y., and Tsushima, K. (1976) Biochim. Biophys. Acta 438, 159-168
12. Itoh, R., and Oka, J. (1985) Comp. Biochem. Physiol. 81B, 159-163
13. Oka, J., Ozasa, H., Itoh, R., and Yokota, S. (1987) Uric Acid Res. (Tokyo) 11, 13-19 (in Japanese)
14. Oka, J., Ozasa, H., and Itoh, R. (1988) Biochim. Biophys. Acta 953, 114-118
15. Itoh, R., and Yamada, K. (1991) Int. J. Biochem. 23, 461-465
16. Berne, R. M. (1980) Circ. Res. 47, 807-813
17. Itoh, R., Oka, J., and Ozasa, H. (1986) Adv. Exp. Med. Biol. 195B, 299-303
18. Itoh, R., Oka, J., and Ozasa, H. (1986) Biochem. J. 235, 847-851
19. Bardenheuer, H., Whelton, B., and Sparks, H. V. (1987) Circ. Res. 61, 594-600
20. Shryock, J. C., Rubio, R., and Berne, R. M. (1988) Am. J. Physiol. 254, H223-H229
21. Yamada, K., and Itoh, R. (1990) Jpn. J. Pharmacol. 52, Suppl. II, 89P (Abstract)

AMPLIFICATION OF T CELL ACTIVATION INDUCED BY CD73 (ECTO-5'NUCLEOTIDASE) ENGAGEMENT

Massimo Massaia, Carmela Attisano, Valter Redoglia, Alberto Bianchi, Linda F. Thompson*, Umberto Dianzani, and Alessandro Pileri

Divisione Universitaria di Ematologia, Universita' di Torino, Ospedale Molinette, via Genova 3, 10126 Torino, Italy
*Immunobiology & Cancer Dept., OMRF, 825 NE 13th St. Oklahoma City, OK 73104, USA

In the last years, the surface 69 kD differentiation antigen CD73 has been demonstrated to play a significant role in the activation of human T cells. Unlike other surface molecules involved in T cell activation, CD73 combines three distinct features: i) it has ecto-5'nucleotidase (5'NT) activity (E.C. 3.1.3.5); ii) it is linked to the cell membrane via a glycosyl phosphatidylinositol (GPI) anchor; iii) it can very effectively amplificate T cell activation even if it is expressed by a minority only (15-25%) of T cells.

Role of CD73 in T cell activation

Since CD73 is expected to regulate the uptake of extracellular purines by converting nontransportable nucleotides into readily transportable nucleosides, earlier studies explored the role of the purine salvage pathway in T cell activation. It was shown that the 5'NT activity was crucial for mitogen-stimulated T cells when their capacity to synthetise purine de novo is lost (1). When subsets studies showed that CD73 expression in T cells was mainly confined to CD8+ CD11b- cells containing CTL precursors (2), functional studies were carried out to investigate whether CD73 was involved in CTL generation. These studies used competitive biochemical inhibitors of 5'NT (AOPCP) and cross-reacting antisera (3). The latter suppressed three different steps of cytotoxicity: recognition of stimulatory cells, activation of the cytotoxic program, and binding of target cells. AOPCP only suppressed the second step of cytoxicity, suggesting that inhibition of the enzyme activity was not the only factor involved in the suppression of T cell activation.

Deeper insights into the role of CD73 in T cell activation were obtained when more specific immunologic tools became available. The first was a polyclonal antiserum against human

CD73 (4). IF studies at the single cell level confirmed the distribution of CD73 in lymphoid subpopulations previously reported by radiochemical and cytochemical assays (4). More importantly, anti-human CD73 serum allowed to purify CD73+ cells and carry on functional analyses (5). It was demonstrated that CD73+ cells have a unique sensitivity to protein kinase C (PKC) activators (5). PKC plays a central role in the cascade of biochemical events initiated by T cell antigen stimulation. Even more interestingly, it was found that CD73 multivalent cross-linking by the polyclonal antiserum delivered activation signals in the presence of submitogenic concentrations of PMA (6). These data demonstrated that CD73 itself could transduce activating signals across the membrane (6). These results were confirmed and extended with monoclonal antibodies (mAbs). So far, at least 5 different mAbs have been produced againts CD73 (1E9, 7G2, 5N2, 27.2, AD2). Functional studies have mostly been carried out with two of them (1E9, 7G2). The latter clearly initiate T cell activation, but they require additional signalling. First, they require cross-linking by a second antibody (6), monocytes (6), or plastic immobilization (7). Second, CD73 engagement requires costimulatory signals like PMA (6) or signals delivered through surface molecules like CD3 and CD2 (7). These data indicate that CD73 itself can transduce activating signals, but this is not enough to fully activate T cells. Rather, CD73 serves as an agonistic molecule to up-regulate T cell activation initiated by antigen recognition (accessory molecule). This function is similar to that of other accessory molecules like CD4, CD8, CD5, CD28, CD44, CD45 etc.. Compared with these molecules, CD73 has some unique features like its enzyme activity and its GPI-linkage to the plasma membrane. Finally, although CD73 is expressed by a minority of T cells and is restricted to a specific subset of CD8+ lymphocytes, its agonistic activity is much more effective than other accessory molecules. CD28 is one of the best characterized accessory molecule which is expressed by more than 70% of T cells. Nevertheless, side by side experiments have shown that CD73 is much more effective than CD28 to promote T cell proliferation induced by CD3 and CD2 stimulations (7). To explain this, it has been proposed that CD73+ cells may act as a specialized subset to amplify immune responses originated by CD3 and CD2 activation pathways: CD73+ cells fully activated by CD3/CD73 costimulation recruit CD73- cells by secreting a number of cytokines (7). Indeed, large amounts of IL-2 are detected in the supernatant of CD3/CD73 costimulated T cells (7). A role of CD73+ cells in the amplification of immune responses is consistent with its decrease in diseases characterized by impaired immune responses (8).

CD73 expression is restricted to naive (CD45RA+) CD8+ cells

In the last years, it has become clear that a major functional distinction among T cells is whether or not they retain the immunological memory of previous antigen encounters. Memory and naive T cells can be discriminated because of a series of phenotypic differences. The most suitable markers to discriminate them are CD45 isoforms (9). Low molecular weight isoforms (CD45R0) are expressed by memory T cells, whereas high molecular weight isoforms are expressed by naive T cells (CD45RA). Beside phenotypic differences, memory and naive cells have different functional properties and activation requirements. Memory cells (CD45R0+) respond much better to recall antigens or CD3 and CD2 stimulations. Their improved response has been ascribed to several causes: i) different profiles of cytokine production (9); ii) increased expression of accessory molecules (9); iii) pre-assembled organization of the CD3/TCR/CD4/CD8/CD45 transducing apparatus (10). Given the unique ability of CD73 to enhance T cell activation and given its heterogenous expression in CD8+ cells, we have investigated CD73 distribution in purified CD8+ CD45RA+ and CD8+ CD45R0+ cells. These subsets were negatively isolated by using the panning technique. Two sequential pannings with CD4 and CD45RA or CD45R0 mAbs were done. Subsets were 85-95% pure as shown by cytofluorometric analyses. CD73 expression was evaluated by radiochemical and cytofluorometric analyses. Unexpectedly, CD73 expression was found to be confined to naive (CD45RA+) cells (Dianzani et al., manuscript in preparation). This finding pointed out once more the unique characteristics of CD73 as an accessory molecule: so far, CD73 is the only accessory molecule overexpressed in naive rather than memory T cells.

The peculiar expression of CD73 in naive cells may help to revise its role in the immune system and its deficiency in some diseases. One can speculate that its deficiency may reflect either an altered distribution of memory and naive cells or an impaired ability of naive cells to develop an effective immunological memory upon the first encounter with the antigen. In the first case, one should expect an enhanced response to recall antigens and/or CD3 and CD2 stimulations (11); in the latter case, one should find that CD73 has a role in the activation of CD45RA+ cells. Indeed, preliminary studies in our lab indicate that CD73 plays a crucial role in the activation of naive CD8+ cells by making these cells able to overcome their hyporeactivity to CD3 engagement.

CONCLUSIONS

It has been proved that CD73 is an important accessory molecule which can highly upregulate T cell activation. CD73 expression is restricted to a small subset of CD3+ CD8+ CD11b- CD45RA+ cells which are naive T cells containing alloreactive CTL precursors. These data may help to restrict the field on which future studies on CD73 should focus on.

ACKNOWLEDGMENTS

This work was partially supported by Associazione Italiana Ricerca sul Cancro (AIRC), Italy, and Ministero Pubblica Istruzione (MPI), Italy

REFERENCES

1. L.F. Thompson, Ecto-5'nucleotidase can provide the total purine requirements of mitogen-stimulated human T cells and rapidly dividing human B lymphoblastoid cells. J. Immunol. 134:3794 (1985).
2. U. Dianzani, M. Massaia, A. Pileri, C.E. Grossi, L.T. Clement, Differential expression of ecto-5'nucleotidase activity by functionally and phenotypically distinct subpopulations of human Leu2+/T8+ lymphocytes. J. Immunol. 137:484 (1986).

3. M. Massaia, A. Pileri, M. Boccadoro, A. Bianchi, A. Palumbo, U. Dianzani, The generation of alloreactive cytotoxic T lymphocytes requires the expression of ecto-5'nucleotidase activity. J. Immunol. 141:3768 (1988).
4. L.F. Thompson, J.M. Ruedi, M.G. Low, L.T. Clement, Distribution of ecto-5'nucleotidase on subsets of human T and B lymphocytes as detected by indirect immunofluorescence using goat antibodies. J. Immunol. 139:4042 (1987).
5. L.F. Thompson, J.M. Ruedi, Functional characterization of ecto-5'nucleotidase positive and negative human T lymphocytes. J. Immunol. 142:1518 (1989).
6. L.F. Thompson, J.M. Ruedi, A. Glass, M.G. Low, A.H. Lucas, Antibodies to 5'nucleotidase (CD73), a glycosyl-phosphatidylinositol-anchored protein, cause human peripheral blood T cells to proliferate. J. Immunol. 143:1815 (1989).
7. M. Massaia, L. Perrin, A. Bianchi, J.M. Ruedi, C. Attisano C, D. Altieri, G.T. Rijkers, L.F. Thompson, Human T cell activation: synergy between CD73 (ecto-5'nucleotidase) and signals delivered through CD3 and CD2 molecules. J. Immunol. 145:1664 (1990).
8. M. Massaia, D.D.F. Ma, M. Boccadoro, F. Golzio, P. Gavarotti, U. Dianzani, A. Pileri, Decreased ecto-5'nucleotidase activity of peripheral blood lymphocytes in human monoclonal gammopathies: correlation with tumor cell kinetics. Blood 65:530 (1985).
9. A.N. Akbar, M. Salmon, G. Janossy, The synergy between naive and memory T cells during activation. Immunol Today 12:184 (1991).
10. U. Dianzani, M. Luqman, J. Rojo, J. Yagi, J.L. Baron, A. Woods, C.A. Janeway, K. Bottomly, Molecular associations on the T cell surface correlate with immunological memory. Eur. J. Immunol. 20:2249 (1990).
11. M. Massaia, A. Bianchi, C. Attisano, S. Peola, V. Redoglia, U. Dianzani, A. Pileri, Detection of hyperreactive T cells in multiple myeloma by multivalent cross-linking of the CD3/TCR complex. Blood, in press.

STUDIES ON THE STRUCTURE AND BIOSYNTHESIS OF THE PHOSPHATIDYL-INOSITOL-GLYCAN ANCHOR AND THE CARBOHYDRATE SIDE CHAINS OF HUMAN PLACENTAL ECTO-5'-NUCLEOTIDASE

R. Burgemeister , S. Reinsch and W. Gutensohn

Institute of Anthropology and Human Genetics
University of Munich
Goethestr. 31, D-8000 Munich 2, Fed.Rep.Germany

The basis of the experimental protocol to study biosynthesis of human ecto-5'-nucleotidase is metabolic labelling of cells in culture followed by an immunoprecipitation using monoclonal antibodies developed previously (Kummer et al., 1984). The cells used are cultured from chorionic villi from placentas of the 7th to the 12th week of gestation. Various methods of immunohistochemistry were used to show that these cells exhibit a strong surface expression of the enzyme.

If these cells are labelled metabolically with ^{35}S-methionine in short 15 min pulses and 5'-N is followed in chase periods, we see a subtle increase in the apparent molecular weight (70 - 72 000 D) within about 60 min. These changes do not represent overall glycosylation but rather the final steps of carbohydrate processing, i.e. the transformation of the high-mannose- to the complex-type sidechains. We had shown previously that using specific inhibitors of the socalled trimming reactions like deoxynojirimycin, deoxymannojirimycin or Swainsonine we do get the expected small molecular weight differences (Burgemeister et al., 1990). The action of these inhibitors could also be shown in a more direct way by a subsequent digestion of the immunoprecipitated 5'-N with endoglycosidase H. It is interesting to note that a small reduction of the molecular weight of 5'-N upon endo H digestion is observed even in uninhibited cells. This suggests that possibly one of the postulated 4 carbohydrate side chains of human 5'-N remains in an endo H-sensitive configuration even in the mature enzyme molecule.

A double immunoprecipitation was employed to distinguish between 5'-N molecules expressed on the cell surface and those in the interior. Chorionic villus cells take about 2 h to establish a steady state distribution of newly synthesized 5'-N between the surface and internal membranes. A block in carbohydrate processing at anyone of the stages mentioned above obviously does not interfere with the establishment of this distribution. Such blocks also do not seem to interfere with the transfer of newly synthesized 5'-N to its phosphatidyl-inositol-glycan (PIG) anchor. This was shown by digesting 5'-N off the surface of the cells with an anchor-specific phospholipase C from Bacillus thuringiensis (Burgemeister et al., 1990).

Biosynthesis of the PIG-anchor proper was studied with the components of the anchor given as precursors. The following compounds were used and were all shown to be incorporated into the PIG-anchor of 5'-N: Palmitate, stearate, myristate, myo-inositol, ethanolamine, phosphate, mannose, galactose and glucosamine. An unequivocal incorporation of the sugars into the anchor-structure can only be shown either in the presence of a block of glycosylation by tunicamycin or by subsequent digestion of the classical carbohydrate side chains with endoglycosidase F. Both methods were used in the experiments summarized in Fig. 1.

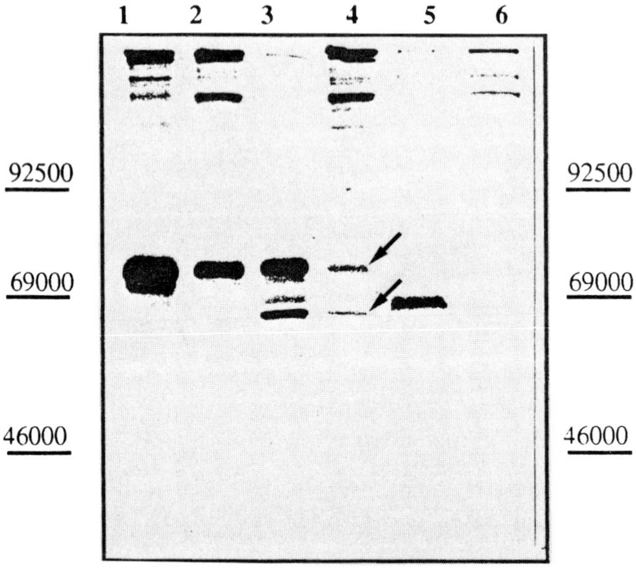

Fig. 1 Chorionic villus cells were labelled in culture with [^{3}H]mannose (lanes 1, 3 and 5) or [^{3}H]galactose (lanes 2, 4 and 6). Control (lanes 1 and 2); presence of tunicamycin during the labelling period (lanes 3 and 4); digestion of the immunoprecipitated 5'-N with endoglycosidase F (lanes 5 and 6). Detection by SDS-PAGE and autoradiography. Arrows point to the position of the fully glycosylated and unglycosylated molecules resp.

Intermediate bands showing up in the tunicamycin experiments (especially Fig. 1, lane 3) are another indication of a total number of 4 carbohydrate side chains, as had been previously predicted by the molecular weight difference between the fully glycosylated and the fully unglycosylated human 5'-N (Burgemeister et al., 1990) and more recently by the total number of N-glycosylation sites in the primary structure (Misumi et al., 1990).

The PIG anchor of ecto-5'-nucleotidase can also be subjected to digestion with a serum-derived phospholipase D. In molecules from cells prelabelled with [^{32}P]phosphate this can be demonstrated by a significant reduction of the phosphate-label. It seems highly probable that this serum-derived activity is responsible for the normal release of 5'-N from many organs into the circulation. The physiological function of this system still remains obscure. Moreover, it seems that the socalled low K_M soluble 5'-nucleotidases isolated from various organs are in effect derived from the membrane-bound ectoenzyme.

The fact of the PIG anchorage of ecto-5'-nucleotidase is meanwhile well established. However, this renders even more difficult the interpretation of results ascribing to 5'-N a role as receptor for stimulatory or costimulatory signals, at least as far as the mechanism of signal transduction is concerned.

Literature

Burgemeister R., Danescu I. and Gutensohn W. (1990)
Glycosylation and processing of ecto-5'-nucleotidase in cultured human chorionic cells.
Biol.Chem.Hoppe-Seyler 371, 355 -361.

Kummer U., Mysliwietz J., Gutensohn W., Buschette S., Jahn H., Neuser D. and Munker R. (1984)
Development and properties of a monoclonal antibody specific for human ecto-5'-nucleotidase.
Immunobiol. 166, 203 - 211.

Misumi Y., Ogata S., Ohkubo K., Hirose S. and Ikehara Y. (1990)
Primary structure of human placental 5'-nucleotidase and identification of the glycolipid anchor in the mature form.
Eur.J.Biochem. 191, 563 - 569.

DIPHOSPHONUCLEOSIDES ARE INDISPENSABLE COFACTORS OF AMP-SPECIFIC CYTOPLASMIC 5'-NUCLEOTIDASE CATALYSED REACTION

Andrzej C Skladanowski, Paola Miscetti, Andrew C Newby and Mariusz M Żydowo

Department of Biochemistry, Academic Medical School Gdansk, Gdansk, Poland and Department of Cardiology, University of Wales College of Medicine, Cardiff, Wales, U.K.

INTRODUCTION

We have proposed that an IMP-selective cytosolic 5'-nucleotidase sets the upper concentration limit of the purine nucleotide pool (Newby & Skladanowski, 1990). An AMP-selective 5'-nucleotidase specifically hydrolyses AMP to form adenosine functioning as a signal to other cells which respond so as to promote energy conservation (Newby, 1984; Newby et al., 1990). To test these hypotheses, more information is needed regarding the molecular structure and binding sites of the 5'-nucleotidases.

We reported earlier (Skladanowski & Newby, 1990) that at near-physiological concentration of AMP, ADP but not ATP can activate AMP-specific 5'-nucleotidase purified from pigeon heart. This means that a fast increase of adenosine formation may occur as a consequence of ATP hydrolysis in this tissue. The activatory effect of other ADP analogues and the protective effect of the enzyme's ligands against heat denaturation have been studied and the conclusions on the binding specificity of activatory site on AMP-specific cytoplasmic 5'-nucleotidase are proposed.

MATERIALS AND METHODS

Materials. ATP, AMP (disodium salt), adenylyl-imidodiphosphate (AMPPNP) and ADP (monopotassium salt) were obtained from Boehringer Corp., Lewes, East Sussex, U.K. ATP and ADP were purified on DEAE-Sephacel (Pharmacia Fine Chemicals, 1974) before use. All other nucleotides, NADH, NADP, adenosine 5'-O(2-thiodiphosphate) (AMPSP), P^1,P^2-di(adenosine-5')tetraphosphate (Ap_4A), α,β-methyleneadenosine 5'-diphosphate (AMPCP), 5-aminoimidazole-4-carboxamide-1-β-D-ribofuranosyl 5'-diphosphate (ZDP) and dithiothreitol were purchased from Sigma Chemical Co., Poole, Dorset, U.K. All chromatographic media, radiochemicals and other reagents were obtained as described previously (Skladanowski & Newby, 1990).

Purification of cytosolic AMP-specific 5'-nucleotidase. The enzyme was isolated essentially as described earlier (Skladanowski & Newby, 1990). Briefly, the procedure comprised of: extraction from the pigeon heart using the buffer with 0.14 M-KCl, salting out with ammonium sulphate (to ca. 70% of saturation), low ionic strenght precipitation, chromatography on phosphocellulose and size-exclusion-chromatography on prepacked column (UltroPak TSK G4000 SW). The enzyme eluted from this column was dialysed to remove added nucleotides, and concentrated in a Minicon block.

Activity of 5'-nucleotidase. The purified 5'-nucleotidase was assayed with use of labelled substrate [2-^{3}H]AMP according to the method described by Newby (1988) and adopted for the purified pigeon heart 5'-nucleotidase (Skladanowski & Newby, 1990). When ZDP was checked as an effector, the progress of reaction was measured by P_i liberation by using the method of Itaya & Ui (1966).

One unit of enzyme activity was defined as the amount needed to convert 1 micromole of substrate per min at 37°C.

Temperature inactivation experiments. The temperature inactivation process was performed at 52°C in the absence or in the presence of enzyme effectors: ADP, TDP and inorganic phosphate. 0.5 ml lots of purified and dialysed AMP-specific 5'-nucleotidase from pigeon heart (0.6 units, 0.05 mg protein) were preincubated in water bath. After different times the aliquots of the preincubation mixture have been withdrawn and immediately placed on ice. The residual activity of 5'-nucleotidase was measured in the optimal conditions at 37°C with 10 mM AMP and 1 mM ADP in the assay mixture.

Estimation of activatory constants. The concentration of compounds giving 50% activation of 5'-nucleotidase ($A_{0.5}$) were computed either by using the linear regression of the transformed Hill equation (Marszalek et al., 1989), V_{max} entry being the experimental value (when ADP was tested), or from either second or third-order polynomic fit to the experimental points (other compounds).

RESULTS AND DISCUSSION

Figure 1 shows how the AMP-selective 5'-nucleotidase responded to the increase of either ATP or ADP at relatively low AMP concentration (100? uM). The sharp increase of the activity in the physiological concentrations of ADP and a virtually flat curve of ATP-dependence indicate that net ATP breakdown might be a trigger for accelerated adenosine production in the heart cytosol.

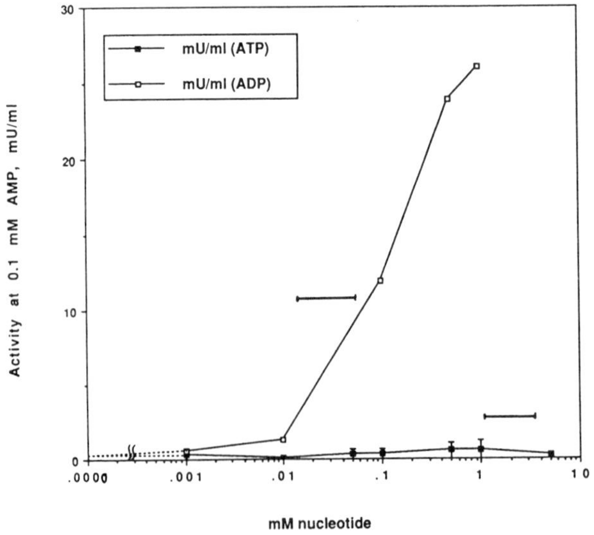

Fig. 1. Effect of ATP and ADP on purified AMP-selective 5'-nucleotidase from pigeon heart.
Each point represents the mean±S.E.M. for 3 incubations.
Horizontal bars represent the ranges of physiological changes.

Table 1 displays the effect of other nucleotides and related compounds on pigeon heart 5'-nucleotidase. It appears that all diphosphonucleosides are able to activate the enzyme with thymidine diphosphate as the most effective when comparing the relative reaction rates and affinity to the enzyme represented by the value of $A_{0.5}$. ZDP was a weak activator probably because its odd base structure. Pirydine nucleotides, NAD and NADP were without effect.

Table 1. Effect of Different Phosphoesters on the Dephosphorylation of AMP by Pigeon Heart Cytoplasmic 5'-Nucleotidase

Effector	Relative reaction rate with 1 mM effector at		$A_{0.5}$ (mM)
	0.1 mM AMP	10 mM AMP	
None	1	1	
ADP	50	2	0.05
ATP	1.7	3.5	-
NADP	1.15	ND[b]	-
NADH	0.72	ND	-
Ap_4A	0.98	1.5	-
AMPCP	1.77	0.77	-
AMPPNP	1.42	1.06	-
AMPSP	42	1.4	0.06
GDP	35	1.63	0.1
CDP	28	1.43	0.3
UDP	36	1.49	0.02
TDP	37[a]	0.86	0.001
ZDP	1.57	ND	ca. 1
2,3-DPG	2.1	1.27	2.9

[a]at 0.1 mM effector
[b]not determined

Neither Ap_4A nor AMPCP influenced the enzyme activity although they activated IMP-selective 5'-nucleotidase from pig lung (Itoh & Yamada, 1990). However other analogue of ADP with modified pyrophosphate moiety, AMPSP, retained its full activating ability. This fact indicates on structural and electrostatic requirements of the part of the binding site proximate to the phosphate residue. Comparing ATP, having visible activating effect at 10 mM AMP concentration with its analogue, AMPPNP, without an activating potential, one may assume that structural requirements are of the primary importance for binding as the electrostatic properties of these compounds are more similar.

2,3-diphosphoglycerate was a weak activator and displayed a cooperative type of binding to the enzyme in contrast to ADP which was bound in a non-cooperative way (Figure 2). 2,3-DPG was reported to be a stronger (reaching 30-fold stimulation) effector for IMP-selective cytoplasmic 5'-nucleotidase from human placenta (Spychala et al., 1988).

The allosteric regulation of the pigeon heart cytoplasmic 5'-nucleotidase may be of a K_M-type as all diphosphonucleosides activated the enzyme only moderately if at all, at high, 10 mM AMP concentration.

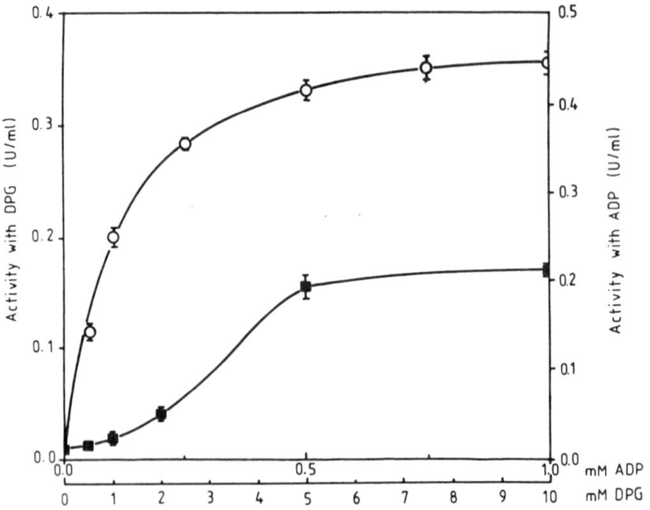

Fig. 2. Dependence of the activity of purified AMP-selective 5-nucleotidase from pigeon heart on the concentration of ADP (-o-) or 2,3-DPG (-■-). Points represent the means±S.E.M. from 3 incubations.

ADP, TDP (activators) and P_i (inhibitor) have been shown to protect AMP-specific 5'-nucleotidase against denaturing factors like heat (Figure 3) or chemical agents. In the presence of TDP the enzyme did not loose its biological activity even after 30 min exposure for 52°C. The dissociation of the enzyme molecule by boiling with sodium dodecylsulphate prior to electrophoresis according to Laemmli (1970) was retarded in the presence of traces of ADP. That was visible in the gel slab as the main band relative to the native mass of the enzyme but not to the subunit (not presented).

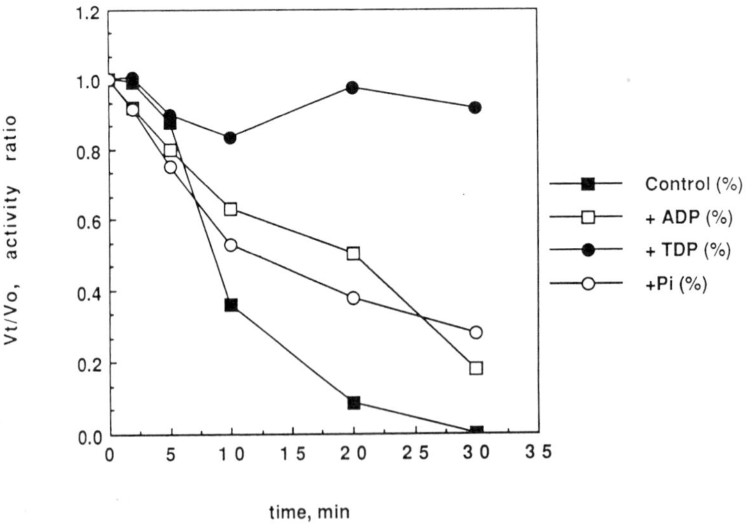

Fig. 3. Inactivation curves of the purified AMP-selective 5'-nucleotidase at 52°C in the absence and in the presence of effectors.
Each point represents the mean±S.E.M. for 3 incubations.

The presence of high affinity but relatively unspecific activatory sites in the molecule of the cytoplasmic AMP-selective 5'-nucleotidase might suggest its more universal role in the maintenance of the total nucleotide homeostasis in myocardial cytoplasm.

ACKNOWLEDGEMENTS

This work was supported by grants from the Medical Research Council (U.K.) and a research grant St-42 from the Polish Committee for Scientific Research.

REFERENCES

Itaya, K., and Ui, M., 1966, A new micromethod for the colorimetric determination of inorganic phosphate., Clin.Chim.Acta, 14:361

Itoh, R., and Yamada, K., 1990, Pig lung 5'-nucleotidase: effect of diadenosine 5',5"'-P^1,P^4-tetraphosphate and its related compounds. Int.J.Biochem., 22:231

Laemmli, U. K., 1970, Cleavage of structural proteins during the assembly of the head of bacteriophage T_4., Nature (London), 227:680

Marszalek, J., Kostrowicki, J., and Spychala, J., 1989, LEHM: a convenient non-linear regression microcomputer program for fitting Michaelis-Menten and Hill models to enzyme kinetic data., CABIOS, 5:239

Newby, A. C., 1984, Adenosine and the concept of 'retaliatory metabolites'. TIBS, 9:42

Newby, A. C., 1988, The pigeon heart 5'-nucleotidase responsible for ischaemia-induced adenosine formation., Biochem.J., 253:123

Newby, A.C.,and Skladanowski, A. C., 1990, 5'-Nucleotidases involved in adenosine formation. in: Abstracts of the 4th International Symposium on Adenosine and Adenine Nucleotides, May 13-17, 1990, Lake Yamanaka, Japan

Newby, A. C., Worku, Y., Meghji, P., Nakazawa, M., and Skladanowski, A. C. 1990, Adenosine: a retaliatory metabolite or not?, News Physiol.Sci., 5:67

Pharmacia Fine Chemicals, 1974, in: "Sephadex Ion Exchangers: A Guide to Ion Exchange Chromatography", p.43, Pharmacia Fine Chemicals, Uppsala

Skladanowski, A. C., and Newby, A. C., 1990, Partial purification and properties of an AMP-specific soluble 5'-nucleotidase from pigeon heart. Biochem.J., 268:117

Spychala, J., Madrid-Marina, V., and Fox, I. H., 1988, High K_M soluble 5'-nucleotidase from human placenta. Properties and allosteric regulations by IMP and AMP., J.Biol.Chem., 263:18759

ADENOSINE METABOLIZING ENZYMES IN BULL AND MAN SPERMATOZOA

A.Minelli, P.Miscetti, M.Moroni R. Fabiani, and I.Mezzasoma

Dip. Med. Sperim. e Sc. Biochim. Univ. di Perugia Italia

INTRODUCTION

A comparative study on the enzymes involved in the metabolism of adenosine in seminal plasma of bull and man (1) has pointed out that hydrolysis of seminal AMP is carried out by enzymes characterized by significant differences in their kinetic parameters. In both mammals , the kinetic parameters of 5'-nucleotidase, the enzyme which specifically cleaves 5'AMP, are similar. In contrast it is stricking that the level of the enzyme in bull seminal plasma is much higher than in man, where non-specific phosphatases predominate. A different involvement of the enzymic activities in regulating seminal adenosine concentration has been consequently proposed.
This report is devoted to the study of adenosine metabolizing enzymes of spermatozoa from man and bull in an attempt to shed light on the physiological role of adenosine in these mammalian semens.

MATERIALS AND METHODS

Bull sperm was provided by Centro Tori (PERUGIA); human ejaculates, provided by Centro di Fisiopatologia della Riproduzione (BOLOGNA), were obtained by normal volunteers.
Bull and man semen was centrifuged at 770Xg X 20 min. at 4° C: the spermatozoa were then washed by suspension of the cell pellet in fresh buffered salt solution (2).
Cell were examined under the light microscope after staining with eosine. All enzyme activities were assayed at 37°C by HPLC, unless otherwise stated. Enzyme activities are expressed in nmoles produced/min/ 10^{6} cells. Determination of spermatozoa adenosine was carried out according to Newby (3). Determination of spermatozoa nucleosides was carried out according to De Korte (4).

RESULTS AND DISCUSSION

Data on AMP hydrolysing activities are reported in Tab.1. The levels of 5'-nucleotidase show the same variability already found for seminal plasma enzymes (1) . Kinetic parameters of AMPase and 5'-nucleotidase are shown in Tab.2.: a Km value of millimolar order is reported for human AMPase with a substrate efficiency roughly 150- fold lower than in bull . As for 5'-nucleotidase, the bull spermatozoa enzyme has a substrate and a catalityc efficiency higher than in man. These data suggest a relatively small involvement of these human enzymic activities in the production of adenosine.

Purine and Pyrimidine Metabolism in Man VII, Part B
Edited by R.A. Harkness *et al.*, Plenum Press, New York, 1991

Table 1. AMP hydrolysing activities in spermatozoa

Enzyme	Bull		Man	
	nmol/min/10^6	%	nmol/min/10^6	%
AMPase	0.402	100	1.03	100
5'nucleotidase	0.336	83.5	0.02	1.9
Acid Phosphatases	0.101	25	0.83	81
Alkaline Phosphatase	-	-	0.18	18

Values are the mean of at least 5 observations.

Another enzymic activity which might be responsible for adenosine release is Ado- Hcy-Hydrolase. Such activity has not been found either in bull or in man spermatozoa, as already stated in seminal plasma (1) .
Enzyme activities involved in adenosine utilization and degradation have also been investigated and the results are reported in Tab. 3. Adenosine seems to be equally deaminated in bull and man , while d-Ado appears to be a better substrate for the enzyme of human spermatozoa.
The kinetic parameters of PNP a show a degree of variability that results in a similar substrate efficiency for d-Ino, while Ino is a better substate for the bovine enzyme by a factor of about 2 .
Xanthine oxidase has not been detected in either mammalian spermatozoa, as already found in seminal plasma (1).
Adenosine kinase seems to be present in both mammalian spermatozoa. Kinetic parameters cannot be given because a precise quantitation of this activity is actually in progress.

Determination of adenosine and its related nucleotides is shown in Tab. 4. It can be seen that the spermatozoa of bull and man contain nearly the same amount of nucleotides while Ado results to be higher in man by a factor of 4. This finding is likely related to a different metabolic regulation of Ado content exerted by the foregoing enzymes .

Table 2. Kinetic paremeters of AMPase and 5'-nucleotidase in spermatozoa from bull and man

Enzyme	Bull		Man	
	Km µM	Vmax nmol/min/10^6	Km µM	Vmax nmol/min/10^6
AMPase	104	1.02	11840	0.822
5'-nucleotidase	84	0.89	31	0.04

Values are the mean of at least 5 observations.

Table 3. Kinetic parameters of ADA and PNP in spermatozoa from bull and man

Enzyme		Bull		Man	
		Km µM	Vmax nmol/min/10^6	Km µM	Vmax nmol/min/10^6
ADA	a	192	0.003	274	0.007
	b	119	0.07	202	0.06
PNP	c	576	0.01	1200	0.01
	d	955	0.005	744	0.004

a:Ado; b: d-Ado; c: Ino; d: d-Ino as substrate. Values are the mean of at least 5 observations.

Tab. 4. Adenosine and nucleotides in spermatozoa from bull and man

	Bull nmol/10^8	Man nmol/10^8
ATP	9.9	8.7
ADP	2.4	2.9
AMP	1.9	1.8
Ado	0.4	1.7

Values are the mean of at least 5 observations.

REFERENCES

1. A.Minelli, R.Fabiani, M.Moroni, and I. Mezzasoma, Adenosine metabolizing enzymes in seminal plasma of bull and man: a comparative study, Comp. Biochem. Physiol. 97B : 675 (1990).
2. K. Brouns, and E. R. Casillas, Partial purification and characterization of an Acetylcarnitine Hydrolase from bovine epididymal spermatozoa, Arch. Biochem. Biophys. 977: 1 (1990).
3. A.C. Newby, Role of adenosine deaminase ecto 5'-nucleotidase in rat poly-morphonuclear leucocytes, Biochem.J. 186 : 907 (1980)
4. D. De Korte, Clin. Chimica Acta, 148: 185 (1985).

CYTOSOLIC 5'-NUCLEOTIDASE/PHOSPHOTRANSFERASE OF HUMAN COLON CARCINOMA

M.G. Tozzi, M. Camici, S. Allegrini, R. Pesi, M. Turriani, A. Del Corso, P.L. Ipata

Dipartimento di Fisiologia e Biochimica, Lab. di Biochimica. Via S. Maria, 55, 56100 Pisa, Italy

INTRODUCTION

Cytosolic 5'-nucleotidase acting preferentially on IMP and GMP has been purified from many sources[1,2,3]. The enzyme resulted to be allosterically activated by ATP, ADP and 2,3-diphosphoglycerate and inhibited by phosphate. Cytosolic 5'-nucleotidase appears to be widely distributed in mammalian tissues suggesting that it may have an essential function in cell metabolism. Furthermore the enzyme activity appears to be higher in tissues in active DNA synthesis or with a high turnover rate of nucleic acids and their precursors[4].

Cytosolic 5'-nucleotidase purified from several sources was also found to catalyze the transfer of the phosphate group of IMP to inosine. The enzyme purified from rat liver was demonstrated to catalyze the phosphate transfer by forming a phosphorylated enzyme as intermediate[5]. Even though the interest in this phosphotransferase activity has been stimulated by the demonstration that purified cytosolic 5'-nucleotidase can phosphorylate 2'-3'-dideoxyinosine and acyclovir in vitro[6,7], the kinetic characteristics and the physiological role of the phosphate transfer catalyzed by 5'-nucleotidase have not been accurately investigated.

In this paper we report the identification and purification of a cytosolic 5'-nucleotidase/phosphotransferase activity from human colon carcinoma and some of its kinetic characteristics which may contribute to the identification of its physiological role.

METHODS

Enzyme assays. Cytosolic 5'-nucleotidase activity was measured by determining the amount of nucleoside formed in the presence of 2 mM labelled nucleoside monophosphate, 20 mM $MgCl_2$, 4.5 mM ATP, 100 mM Tris-HCl pH 7.4, 1 mM inosine in a total volume of 50 or 100 μl. The nucleoside produced was separated by PEI-cellulose chromatography developed with water. The phosphate formed during the reaction catalyzed by 5'-nucleotidase/nucleoside phosphotransferase was determined according to Chen et al.[8]

Nucleoside phosphotransferase activity was measured by determining the nucleoside monophosphate synthesized in the presence of labelled nucleoside. Incubation medium was the same described for 5'-nucleotidase assay with the exception of the labelled compound; in this case 1 mM 8^{14}C inosine was used. The reactions were stopped by spotting 10 μl of the incubation medium on DE-81 paper disks followed by washing 15 min in 1 mM ammonium formate and 10 min in two changes of water.

Preparation of tissue extracts. Tissue samples were obtained from patient operated upon for intestinal cancer. Tissue extracts were prepared as previously described[9].

Purification of 5'-nucleotidase. The soluble fractions of tumoral tissue extracts were pooled. Thirty ml of the pooled fractions (1.65 mg/ml of protein) were applied on DEAE-cellulose column (Whatman, DE-52) previously equilibrated with 10 mM Tris HCl pH 7.4 containing 1 mM dithiothreitol (standard buffer). Elution was carried out with a linear gradient of NaCl from 0 to 0.4 M NaCl in the standard buffer. Active fractions were pooled and dialyzed against standard buffer. The sample was then loaded on a Matrex Green A column equilibrated with standard buffer. After washing with standard buffer supplemented with 0.4 M KCl, proteins were eluted with a linear gradient from 0.4 to 1.6 M KCl in standard buffer. Active fractions were pooled, concentrated and applied on a precalibrated Sephacryl S-300 column equilibrated with standard buffer supplemented with 0.2 M KCl. The enzyme characterization was performed using partially purified enzyme preparation obtained after Matrex Green A chromatography supplemented with 1 mg/ml bovine serum albumin.

RESULTS AND DISCUSSION

The rate of inosine formation from IMP in the presence of ATP (namely 5'-nucleotidase activity) and the rate of IMP formation from inosine in the presence of ATP and IMP (namely phosphotransferase activity) were measured in human colon carcinoma extracts and in peritumoral tissues apparently normal from the same patients. Both specific activities were significantly increased in tumor with respect to normal tissues (about 2.5 times). We have recently demonstrated that in samples of human colon carcinoma some purine salvage enzymes, such as hypoxanthine-guanine phosphoribosyltransferase and purine nucleoside phosphorylase display an increased specific activity with respect to the control[9], thus indicating that the metabolic program in tumor cells is in favor of IMP accumulation. Our results on cytosolic 5'-nucleotidase indicate that this enzyme behave as a purine salvage enzyme and are in line with the observation that the enzyme displays a higher specific activity in tissues in active nucleic acids synthesis. The enzyme was purified from human colon carcinoma extracts in three steps with a yield of about 10%. The final preparation had a specific activity of 1,6 μmol of IMP synthesized per min per mg of protein.

Fig 1 shows Matrex Green-A elution profile. Both enzyme activities coelute at a KCl concentration of about 1 M. Fig.2 shows the elution profile of Sephacryl S-300 gel filtration, 5'-nucleotidase and nucleoside phosphotransferase activities coelute also in this chromatographic system and indicate a molecular weight of about 250000 which is in good agreement with the enzyme purified from other sources[3].

Owing to the instability of the purified enzyme, the Matrex pooled fractions were used for the kinetic characterization. The following enzymatic activities were found to be absent in this preparation: HGPRT, PNP, adenosine kinase, inosine kinase, guanosine kinase, deoxyguanosine kinase, adenosine deaminase, alkaline phosphatase, acid phosphatase. Table 1 shows the nucleoside monophosphates as substrates of 5'-nucleotidase activity (rate of nucleoside production) and as phosphate donors in the nucleoside phosphotransferase assay (rate of IMP formation from inosine). In both assays IMP, GMP and their deoxycounterparts were the best substrates, while AMP was a poor substrate and the enzyme activity on deoxyAMP was almost undetectable. At a final concentration of 1 mM and in the presence of IMP and ATP, inosine and deoxyinosine appeared to be the best substrates for the phosphotransferase reaction.

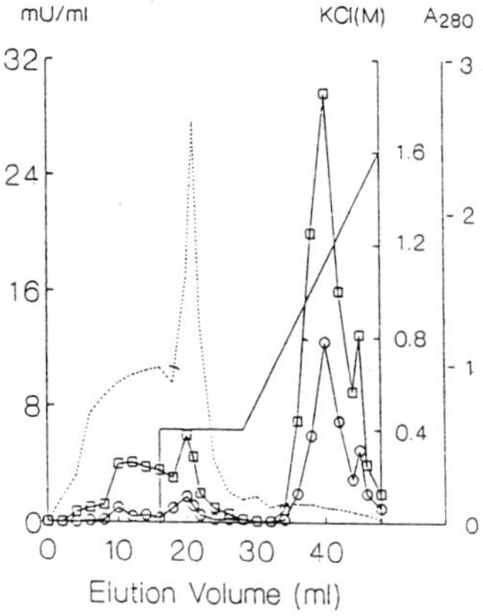

Fig.1. Matrex Green A chromatography. (·····):absorbance at 280 nm; (—):KCl; (□):5'N activity; (○): PhT activity.

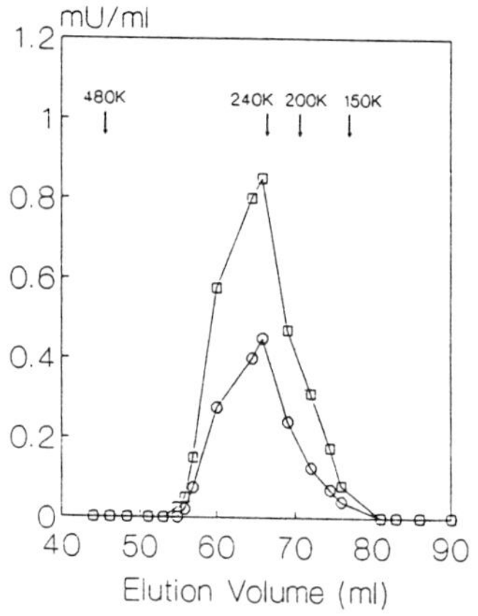

Fig.2. Sephacryl S-300 chromatography. (□): 5'N activity ; (○); PhT activity.

TAB.1. NUCLEOSIDE MONOPHOSPHATE AS SUBSTRATES OF 5'-NUCLEOTIDASE (5'-N) AND PHOSPHOTRANSFERASE (PhT) ACTIVITIES

	none	IMP	dGMP	AMP	dAMP	dIMP	GMP	CMP
5'N	0	100	68	18	0.4	N.D	N.D	N.D
PhT	0	78	76	11	3.0	100	82	10

N.D = Not determined ; Enzyme activities are expressed as %.

The Km measured in the presence of ATP was 830 μM for both inosine and deoxyinosine. The enzyme was also able to transfer the phosphate from IMP to 8-azaguanosine and 2',3'-dideoxyinosine, two nucleoside analogs displaying antimetabolite and antiviral effects, respectively, at a rate comparable to that measured with natural substrates. As previously reported for cytosolic 5'-nucleotidase purified from other sources[1,2,3], the enzyme purified from human colon carcinoma was activated by ATP, ADP and 2,3-diphosphoglycerate; the same compounds activated also phosphotransferase activity (Table 2). The effect of ATP on 5'-nucleotidase activity was an increase of about 3 times of Vmax and a decrease of the Km for IMP from 80 μM to 20 μM. Phosphotransferase activity displayed an absolute requirement for some effectors. In fact, the value referred to as the activity measured in the absence of effectors, was measured at an actual KCl concentration of 200 mM. This salt

TAB.2. LIST OF EFFECTORS FOR CYTOSOLIC 5'-NUCLEOTIDASE (5'N) AND NUCLEOSIDE PHOSPHOTRANSFERASE (PhT) ACTIVITIES

	none	ATP	DPG	dATP	ADP	GTP
5'N	11.8	39.0	31.2	N.D	N.D	N.D
PhT	3.7	25.8	22.1	27.5	25.6	22.8

N.D = Not determined ; Enzyme activities are expressed as nmol/min/mg .

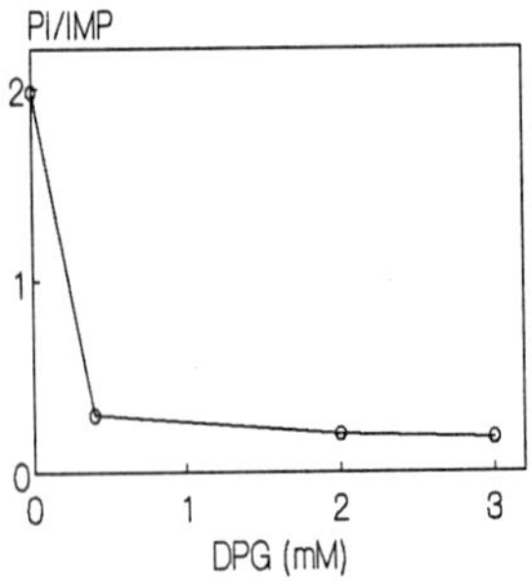

Fig.3. Ratio between IMP and Pi formation during the assays performed at different concentrations of DPG.

is a stabilizer of enzymatic activity and at high concentration can substitute for ATP as activator. The phosphotransferase activity was completely undetectable if measured in the absence of KCl and any other activator. Furthermore both 5'-nucleotidase and phosphotransferase activities displayed a complete requirement for $MgCl_2$. 5'-nucleotidase activity is usually measured as the formation of inosine from IMP which is the step in common between the hydrolytic and phosphotransferasic activity of the enzyme. In fact the enzyme-phosphate intermediate can transfer the phosphoryl moiety either to water or to a nucleoside present in the mixture. In this view, the rate of inosine formation from IMP can be taken as the sum of the rate of IMP hydrolysis and the rate of the phosphate transfer. Table 2 shows that while the rate of inosine formation increased, approximately 3 fold, in the presence of activators, the rate of phosphate transfer increased of about 7 fold in the same condition. On the base of these results, the presence of activator seems to favor the transfer of the phosphate to a nucleoside instead of the water. In fact, the presence of increasing concentrations of DPG caused a decrease of the ratio between the phosphate and IMP production during the assays (Fig 2). This finding might throw a new light on the function of cytosolic 5'-nucleotidase, which cannot be considered as a pure catabolic enzyme, but in appropriate metabolic conditions, might function preferentially as a phosphotransferase. Finally, since this enzyme activity is significantly enhanced in human colon carcinoma, and acts on purine analogs as well, it should be taken into account in the study of purine pro-drugs activation.

REFERENCES

1. Pinto, R.M., Canales, J., Faraldo, A., Sillerio, A., and Gunther Sillerio, M.A. (1987) Comp. Biochem. Physiol. 86B, 49-53.
2. Truong,Vu L., Collins, A. R., and Lowenstein, J.M. (1988) Biochem. J. 253, 117-121.
3. Bontemps, F., Van Den Berghe, G., and Hers, H.G. (1988) Biochem. J. 250, 687-696.
4. Itoh, R., and Yamada, K. (1991) Int. J. Biochem. 23, 461-465.
5. Worku, J., and Newby A.C. (1982) Biochem. J. 205, 503-510.
6. Keller, P.M., McKee, S.A., and Fyfe, J.A. (1985) J. Biol. Chem 260, 8664-8667.
7. Johnson, M.A., and Fridland, A. (1989) Mol. Pharmacol. 36, 291-295.
8. Chen, P.S., Toribara, Jr.,T.,Y., and Warner, H. (1956) Anal. Chem. 28, 1756-1758.
9. Camici, M., Tozzi, M.G., Allegrini, S., Del Corso, A., Sanfilippo, O., Daidone, M.G., De Marco, C., and Ipata P.L. (1990) Cancer. Biochem. Biophys. 11, 201-209.

PURINE NUCLEOSIDE PHOSPHORYLASE: ALLOSTERIC REGULATION OF A DISSOCIATING ENZYME

Thomas W. Traut, Patricia A. Ropp, and Allen Poma

Dept. of Biochemistry and Biophysics
University of North Carolina School of Medicine
Chapel Hill, NC 27599 U.S.A.

INTRODUCTION

Purine nucleoside phosphorylase (PNP; EC 2.4.2.1) catalyzes the phosphorolysis of (d)guanosine or inosine to form the base (guanine or hypoxanthine) and (d)ribose-1-P; the enzyme has therefore been defined as catabolic (1,2). The enzyme may also be considered as biosynthetic, since the salvage of purines to form nucleotides is largely via a one step phosphoribosyltransferase reaction. Reports characterizing this enzyme have not been consistent. Based on the native M_r, the enzyme has been reported as a trimer, a dimer, or a monomer (references in #3). Also, some studies have reported negative cooperativity with phosphate (references in #4), or nucleosides (references in #4). Some ambiguity about the above results arose when other studies reported no cooperativity with phosphate, or with nucleosides (references in #4).

The three-dimensional structure for human PNP was recently reported (5), with some evidence for two distinct phosphate binding sites. The present work shows physical and kinetic studies that are also best explained by two phosphate binding sites: the catalytic site plus a potential regulatory site. The allosteric effects of ligands, plus the possible oxidation of enzyme cysteines are suggested to account for many of the varying results reported for this enzyme.

MATERIALS AND METHODS

Crystalline PNP (bovine spleen) was from Sigma; all other reagents were the best grade available. The standard spectrophotometric assay measured the appearance of uric acid, at 293 nm, in the presence of the coupling enzyme, xanthine oxidase. A 1 ml reaction mixture contained 50 mM citrate (pH 6.5), 0.2 units xanthine oxidase, and PNP at concentrations in figure legends. The substrates were normally at fixed concentrations (10 mM P_i and 0.5 mM inosine) or at concentrations shown in figures.

The native M_r was determined by gel permeation chromatography with a Sephacryl S-300 column (1 x 110 cm), equilibrated in standard buffer $\pm$ P_i. A more detailed description has been reported for the various experiments (3, 4). For stability studies enzyme was freshly prepared $\pm$ dithiothreitol, and stored at 4°C for the times indicated in figure legends.

RESULTS

The activity of PNP is very sensitive to enzyme concentration, and continued

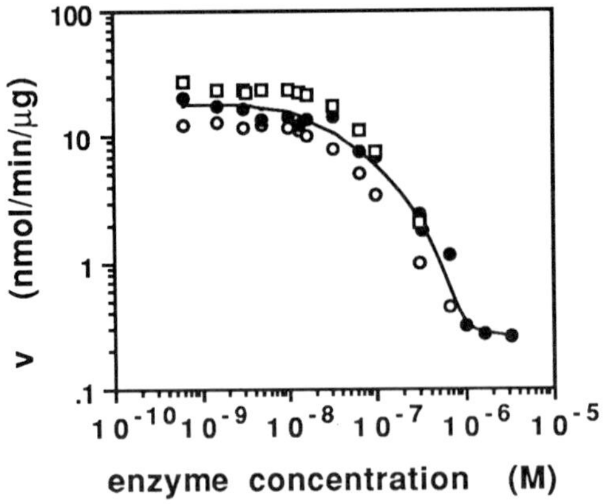

Fig. 1. Change in specific activity of PNP with dilution of enzyme in buffer only (○), or in buffer with 50 μM inosine (q), or 50 mM P_i (●).

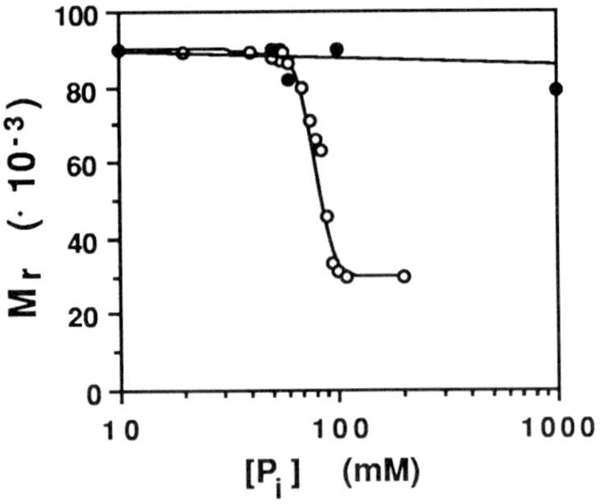

Fig. 2 Dissociaition of PNP in response to P_i. Gel permeation chromatography at 24°C (○), or at 4°C (●).

dilution led to a significant increase in specific activity (Fig. 1). Dilution of enzyme in the presence of one of the substrates produced the maximum specific activity. Since dilution frequently leads to dissociation of polymeric enzymes, the ability of PNP to dissociate was tested. Fig. 2 shows that the enzyme behaves as a stable trimer in standard buffer, ± P_i up to 50 mM. At higher phosphate concentrations dissociation became evident, and the enzyme was completely in the monomeric form (M_r = 30 kDa) at P_i concentrations ≥ 100 mM. Inosine at a concentration of 10 mM had no influence on dissociation. Although each substrate stabilized the enzyme during dilution (Fig. 1), they must stabilize different conformations.

Kinetic studies support this interpretation. With inosine at low concentrations, the enzyme exhibits positive cooperativity (Fig. 3); the diagonal reference line in the Hill plots has a slope of 1.0. Fairly comparable positive cooperativity was observed when PNP was very dilute (Fig. 3A) or 30 fold more concentrated (Fig. 3B). At inosine concentrations above 60 μM, negative cooperativity became very evident. However, with the alternative nucleoside substrate guanosine, no cooperativity was observed at any concentration of the substrate (Fig. 3C).

The activity of PNP towards the substrate inosine was sensitive to the presence of reducing thiols. Storing PNP in solution led to certain specific losses in enzyme function. After 21 days storage ± dithiothreitol, positive cooperativity was difficult to observe (Figs. 4A, 4B). In the presence of DTT, negative cooperativity was retained (Fig. 4A), but in the absence of DTT negative cooperativity could not be observed (Fig. 4B). Thus the presence of DTT had no effect on this decline in positive cooperativity, but did maintain the expression of negative cooperativity. A brief incubation (1 h) of freshly prepared enzyme with H_2O_2 gave a comparable change in the kinetics (not shown). Overall, stored enzyme showed a significant change in affinity for inosine. For freshly prepared enzyme the $S_{0.5}$ with inosine was 23 μM; for enzyme stored 21 days the affinity decreased about 10 fold, with an $S_{0.5}$ of 242 μM. When comparable experiments were done with the substrate P_i, there was no observable effect of storage, or of the absence of reducing thiols (Fig. 5).

DISCUSSION

The dilution curves (Fig. 1) clearly show that at concentrations of PNP below 10^{-7} M, the enzyme exhibited a significant increase in specific activity, consistent with a conformational change stabilized by dissociation. Since the assay mixture was always adjusted to contain a constant concentration of each substrate, then the effect of substrate in the dilution experiment represents the stabilization of a more optimal conformation before enzyme activity was measured. Although the enzyme

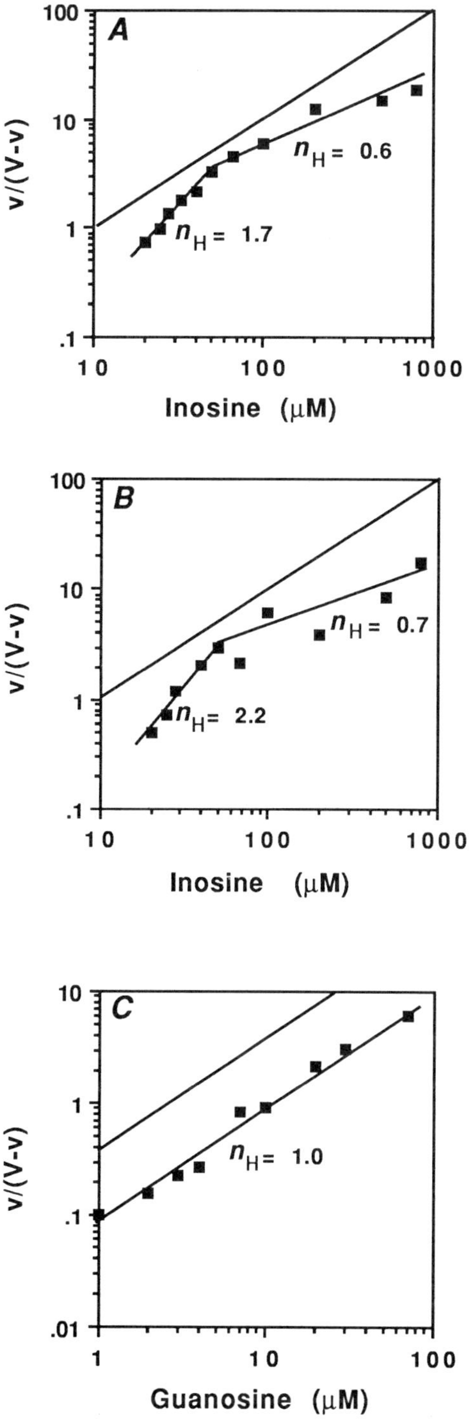

Fig. 3. Activity of PNP with the nucleoside substrates inosine or guanosine. Enzyme concentration was 0.63 μg/ml ((A, C) or at 20 μg/ml (B).

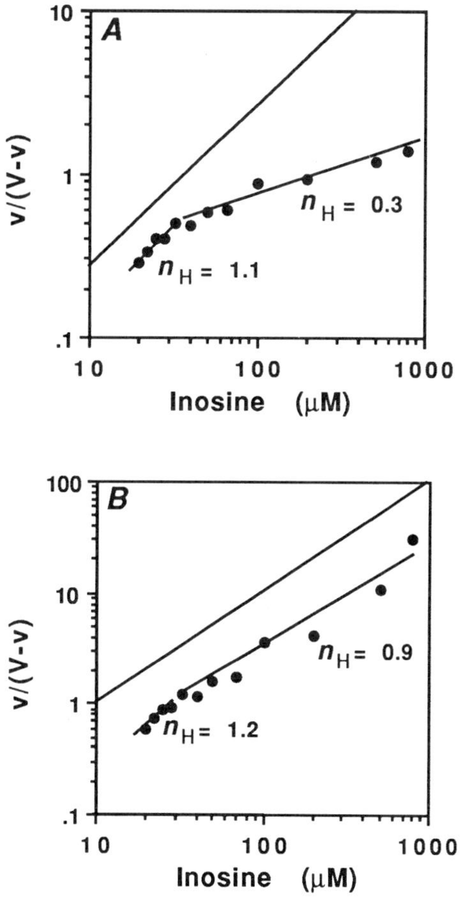

Fig. 4. Stability of PNP as a function of storage. Enzyme was stored at 4°C for 21 days with (A) or without (B) dithiothreitol.

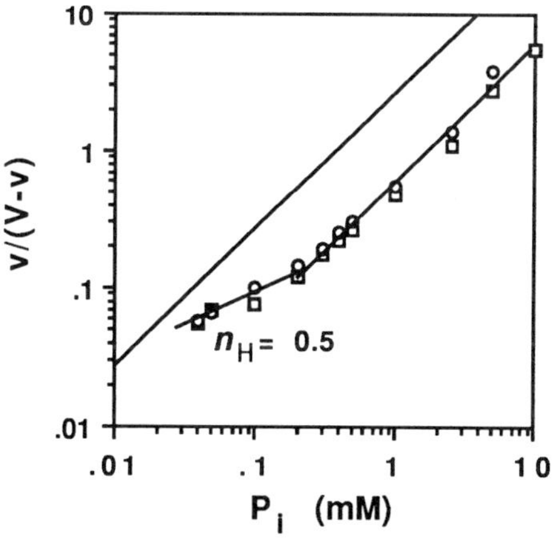

Fig. 5. Activity of PNP with phopsphate. Enzyme was prepared fresh (m), or stored for 30 days (q); 0.63 μg PNP/ml.

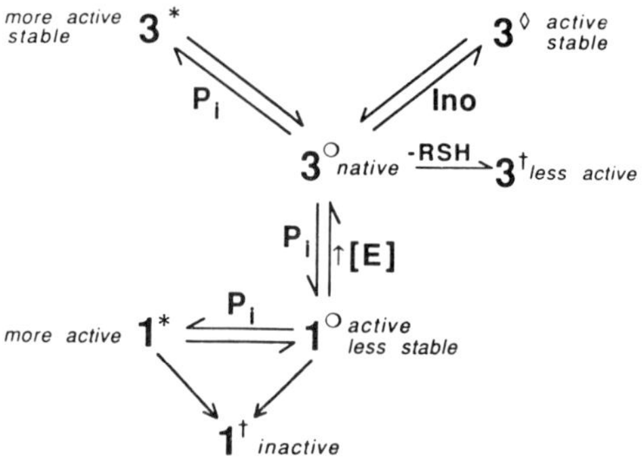

Fig. 6. Summary model for different polymeric and conformational states of purine nucleoside phosphorylase.

appears most active as a diluted, and therefore dissociated monomer, in most eucaryotic sources this enzyme, based on the final specific activity of different purification schemes, and the purification fold, is generally at a concentration of 20-100 μg/ml (0.7 - 3.5 μM), and should therefore exist as a stable trimer.

Physiological concentrations for inosine are in the range (1-20 μM), so that the positive cooperativity observed may be physiologically relevant. The negative cooperativity occurs at higher concentrations of inosine that are less likely to occur. In contrast, since physiological concentrations of P_i are in the range (0.4 - 6 mM), the observed negative cooperativity with this substrate may have a physiological meaning. The affinity for phosphate as measured by catalytic activity ($S_{0.5}$ = 3.1 mM) is substantially different from a calculated $L_{0.5}$ (3) for the dissociation process in response to phosphate ($L_{0.5}$ = 88 mM). In addition, the Hill coefficient for activity was 0.5, while a similar coefficient calculated from the data in Fig. 2 has the value of 10.8, indicating unusual positive cooperativity (3). Clearly, 2 separate sites on the enzyme are required to explain such different affinities and cooperativity. Thus, these results are entirely consistent with the suggestion of 2 P_i sites on this protein based on the crystal structure (5). The affinity of P_i at the second site is too weak for P_i to regulate the enzyme at this site. It is more reasonable that P_i acts as a modest analog of some nucleotide that is the true regulatory effector for PNP in vivo.

Studies described here, as well as additional experiments reported (3, 4) are best summarized by the model in Fig. 7. This describes the various properties exhibited by the enzyme in vitro. It is important to distinguish such a model from conditions in vivo, where reducing thiols (RSH) are always present, and where enzyme concentration is not likely to decrease by orders of magnitude. While the dissociated enzyme is more active, it is also less stable. This may explain why this enzyme is so abundant in many tissues or microorganisms.

REFERENCES

1. Parks, R.E., Jr. and Agarwal, R.P., Purine nucleoside phosphorylase, in "The Enzymes", (Boyer, P., Ed.) Vol. 7, pp. 483-514, Academic Press, New York (1976).

2. Stoeckler, J.D., Ealick, S.E., Bugg, C.E., and Parks, R.E., Jr., Design of purine nucleoside phosphorylase inhibitors, Fed. Proc. 45: 2773 (1986).

3. Ropp, P.A. and Traut, T.W., Purine nucleoside phosphorylase: allosteric regulation of a dissociating enzyme, J. Biol. Chem. 266: 7682 (1991).

4. Ropp, P.A. and Traut, T.W., Allosteric regulation of purine nucleoside phosphorylase, Arch. Bioch. Biophys. 288: (in press) (1991).

5. Ealick, S.E., Rule, S.A., Carter, D.C., Greenhough, T.J., Babu, Y.S., Cook, W.J., Habash, J., Helliwell, J.R., Stoeckler, J.D., Parks, R.E., Jr., Chen, S., and Bugg, C.E., Three-dimensional structure of human erythrocytic purine nucleoside phosphorylase at 3.2 Å resolution, J. Biol. Chem. 265: 1812 (1990).

PURINE NUCLEOSIDE PHOSPHORYLASE OF BOVINE LIVER MITOCHONDRIA

Roger A. Lewis and Robert K. Haag

Biochemistry Department
University of Nevada
Reno, Nevada 89557

INTRODUCTION

Purine nucleoside phosphorylase (PNPase, purine nucleoside : orthophosphate ribosyltransferase E.C. 2.4.2.1) in mammals reversibly catalyzes the phosphorolysis of the purine ribonucleosides (deoxy)inosine and (deoxy)guanosine to (deoxy)ribose-1-phosphate and the respective base. This enzyme has been isolated and purified from various cytosolic sources (1-4). This report describes the partial characterization of a newly discovered PNPase activity isolated from the mitochondria of bovine liver.

The mitochondrial enzyme appears to possess characteristics common to other mammalian PNPases, however its mitochondrial location suggests that this PNPase is separate from the cytosolic proteins. Other reports (in preparation) document that mitochondrial PNPase has both physical and kinetic characteristics which are different from those reported for the cytosolic enzyme.

That mitochondria possess a deoxyguanosine kinase separate from the cytosolic dC/dG/dA kinase activity (5) and that a mitochondrial PNPase exists with unique enzymatic parameters, suggest that mitochondrial purine metabolism is separate from that of the cytosol. These differences may prove useful in the design of therapies for certain disorders caused by alterations of purine metabolizing enzymes or for those diseases caused by alterations of mitochondrial nucleic acid metabolism.

METHODS

Isolation of PNPase

Mitochondria were isolated from beef liver by the method of Heisler (6). After lysis of the mitochondria by sonication, PNPase was partially purified by a procedure to be published elsewhere.

Assay for PNPase Activity

The coupled enzymatic assay of Kalcker (7) was used to determine catabolic activity of PNPase. The assay was run in one ml aliquots containing 50 mM potassium phosphate buffer (pH 7.4), 1.0 mM inosine, 0.023 U/ml xanthine oxidase, and 10-50 μl of mitochondrial PNPase.

Purine and Pyrimidine Metabolism in Man VII, Part B
Edited by R.A. Harkness *et al.*, Plenum Press, New York, 1991

Mitochondrial PNPase activity in the anabolic direction was assayed using radiolabeled purine bases. A total assay volume of 600 μl consisted of 500 μM hypoxanthine, 1.67 mM ribose-1-phosphate, 6 μCi [8-^{3}H]hypoxanthine (spec. act. 20 mCi/mmole), 25 mM 3-(N-morpholine) propanesulfonic acid (MOPS) buffer (pH 7.4). The enzymatic reaction was initiated by the addition of 25 μl of enzyme. Reactions were run for 3 to 8 minutes. The reaction was stopped by the addition of 1.0 ml of 95% ethanol. Protein was removed by centrifugation and 40 μl of supernatant, along with standards of hypoxanthine and inosine, were spotted on polygram cellulose 300 MN thin-layer chromatography plates. Two-dimensional chromatography was carried out using a solution of acetonitrile-ammonium acetate (0.10 M pH 7) - ammonia (60:30:10) in the first direction followed by chromatography using a solution composed of 1 - butanol - methanol - water - ammonia (60:20:20:1) in the second direction. After drying, the hypoxanthine and inosine spots were visualized by UV light, cut-out and quantified by liquid scintillation counting.

RESULTS

Mitochondrial PNPase was purified 63 fold (details are to be published elsewhere). This degree of purification produced an activity in which the nucleoside substrate was converted to its corresponding free base and no competing reactions for the substrate were detected. Also no competing reactions were observed in the anabolic reaction when hypoxanthine was used as substrate. Whereas the activity of PNPase in the crude mitochondrial lysate or the supernatant of this sample was stable when stored at -20°C for up to three months, the most purified preparation was relatively unstable when stored at either -20°C or -70°C in Tris-HCl (pH 7.4), 15% glycerol and 0.01 M mercaptoethanol. These samples were used within a week of storage.

The enzyme showed maximum activity over a range of pH from 6.5-9.0 (Figure 1). Table 1 demonstrates the substrate specificity of mitochondrial PNPase. In the catabolic direction, inosine and deoxyinosine were good substrates. When challenged with competing substrates, adenosine and deoxyadenosine proved to be poor competitors, while guanosine and deoxyguanosine were effective at inhibiting the reaction. This suggests that these latter two compounds may also serve as substrates. In the anabolic direction, hypoxanthine was a good substrate, whereas adenine was not. These results are consistent with those reported for mammalian PNPase.

Table 1. Substrate Specificity of Mitochondrial PNPase

A. Catabolic Reaction

Substrate	Competing Substrate	% Activity
Inosine	_	100
Deoxyinosine	_	96
Inosine	Adenosine	90
Inosine	Deoxyadenosine	86
Inosine	Guanosine	63
Inosine	Deoxyguanosine	68

B. Anabolic Reaction

Substrate	% Activity
Hypoxanthine	100
Adenine	3

A. Assays were conducted using equal concentrations of substrate and competing substrate (1mM).

B. Hypoxanthine and adenine concentrations were 0.5mM.

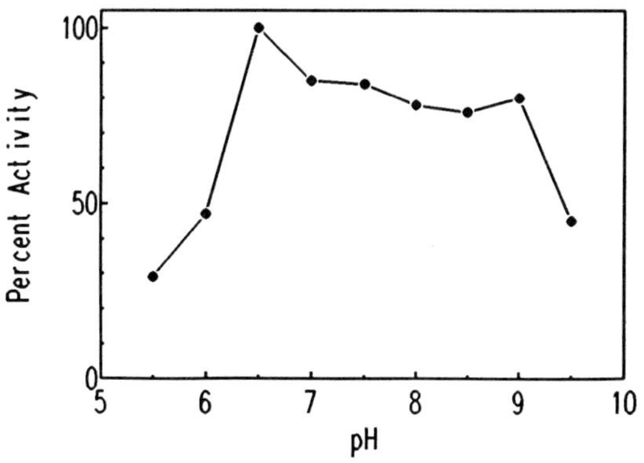

Figure 1. pH Profile

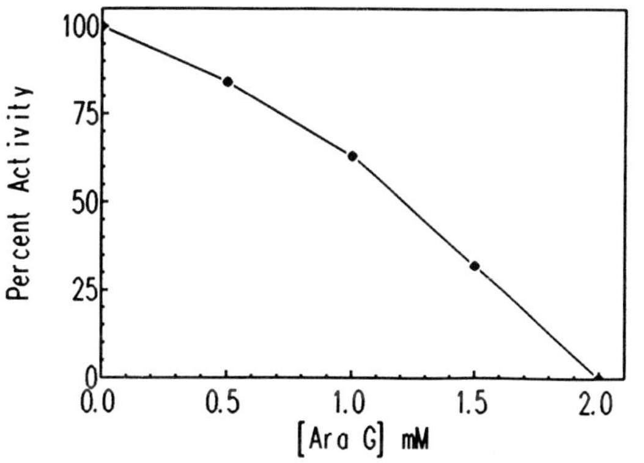

Figure 2. Inhibition of PNPase by Arabinosylguanine

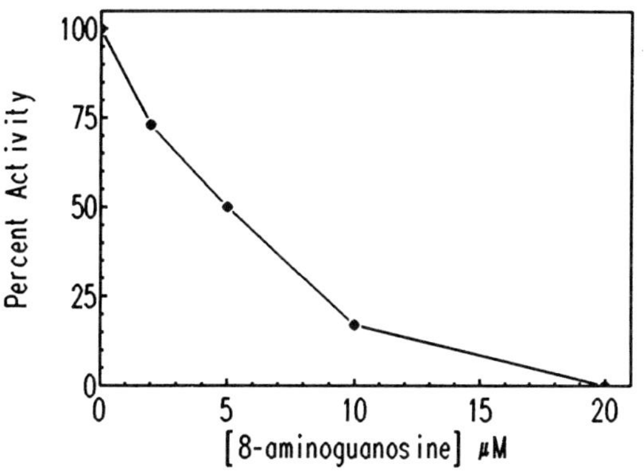

Figure 3. Inhibition of PNPase by 8-Aminoguanosine

When arabinosyl guanine (araG) and 8-aminoguanosine were tested, both were found to be inhibitors of mitochondrial PNPase (Figure 2 and 3). AraG exhibited an $I.D._{50}$ value of 1.2mM and 8-aminoguanosine revealed one of 5 μM.

DISCUSSION

Earlier reports have demonstrated that mitochondria contain a deoxyguanosine kinase (8,9) and herein is describe a mitochondrial PNPase. Furthermore, the addition of deoxyguanosine to intact mitochondria produce several metabolites including dGMP, dGDP, dGTP and guanine (10). Collectively, this suggests that mitochondria can both catabolically and anabolically transform deoxyguanosine by the following scheme.

$$\begin{array}{l} dG \rightarrow dGMP \rightarrow dGDP \rightarrow dGTP \\ \downarrow \\ gua \end{array}$$

The use of the demostrated inhibitors, 8-aminoguanosine and araG will be useful in the study of the two observed mitochondrial enzymes and in the metabolism of deoxyguanosine in intact mitochondria.

REFERENCES

(1) A.S. Lewis and M.D. Glantz, J. Biol. Chem. 251:407-413 (1976).
(2) K. Murakami and K. Tsushima, Biochim. Biophys. Acta 453:205-210 (1976).
(3) A.S. Lewis and M.D. Glantz, Biochem. 15:4451-4456 (1976).
(4) B.K. Kim, S. Cha and R.E. Parks Jr., J. Biol. Chem. 243:1763-1770 (1968).
(5) J.P. Durham and D.H. Ives, J. Biol. Chem. 245:2276-2284 (1970).
(6) C.R. Heisler, Biochem. Ed. 19:35-38 (1991).
(7) H.M. Kalckar, J. Biol. Chem. 167:429-443 (1947).
(8) W.R. Gower, Jr., M.C. Carr and D.H. Ives, J. Biol. Chem. 254:2180-2183 (1979).
(9) R.A. Lewis and L.F. Watkins In: Purine Metabolism in Man, (C.H. DeBruyn, H.A. Simmons and M.M. Muller, eds), vol. IV, Part B., pp. 7:9-82, Plenum, New York (1984).
(10) L.F. Watkins and R.A. Lewis, Molec. Cell. Biochem. 77:153-160 (1987).

URIDINE AND PURINE NUCLEOSIDE PHOSPHORYLASE ACTIVITY IN HUMAN AND RAT HEART

J.W. de Jong[a], R.T. Smoleński[b], M. Janssen[a], D.R. Lachno[c], M.M. Żydowo[b], M. Tavenier[a] and M.H. Yacoub[c]

[a]Thoraxcenter, Erasmus Univ. Rotterdam, NL; [b]Dept. of Biochem., Acad. Med. School, Gdańsk, PL; [c]Thoracic & Cardiac Surg. Unit, Harefield Hosp., Harefield, UK

Literature data on cardiac pyrimidine metabolism are scarce, contrasting our knowledge on myocardial purine metabolism. However, degradation of pyrimidine nucleotides in the ischemic myocardium seems to parallel catabolism of adenylates[1]. More knowledge about pyrimidine nucleotide breakdown is useful because of 1) the regulatory role of pyrimidine derivatives[2]; 2) the interconversions of anticancer drugs by pyrimidine metabolizing enzymes[3,4]; 3) the potential diagnostic value of pyrimidine catabolites.

During cardiac surgery we observed the accumulation and release of uridine - a pyrimidine nucleotide catabolite in ischemic heart. We evaluated differences in degradation of this compound by uridine phosphorylase (E.C. 2.4.2.3) between human and rat myocardium. For comparative reasons, we also assayed purine nucleoside phosphorylase (E.C. 2.4.2.1).

MATERIALS AND METHODS

Uridine assay in human heart and blood

For nucleotide catabolite determination[5,6], blood-free coronary effluent was collected during corrections of congenital or acquired heart defects at the time of subsequent infusions of cardioplegic solution. Biopsy and blood specimens were collected during heart or heart-lung transplantation. Tissue samples were taken before collection of the heart, before starting the implantation procedure, and 0.5 h after reperfusion.

In vitro ischemia

Human heart papillary muscle, collected during mitral valve surgery, was divided into 3-5 mm size cubes. Tissue slices were incubated at 37°C for 0.5 h. Rat-heart left ventricle, flushed with cold cardioplegic solution, received the same treatment. After incubation, tissue was powdered in liquid N_2 and extracted in 0.8 ml 0.6 M $HClO_4$. The neutralized supernatant fluid was assayed for uridine by HPLC[6], the pellet for protein.

Uridine and purine nucleoside phosphorylase assays

Explanted human ventricle or papillary muscle and flushed rat ventricle were homogenized in 50 mM Na-phosphate (pH 6.9), 1 mM EDTA and 0.1 mM dithiothreitol. Incubation was started by addition of 0.1 ml homogenate to 0.1 ml Na-phosphate, 2 mM uridine or inosine. Incubations carried out at 37°C were terminated after 15 min with 0.2 ml 8% $HClO_4$. Conversions were measured by HPLC[6] in neutralized supernatant fluids.

RESULTS

Uridine concentrations during heart surgery

Before harvesting for transplantation, the uridine content of donor hearts was 22 ± 6 nmol/g protein (mean ± SE, n=7). We found a threefold increase ($P<0.05$) during cold storage of the organs which was not reversed upon reperfusion.

Following implantation, immediately after aortic declamping, the arterial [uridine] was 2.8 ± 0.5 μM (n=6). Reperfused myocardium released significant amounts of uridine for at least 10 min, with a coronary sinus peak value of about 6 μM.

We also observed washout of uridine formed in human hearts during corrections of congenital or acquired heart defects. The coronary effluent concentration increased from 0.82 ± 0.09 μM after the first infusion of cardioplegic solution to 6 ± 2 μM after the third ($P<0.05$, n=5-7), indicating a continuous degradation of pyrimidine nucleotides in the ischemic myocardium.

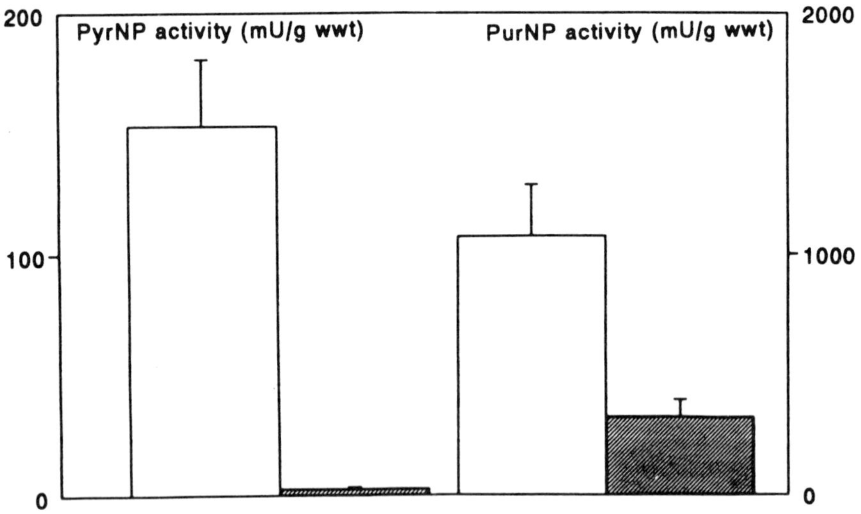

Fig. 1. Activity of uridine phosphorylase (PyrNP, n=3-5) and purine nucleoside phosphorylase (PurNP, n=5) in hearts of rats (open bars) and humans (hatched bars).

In vitro uridine formation by ischemic heart

During in vitro ischemia of human myocardium, uridine increased in 15 min from 0.1 to 1.5 μmol/g protein with little increase thereafter. Uridine in rat myocardium rose transiently to only 0.5 μmol/g. In contrast, the rate of ATP degradation and purine catabolite formation was comparable in human and rat myocardium (data not shown).

Cardiac uridine phosphorylase and purine nucleoside phosphorylase

Fig. 1 shows that uridine phosphorylase in human heart was about 60x less active than that in rat heart ($P<0.01$). Purine nucleoside phosphorylase was substantially more active in both species. It differed only a factor 3 ($P<0.02$) between human and rat heart.

DISCUSSION

Low uridine phosphorylase in human heart

The major finding of this study is the demonstration that ischemic human heart hardly catabolizes uridine, contrasting rat heart. We observed this in the clinical setting (heart transplantation, correction of heart defects) and during in vitro experiments. Uridine phosphorylase, the enzyme responsible for uridine breakdown, proved to be virtually absent in human heart.

Implications

There are several possible implications of the above observation. The first concerns the metabolism of pyrimidine analogs used in anticancer therapy[3,4]. The prominent difference between the activity in human and rat heart makes it difficult to predict cardiotoxicity of these compounds on the basis of animal experiments. In heart purine and pyrimidine nucleoside phosphorylase activity is localized in the endothelial cells[7]. (Recent data indicate that the former is also present in the cardiomyocytes[8].) This strategic localization can lead to a complete change in pattern of pyrimidine metabolites reaching myocardial cells from the blood. Human neoplastic tissues have also a relatively low activity of uridine phosphorylase in comparison to those in rodents[9].

The next implication concerns the potential application of uridine to evaluate ischemic myocardial injury[10]. The concentration of this compound in ischemic heart or coronary effluent is several times lower than that of purine catabolites. Nevertheless, uridine determination may be important where one evaluates the cardioprotective action of purine compounds and determination of purine catabolites has limited value[11]. It is also important that uridine metabolism in human blood is slow[12], which contrasts purine catabolism[13].

Conclusion

Human myocardium possesses an active purine nucleoside phosphorylase, but very little uridine phosphorylase activity. The latter explains why uridine accumulates in heart and blood due to ischemia. Rat heart shows considerable activity of both enzymes. Uridine can potentially indicate myocardial ischemia.

ACKNOWLEDGEMENTS

We gratefully acknowledge the Polish Academy of Sciences and the British and Netherlands Heart Foundation for financial support, and Ms. A.S. Nieukoop for excellent analytical help.

REFERENCES

1. J.L. Swain, R.L. Sabina, P.A. McHale, J.C. Greenfield & E.W. Holmes, Prolonged myocardial nucleotide depletion after brief ischemia in the open-chest dog, Am. J. Physiol. 242:H818 (1982)
2. S. Lortet, J. Aussedat & A. Rossi, Synthesis of pyrimidine nucleotides in the heart: uridine and cytidine kinase activity. Arch. Int. Physiol. Biochim. 95:289 (1987)
3. P.M. Schwartz, R.D. Moir, C.M. Hyde, P.J. Turek & R.E. Hanschumacher, Role of uridine phosphorylase in the anabolism of 5-fluorouracil, Biochem. Pharmacol. 34:3585 (1985)
4. G.J. Peters, E. Laurensse, A. Leyva, J. Lankelma & H.M. Pinedo, Sensitivity of human, murine and rat cells to 5-fluorouracil and 5'-deoxy-5-fluorouridine in relation to drug metabolizing enzymes, Cancer Res. 46:20 (1986)
5. R.T. Smoleński, A.C. Skladanowski, M. Perko & M.M. Żydowo, Adenylate degradation products release from the human myocardium during open heart surgery, Clin. Chim. Acta. 182:63 (1989)
6. R.T. Smoleński, D.R. Lachno, S.J.M. Ledingham & M.H. Yacoub, Determination of sixteen nucleotides nucleosides and bases using high-performance liquid chromatography and its application to the study of purine metabolism in hearts for transplantation, J. Chromatogr. 527:414 (1990)
7. R. Rubio & R.M. Berne, Localisation of purine and pyrimidine nucleoside phosphorylases in heart, kidney and liver, Am. J. Physiol. 239:H721 (1980)
8. J.W. de Jong, E. Keijzer, T. Huizer & B. Schoutsen, Ischemic nucleotide breakdown increases during cardiac development due to drop in adenosine anabolism/catabolism ratio, J. Mol. Cell. Cardiol. 22:1065 (1990)
9. Y. Maehara, Y. Sakaguchi, T. Kusumoto, H. Kusumoto & K. Sugimachi, Species differences in substrate specificity of pyrimidine nucleoside phosphorylase, J. Surg. Oncol. 42:184 (1989)
10. R.A. Harkness, Hypoxanthine, xanthine and uridine in body fluids, indicators of ATP depletion, J. Chromatogr. 249:255 (1988)
11. J.W. de Jong, P. van der Meer, H. van Loon, P. Owen & L.H. Opie, Adenosine as an adjunct to potassium cardioplegia: effect on function, energy metabolism and electrophysiology, J. Thorac. Cardiovasc. Surg. 100:445 (1990)
12. J. Tseng, J. Barelkovski & E. Gurpide, Rates of formation of blood-borne uridine and cytidine in dogs, Am. J. Physiol. 221:869 (1971)
13. E. Harmsen, J.W. de Jong & P.W. Serruys, Hypoxanthine production by ischemic heart demonstrated by high pressure liquid chromatography of blood purine nucleosides and oxypurines, Clin. Chim. Acta. 115:73 (1981)

A COMPARATIVE STUDY OF THE SMALL FORMS OF ADENOSINE DEAMINASE FROM VARIOUS ORGANISMS

Cynthis Ma and Pang Fai Ma

Center for Medical Education
Ball State University
Muncie, IN 47306 U.S.A.

INTRODUCTION

Adenosine deaminase (ADA) (EC 3.5.4.4) catalyzes the hydrolytic deamination of adenosine to inosine. Three forms of ADA have been reported in tissues of various sources. These enzyme forms are designated as A, B and C forms, corresponding to molecular weights of 200,000, 100,000 and 34,000 respectively (1). In higher mammals, the B form was found absent in those tissues studied; whereas the prevalent A and C form enzymes are interconvertible (2,3). It was reported that the A and C form ADA from mammalian sources showed extensive cross-reactivity immunochemically (4,5,6) indicating the similarity of the two enzyme forms. It was later shown that the large A form ADA was a complex of the small C form enzyme and a nonenzymatic binding protein, the complexing protein, which may be a soluble (7.8) or membrane-bound glycoprotein (9,10).

In a study with normal rats, only the C form of ADA was found in all the tissues examined (11). Two isozymes of the C type have been demonstrated in the epithelial cells of rat intestinal mucosa (12). A C form ADA was found in rat brain, which resembled the other mammalian C form ADA in all the kinetic parameters compared (13). Molecular forms distribution patterns of adenosine deaminase vary among different tissues, showing tissue and species specific differences. In this report, the small C form adenosine deaminases from human, calf, rat and frog are compared in the interaction with the complexing protein extracted from human liver. This study shows species specific difference in the small C form ADA, even among the mammalian enzymes which are similar in many enzymic properties.

MATERIAL AND METHODS

Materials

Adenosine and adenosine deaminase from calf intestine were obtained from Sigma Chemical Co. Sephadex G-150 was bought from Pharmacia. Human tissues were obtained from the Ball Memorial Hospital, Muncie, Indiana. Tissues were immediately frozen after excision in autopsy, and kept in a deep freeze at -20°C.

Purine and Pyrimidine Metabolism in Man VII, Part B
Edited by R.A. Harkness *et al*., Plenum Press, New York, 1991

The procedure for extraction of enzymes and the complexing protein was the same as previously reported (14). Tissues were homogenized in glass-distilled water at a concentration of 1 g. of tissue per 5 ml. After centrifugation at 18,000g in a refrigerated centrifuge for 1 hour, the supernatant was fractionated with solid ammonium sulfate. Successive 10% saturation fractions were made from 40 to 90%. The precipitated protein of each fraction was dissolved in 0.05 M phosphate buffer (pH 7.0) and assay for enzyme activity. All active fractions were combined and used in the aggregation study.

Enzyme Assay

Enzyme activity was measured by following the decrease in absorbance at 265 nm (15). A unit of enzyme activity is defined as that amount of enzyme which catalyzes the hydrolysis of 1 umole of adenosine/min under standard conditions; i.e., with a 1-cm path-length cell containing 3.0 ml of $1.0 * 10^{-4}$ M adenosine in 0.2 M phosphate buffer (pH 7.0) at 38°C.

Gel Filtration Chromatography

Sephadex G-150 was suspended in 0.05 M phosphate buffer (pH 7.0). Columns (0.6 cm x 30 cm) were poured in the cold and equilibrated overnight with flowing buffer. The void volume was determined with blue dextran. In all experiments, 50 ul of sample was applied to the column, and fractions of 0.25 ml each were collected. The ADA molecular forms were identified based on the order of appearance in the elution profiles.

Enzyme And Complexing Protein Interaction

Equal portions of the C form ADA from different sources and extract of the complexing protein were mixed in a small test tube, and left at 30°C for 60 minutes. Immediately, the mixture was subjected to gel filtration column chromatography to determine quantitatively the two molecular forms. For controls, the same volume of an enzyme or complexing protein extract was incubated with equal volume of phosphate buffer, and the mixture was subjected to the same procedures.

RESULTS

The complexing protein was isolated from human liver. The supernatant of the liver homogenate was first made 40% saturation with solid ammonium sulfate. Successive 10% saturation fraction was made up to 90%. Each saturation cut yielded a protein precipitate which was resuspended in phosphate buffer, and used to measure for ADA, or the complexing protein activity. Fractions over 60% saturation with ammonium sulfate contain relatively small amount of ADA activity. The relative amounts of the complexing protein in different fractions are listed in Table 1.

Table 1. Relative Amounts of the Complexing Protein Activity in Fractions of Successive 10% Saturation of Ammonium Sulfate

Fraction (%)	Relative Amount of Complexing Protein
70	2.0
80	1.2
90	1.0

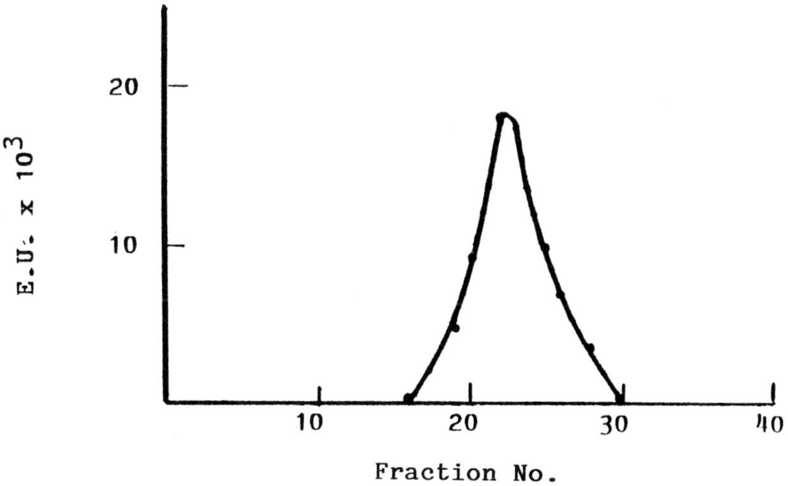

Fig. 1. A gel filtration column chromatographic pattern of the mixture containing the C form ADA from calf intestine and phosphate buffer, incubated at 30°C for 60 minutes.

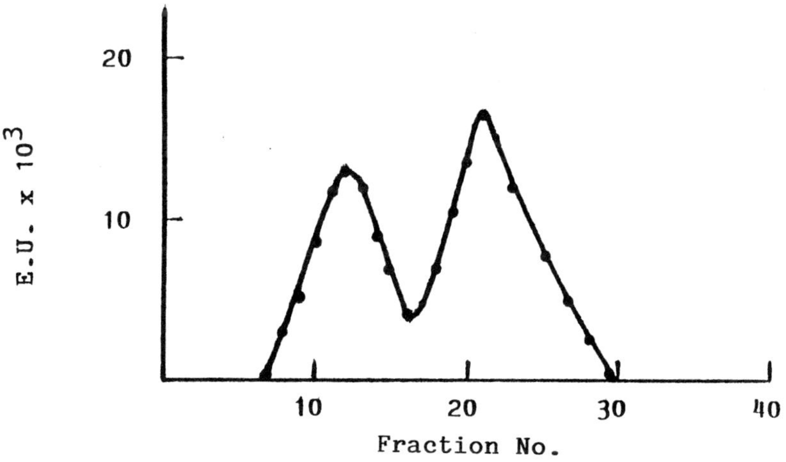

Fig. 2. A gel filtration column chromatographic pattern of the mixture containing the C form ADA from calf intestine and the complexing protein extract from human liver, incubated at 30°C for 60 minutes.

Table 2 summarizes the results of the experiments to establish the minimum incubation time required for the conversion of the C to A form ADA. Equal portions (25 ul) of the C form ADA and the complexing protein extract were mixed. The mixture was left at 30°C for varying amount of time. At the end of the incubation period, the sample was immediately passed through a gel filtration column. The eluent was collected as 0.25 ml fractions which were assayed for ADA activity.

Table 2. An Incubation Time Study of the Reaction of Converting the C form ADA and the Complexing Protein to the A Form ADA

Time (Min.)	A Form ADA Recovered (E.U. x 10^3)
20	0.0
30	2.5
45	9.0

In order to study the ability of the C form ADA from various sources to form an aggregate with the complexing protein, the following control experiments were performed. In separate experiments, equal portions of the C form ADA or the complexing protein extract, and phosphate buffer were mixed. The mixture was left at 30°C for 60 min before it was passed through a gel filtration column. Enzyme assay of the eluent fractions of the complexing protein extract revealed no traces of the A form ADA activity. Results of a representative set of experiment are shown in Figure 1 and Figure 2.

DISCUSSION

In human liver, the complexing protein appears to be concentrated in the ammonium sulfate saturation range of 60 - 70%. In this fraction, the presence of ADA activity is not detectable. The minimum incubation time for the aggregation of the C form ADA and the complexing protein is at least 30 minutes. The optimal temperature for this interaction is 36°C. In this study, the incubation of the reaction mixture is 30°C so as to avoid denaturation of the enzyme or the complexing protein. In all the aggregation study, the incubation mixture was performed at 30°C for 60 minutes.

The C form ADA from human, calf, rat and frog were used to study the ability of each in forming an aggregate with the complexing protein isolated from human liver. The human and the calf C form ADA are readily converted to the A form as illustrated in Figure 1 and Figure 2; whereas the rat and the frog C form ADA's do not form aggregates with the human complexing proteins to produce the A form ADA's. This enzyme property characterizes one of the species specific differences among the mammalian C form ADA isozymes.

REFERENCES

1. P. F. Ma and J. R. Fisher, Com. Biochem. Physiol., 27:105 (1968).
2. P. F. Ma and J. R. Fisher, Comp. Biochem. Physiol. 31:771 (1969).

3. H. Akedo, H. Nishihara, K. Shinkai and K. Komatsu, Biochim. Biophys. Acta, 212:189 (1970).
4. P. C. Lee, J. R. Fisher and P. F. Ma, Comp. Biochem. Physiol., 40B1071 (1971).
5. P. C. Lee, J. R. Fisher and P. F. Ma, Comp. Biochem. Physiol., 46B:483 (1973).
6. H. Akedo, H. Nishihara, K. Shinkai, K. Komatsu and S. Ishikawa, Biochim. Biophys. Acta, 276:257 (1972).
7. P. E. Daddona and W. N. Kelley, J. Biol. Chem., 253:4617 (1978).
8. W. P. Schrader and A. R. Stacy, J. Biol. Chem., 252:6409 (1977).
9. W. P. Schrader and P. J. Bryer, Arch. Biochem. Biophys., 215:107 (1982).
10. P. P. Trotta, Biochemistry, 21:4014 (1982).
11. K. J. Collier Jr. and P. F. Ma, Int. J. Biochem., 12:659 (1980).
12. P. P. Trotta and M. E. Balis, Cancer Res., 37:2297 (1977).
13. J. J. Centelles, R. Franco and J. Bozal, J. Neurosci. Res., 19:258 (1988).
14. P. F. Ma and T. A. Magers, Int. J. Biochem., 6:281 (1975).
15. H. M. Kalckar, J. Biol. Chem., 167:461 (1947).

ADENOSINE DEAMINASE IN THE DIAGNOSIS OF PLEURAL EFFUSIONS

Manuel López Jiménez, Angela Rodríguez-Piñero, Manuel A. Carnicero, Antonio Zapatero, José Perianes, Luis Vigil and Julian Ruiz Galiana

Departments of Medicine and Clinical Biochemistry
Hospital General de Móstoles
Madrid, Spain

INTRODUCTION

Adenosine deaminase (ADA; EC 3.5.4.4) catalyzes the irreversible deamination of adenosine to inosine and 2'-desoxyadenosine to 2'-desoxyinosine (1). ADA is essential for the differentiation of lymphoid cells, the enzyme activity is greater in T lymphocytes than in B lymphocytes and it is also inversely proportional to the degree of T cell differentiation (2,3).

A hereditary deficiency of ADA results in severe combined immunodeficiency (4). In another hand, an increase of serum ADA activity has been described in several diseases where cellular immunity is stimulated, such as infectious mononucleosis, typhoid fever, brucellosis and boutonneuse fever (5,6). Also, ADA activity is higher in cerebrospinal, pleural, pericardial and peritoneal fluid of patients with tuberculous meningitis and pleural, pericardial and peritoneal tuberculosis than in other diseases (7-10).

The aim of this study was to evaluate the usefulness of determining ADA activity in the diagnosis of pleural effusions.

PATIENTS AND METHODS

We have determined the concentration of ADA in pleural fluid of 157 patients with pleural effusions of different causes. The patients were subdivided into seven groups: group 1, tuberculosis (32 cases); group 2, malignant effusions (27); group 3, empyemas (12); group 4, parapneumonic effusions (23); group 5, transudates (32); group 6, other etiologies (7 with pulmonary embolism, 4 with pancreatitis and 1 with splenic abscess); group 7, unknown cause (19).

The activity of ADA was determined according to the colorimetric method of Blake and Berman (11) using adenosine sustrate and amonium kit from Boehringer Mannheim. Pleural fluid samples were collected by needle punture and stored at -20°C after centrifugation to discard the cell pellet.

Groups were compared according to an unmatched Student's *t* Test.

RESULTS AND DISCUSSION

Individual results of ADA activity are showed in the figure 1. The mean pleural fluid ADA activity of the patients with tuberculous pleuritis was 60.2±29.6 U/L significantly higher than that of the patients with other causes of pleural effusion ($P<0.001$), excepting the group 3 (empyemas). In group 2 the mean±SD value was 18,8±14.5 U/L, in group 3 55.6±36.4 U/L, in group 4 17.2±8.4 U/L, in group 5 9.5±3.3 U/L, in group 6 13.9±4.7 U/L and in group 7 11.7±3.4 U/L.

At a cutoff level of 32 U/L the sensitivity, specificity and overall accuracy in detecting tuberculous pleuritis were of 0.84, 0.91 and 0.90, respectively. Eleven patients showed false-positive results (7 with empiema, 2 with parapneumonic pleural effusion, 1 with lung adenocarcinoma and 1 with Burkitt's lymphoma), but pleural fluid ADA activities in these patients were relatively low compared with those in tuberculous pleural effusion, so if cutoff level was raised to 56 U/L there were only two false-positive results (1 with empyema and 1 with Burkitt's lymphoma).

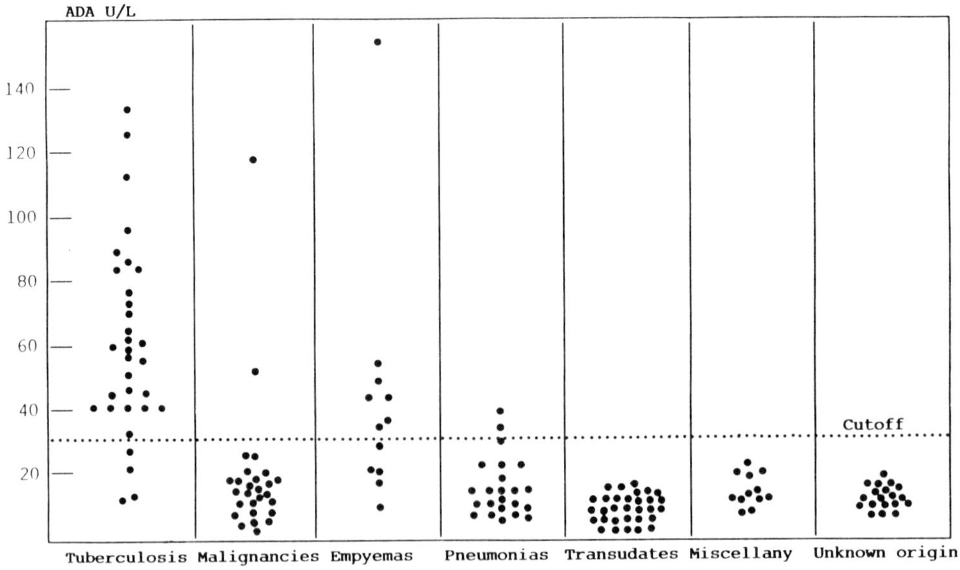

Fig. 1 ADA activity in pleural fluid of 157 patients.

In clinical practice, pleural effusions are frequent and often constitute difficult diagnostic problems. Pleural fluid cultures to detect the presence of Mycobacterium tuberculosis are positive only in 20-30% of patients with tuberculosis pleuritis (12), and biopsy of the parietal pleura shows typical epithelioid cell granulomas in 50-80% of patients with this disease (13,14). The present study showed that determination of ADA activity in pleural fluid help to distinguish tuberculous pleural effusions from others etiologies and confirmes the results of previous studies (9,15).

In summary, pleural fluid ADA activity is useful for the diagnosis of tuberculous pleural effusions because of its high sensitivity and specificity.

REFERENCES

1. N. M. Kredich, M. S. Hersfield, Immunodeficiency diseases caused by adenosine deaminase deficiency and purine nucleoside phosphorylase deficency, in: "The Metabolic Basis of Inherited Diseases", C. R. Scriver, A. L. Beaudet, W. S. Sly and D. Valle, eds., McGraw-Hill, New-York (1989).
2. J. L. Sullivan, W. R. A. Osborne, W. R. J. Wedgwood, Adenosine deaminase activity in lymphocytes, Br. J. Hematol. 37:157 (1977).
3. A. Shore, H. M. Dosch, E. W. Gelfand, Role of adenosine deaminase in the early stages of precursor T cell maturation, Clin. Exp. Immunol. 44:152 (1981).
4. E. R. Giblett, J. E. Anderson, F. Cohen, B. Pollara, H. J. Meuwissen, Adenosine deaminase deficiency in two patients with severely impaired cellular immunity, Lancet 2:1067 (1972).
5. B. Galanti, S. Nardiello, M. Russo, F. Fiorentino, Increased lymphocyte adenosine deaminase in typhoid fever, Scand. J. Infectious Dis. 130:4750 (1981).
6. M. A. Piras, C. Gakis, M. Brudoni, G. Andreoni, Immunological studies in Mediterranean spotted fever, Lancet 1:1249 (1982).
7. M. A. Piras, C. Gakis, Cerebrospuinal fluid adenosine deaminase activity in tuberculous meningitis, Enzyme 14:317 (1973).
8. J. M. Martínez Vázquez, I. Ocaña, E. Ribera, R. M. Segura, C. Pascual, Adenosine deaminase in the diagnosis of tuberculous peritonitis, Gut 27:1049 (1986).
9. I. Ocaña, J. M. Martínez Vázquez, R. M. Segura, T. Fernández de Sevilla, J. A. Capdevila, Adenosine deaminase in pleural fluids, test for diagnosis of tuberculous pleural effusion, Chest 84:51 (1983).
10. J. M. Martínez Vázquez, E. Ribera, I. Ocaña, R. M. Segura, R. Serrat, J. Sagrista, Adenosine deaminase activity in tuberculous pericarditis, Thorax 41:888 (1986).
11. J. Blake, P. Berman, The use of adenosine deaminase assays in the diagnosis of tuberculosis, S. Afr. Med. J. 62:19 (1982).
12. H. W. Berger, E. Mejía, Tuberculous pleurisy, Chest 63:88 (1973).

13. H. Levine, W. Metzger, D. Lacera, L. Kay, Diagnosis of tuberculous pleurisy by culture of pleural biopsy specimen, Arch. Intern. Med. 126:269 (1970).
14. L. Scharer, J. H. McClement, Isolation of tubercle bacilli from needle biopsy specimens of parietal pleura, Am. Rev. Respir. Dis. 97:466 (1968).
15. T. Petterson, K. Ojala, T. H. Weber, Adenosine deaminase in the diagnosis of pleural effusions, Acta Med. Scand. 215:299 (1984).

MOLECULAR FORMS OF HUMAN KIDNEY AMP-DEAMINASE

Grzegorz Nowak and Krystian Kaletha

Department of Biochemistry
Academic Medical School
Gdańsk, Poland

INTRODUCTION

AMP-deaminase (EC 3.5.4.6) - the enzyme catalyzing hydrolytic deamination of adenylic acid is widely distributed in vertebrate tissues.

The physiological significance of AMP-deaminase in cellular metabolism is still unclear. Among many postulated roles, stabilization of the adenylate energy charge (1), regulation of the purine nucleotide pool size (2), and replenishment of the citric acid cycle intermediates (3) are most frequently quoted. In the kidney, the role of AMP-deaminase (throughout the purine nucleotide cycle) in ammoniagenesis has been also proposed (4).

It is known since recently (5), that at least two, different genes code AMP-deaminase production in various human tissues. One of them (ampd1) is specific for mature skeletal muscle, whereas the second one (ampd2) is expressed in other tissues, including that of kidney.

In this paper some physicochemical properties of AMP-deaminase isolated from human kidney are described.

MATERIALS AND METHODS

Human kidneys of autopsied young, healthy subjects were taken 12-24 hr after death, washed and weighed. After removing fat and membranes, the tissue was homogenized in three volumes (v/w) of the extraction buffer (0.089 M phosphate buffer, pH 6.5, containing 0.18 M KCl and 1 mM thioethanol), twice centrifuged, and AMP-deaminase was subsequently isolated by chromatography on phosphocellulose, as described previously (6). The most active fractions from the two activity peaks obtained (see Fig. 1) were pooled, and used for kinetic studies. For electrophoretic analysis and molecular weight determinations, the partially purified AMP-deaminase (the enzyme form present in the second activity peak) was rechromatographed on phosphocellulose, concentrated, applied onto Sepharose CL-6B column, and then eluted with 0.1 M potassium-cacodylate buffer, pH 7.0 with the addition of 0.5 M KCl.

Activity of AMP-deaminase was estimated according to the

Purine and Pyrimidine Metabolism in Man VII, Part B
Edited by R.A. Harkness *et al.*, Plenum Press, New York, 1991

phenol-hypochlorite method of Chaney and Marbach (7). The incubation medium, in the final volume of 0.5 ml, contained 0.1 M potassium-cacodylate buffer, pH 6.6, various concentrations of the substrate and effectors.

The kinetic parametrs of the reaction ($S_{0.5}$, V_{max} and n_H) were calculated, as described previously (8).

SDS PAG electrophoresis was performed according to method of Weber and Osborn (9).

RESULTS

The activity of AMP-deaminase in human kidney extract amounted about 0.008 µmoles/min per mg of extractable protein and was the same in both parts (cortex and medulla) of the organ.

Figure 1 presents a typical elution profile of human kidney AMP-deaminase from the phosphocellulose column. As may be seen from the Figure, the enzyme eluted in the form of two well separated activities, indicating that two, chromatographically different forms of AMP-deaminase were present in the human kidney extract. The activity eluted with 0.75 M potassium chloride constituted less than 10 percent of the total activity released, and the enzyme form present in this eluate (isoform A) manifested a regular, hyperbolic (n_H=1.1) type of substrate saturation kinetics (see Fig.2) with the $S_{0.5}$ parameter value about 3 mM. The presence of 1 mM ATP (not shown here) decreased this value to 0.9 mM. The main amount of AMP-deaminase activity eluted from the column at 1.1-1.2 M KCl concentration (Fig.1). The isoform eluted in this way (isoform B) manifested distinctly sigmoidal (n_H = 1.8) substrate saturation kinetics (see Fig.2), with the $S_{0.5}$ value as high as 10 mM. The presence of 1 mM ATP (not shown here) decreased this value to 1.1 mM.

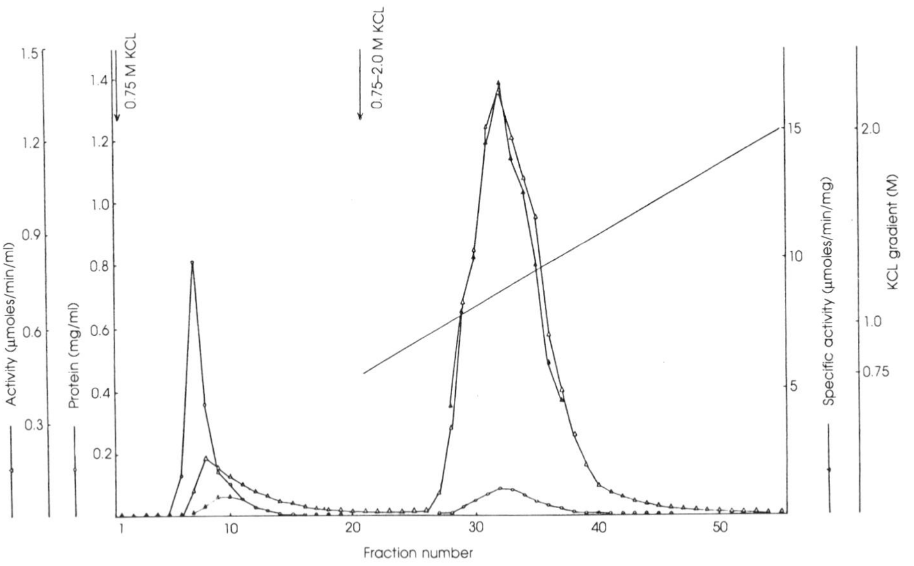

Fig.1 Elution profile of human kidney AMP-deaminase from a phosphocellulose column.

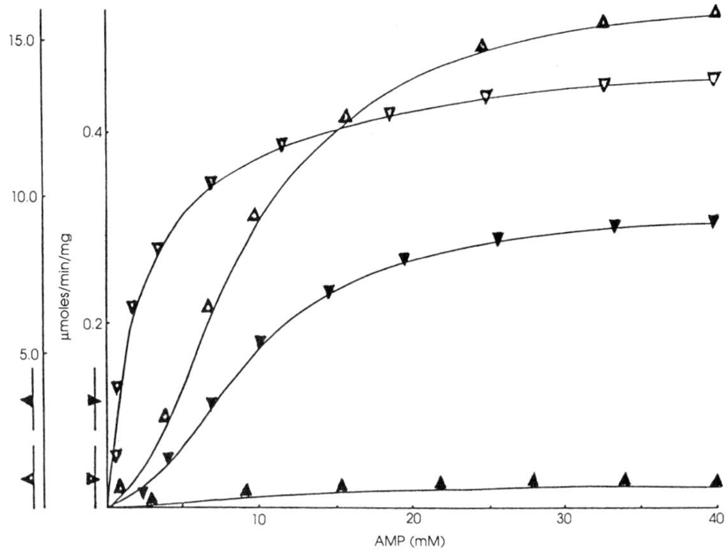

Fig.2 Effect of substrate concentration on velocity of the reaction catalysed by two forms - form A (▽,▼) and form B (△,▲) - of human kidney AMP-deaminase in the absence (open symbols) and in the presence (close symbols) of 30 µM stearyl-CoA.

Activated stearyl acid inhibited distinctly the reaction catalysed by the two forms of kidney AMP-deaminase (Fig.2), influencing mainly either maximum velocity of the reaction (in the case of isoform B), or $S_{0.5}$ parameter value (in the case of isoform A).

The results of gel filtration on Sepharose CL-6B column and PAG electrophoresis in the presence of SDS (not shown here) allowed to estimate molecular weight of the native enzyme and its subunit as close to 140 kDa and 35 kDa, respectively.

DISCUSSION

Chromatography on phosphocellulose revealed the presence of two kinetically different activities of AMP-deaminase in human kidney extract. The result obtained was not an artefact because other chromatographic separations performed gave the same results. Moreover, a similar results have been obtained during chromatography of human uterine (6) and human heart (10) enzyme.

Several possible reasons may be responsible for the permanent presence of specific AMP-deaminase in the 0.75 M KCl eluate. The isoform A may simply represent the proteolitically degraded form of the native enzyme (isoform B); it may represent also the native enzyme combined with other protein(s) released by 0.75 M potassium chloride, or alternatively, it may represent the product of the abortive expression of different (ampd1 or other not identified yet) AMP-deaminase gene. The experimental background supporting each of the above mentioned explanations do exist (11,12,5).

The results presented here indicate that physicochemical properties of human kidney AMP-deaminase are very similar to these desribed for AMP-deaminases from human heart and uterus (6,10). This finding should not be surprising in view of their common origin (5).

REFERENCES

1. Chapman A.G. and Atkinson D.E. (1973) J. Biol. Chem. 248, 8309-8312
2. Setlow B. and Lowenstein J.M. (1967) J. Biol. Chem. 242, 607-615
3. Aragon J.J. and Lowenstein J.M. (1980) Eur. J. Biochem. 110, 371-377
4. Bogusky R.T., Lowenstein L.M. and Lowenstein J.M. (1976) J. Clin. Invest. 58, 326-335
5. Morisaki T., Sabina R.L. and Holmes E.W. (1990) J. Biol. Chem. 265, 11482-11486
6. Nagel-Starczynowska G., Nowak G. and Kaletha K. (1991) Biochim. Biophys. Acta 1073, 470-473
7. Chaney A.L. and Marbach E.P. (1962) Clin. Chem. 8, 130-132
8. Kaletha K. and Nowak G. (1988) Biochem. J. 249, 255-261
9. Weber K. and Osborn M. (1969) J. Biol. Chem. 244, 4406-4412
10. Nowak G. and Kaletha K. (1991) Biochem. Med. Metab. Biol. (in press)
11. Ranieri-Raggi M. and Raggi A. (1990) Biochem. J. 272, 755-759
12. Shiraki H., Ogawa H., Matsuda Y. and Nakagawa H. (1979) Biochim. Biophys. Acta 566, 345-352

MOLECULAR ANALYSIS OF ACQUIRED MYOADENYLATE DEAMINASE DEFICIENCY IN POLYMYOSITIS (IDIOPATHIC INFLAMMATORY MYOPATHY)

R. L. Sabina, A. R. Sulaiman, and R. L. Wortmann

Medical College of Wisconsin, Departments of Cellular Biology and Anatomy Neurology, and Medicine, Clement J. Zablocki VA Medical Center, Milwaukee Wisconsin, U.S.A.

Myoadenylate deaminase (mAMPD) deficiency occurs in primary (inherited) and secondary (acquired) forms (1). Clinically, inherited mAMPD deficiency is accompanied by exercise intolerance, muscle cramping, and myalgia. At the molecular level inherited mAMPD deficiency is characterized by normal transcript abundance without any immunoreactive mAMPD peptide (2). Recently, inherited mAMPD deficiency has been defined as a single point mutation in the human gene for myoadenylate deaminase (Ampd1) and is estimated to occur in 2% of the normal population (3).

Secondary deficiencies occur in association with a wide variety of neuromuscular diseases. These include Duchenne, Becker, and Fukuyama-type muscular dystrophies, Werdnig-Hoffman disease, myotonic dystrophy, hypothyroidism, periodic paralysis, scleroderma, systemic lupus erythematosus, and a variety of inflammatory myopathies (4,5). A high percentage of the reported cases associated with a secondary deficiency have inflammatory muscle disease (6). Inverse relationships between the relative severity of the pathologic change of the primary neuromuscular disease and mAMPD activity have been observed. In addition creatine phosphokinase (CPK) activity is affected in these individuals, although not as dramatically as mAMPD (4). The molecular abnormalities that occur with these acquired mAMPD deficiency states are not as well understood. Investigation of the relative expression of the Ampd1 and CPK (M-CPK) genes in a patient population with the primary diagnosis of the idiopathic inflammatory myopathy (polymyositis) was undertaken to characterize the molecular events that underlie this type of secondary deficiency state.

Patients with idiopathic inflammatory myopathies met the criteria for the diagnosis of polymyositis proposed by Bohan and Peter (7). Muscle biopsies from patients with polymyositis were compared to controls whose condition was characterized by diffuse myopathic symptoms, normal muscle histopathology, and normal histochemical staining for mAMPD. In a double-blinded fashion, each specimen was evaluated for relative myofiber necrosis and relative muscle-specific gene expression by a) mAMPD activity, b) Ampd1 transcript abundance, and c) M-CPK transcript abundance.

Relative myofiber necrosis was assessed in fixed skeletal muscle tissue sections by analyzing eight micron sections prepared from each biopsy and stained with Hematoxylin/Eosin.. None of the 13 control cases exhibited fiber necrosis and the cases of polymyositis were grade as having mild-to-moderate versus severe necrosis.

Table 1

Group	N	mAMPD Activity[a]	Ampd1 Transcript[b]	M-CPK Transcript[b]
Controls	13	481	100	100
Polymyositis-necrosis				
Mild to moderate	3	376	58	75
Severe	3	137	24	11

a=enzyme activity in nmol/min/mg protein
b=control results assigned value of 100%

Activity of mAMPD was quantitated by a radiochemical assay (8) performed on skeletal muscle extracts. Activity decreased with increasing severity of myofiber necrosis in muscle form patients with polymyositis, reaching 28% of control (mean 481 nmol/min/mg protein) in severely necrotic samples. The activity of mAMPD was significantly decreased compared to control values ($P<0.05$, control vs. severely necrotic).

Transcript was detected by Northern blot analysis performed on total skeletal muscle RNA utilizing radiochemically-labelled cDNA inserts as probes, quantitated by densitometry, and normalized to the abundance of alpha-tubulin (a-TUB) transcript in the same sample (9). The control group was set at 100%. This value had an associated 22% standard error of the mean(S.E.M.). Similar to mAMPD activity, Ampd1 transcript abundance tended to decrease with increasing severity of myofiber necrosis in myositic muscle, reaching 24% of control mean in severely necrotic patients. This value was 58% of control for the the mild to moderate group. However, due to the greater variability in estimating relative transcript abundance and the small patient sample, this difference did not reach statistical significance.

M-CPK transcript abundance was also set at 100% using the control group. This estimate had an associated S.E.M. of 18%. M-CPK transcript abundance was significantly reduced in severely necrotic myositic skeletal muscle, reaching 11% of control mean. An intermediate but not significant value of 75% was observed for the mild to moderate group.

The results of these investigations confirm and extend previous studies related to an acquired deficiency of mAMPD in human skeletal muscle by demonstrating decreased mAMPD activity in skeletal muscle from patients with polymyositis as a function of relative myofiber necrosis. This change occurs in conjunction with a parallel decrease in Ampd1 transcript abundance and in severely necrotic skeletal muscle. Thus the molecular events that underlie the secondary deficiency of AMPD associated with idiopathic inflammatory myopathy differ from those in primary conditions. Where decreased enzyme protein is found in the presence of adequate or increased transcript abundance. In addition these results indicate that fiber necrosis is most certainly a major contributing factor to the observed decrease in muscle-specific gene expression in polymyositis. Additional studies are needed to see if this pattern holds for other neuromuscular diseases associated with secondary mAMPD deficiency and whether factors other than myonecrosis play a role.

References

1. W. N. Fishbein, Myoadenylate deaminase deficiency: inherited and acquired forms. Biochemical Medicine 33:158-169 (1985).
2. R. L. Sabina, W. N. Fishbein, G. Pezeshkpour, P.R.H. Clark, and E. W. Holmes, Molecular analysis of the myoadenylate deaminase deficiencies. Neurology, in press.
3 M. Gross, H. Morisaki, T. Morisaki and E. Holmes, A common point mutation is responsible for AMP deaminase deficiency. International Journal of Purine and Pyrimidine Research 2 (supp 1):50 (1991).
4. N. C. Kar and C. M. Pearson, Muscle adenylic acid deaminase activity: selective decrease in early-onset Duchenne muscular dystrophy. Neurology 23:478-482 (1973).

5. H. Nagao, AMP deaminase activity of skeletal muscle in neuromuscular disorders in childhood: histochemical and biochemical studies. Neuropediatrics 17:193-198 (1973).
6. A. Bohan and J. B. Peter, Polymyositis and dermatomyositis (first of two parts) New England Journal of Medicine 292:344-347 (1975).
7. R. L. Sabina, J. L. Swain, and E. W. Holmes, Myoadenylate deaminase deficiency, in: The Metabolic Basis of Inherited Disease, 6th Edition. C. Scriver, A. L. Beaudet, W. S. Sly, and D. Valle, editors. McGraw-Hill, New York, 1077-1084 (1989).
8. R. L. Sabina, et al, Myoadenylate deaminase deficiency: functional and metabolic abnormalities associated with disruption of purine nucleotide cycle. Journal of Clinical Investigation 73:720-730 (1984).
9. R. L. Sabina, R. Marquetant, N. M. Desai, K. Kaletha, and E. W. Holmes, Cloning and sequence of rat myoadenylate deaminase cDNA: evidence for tissue-specific and developmental regulation. Journal of Biological Chemistry 262:12397-12400 (1987).

MOLECULAR GENETIC ANALYSIS OF CHROMOSOME 9p IN METHYLTHIOADENOSINE PHOSPHORYLASE DEFICIENT GLIOMA CELL LINES

David J. Wu, Linda Reynolds, Dennis A. Carson, Tsutomu Nobori

Department of Medicine, University of California, San Diego
La Jolla, California 92093-0945 USA

INTRODUCTION

Methylthioadenosine (MTA) phosphorylase is the enzyme involved in the metabolism of polyamines and purines.[1] MTA, the substrate for this enzyme, is produced during the synthesis of spermine and spermidine and is cleaved to methylthioribose 1-phosphate and adenine, which are recycled to methionine and adenine nucleotide, respectively.[1] This enzyme has been known to be deficient in human leukemias, lymphomas, and solid tumors.[2,3,4] The gene locus for this enzyme (designated MTAP) has been mapped to human chromosome 9p.[5] Enzyme deficient malignant cell lines are frequently found to have chromosome 9p abnormalities.[6] Although cytogenetic analysis of human gliomas has demonstrated a very prevalent chromosomal abnormality involving deletions or translocations of chromosome 9p,[7] these tumors have not been tested for MTAP deficiency. In this report, we screened eight human glioma cell lines and six primary brain tumors for MTA phosphorylase activities using the radiochemical method and immunoblot analysis. Of 14 cell lines and primary tumors, five cell lines and five primary tumors (70%) were found to be enzyme-deficient. To elucidate the molecular mechanism of this enzyme deficiency and its relevance to the genesis of human brain tumors, DNA analysis and pulsed field gel analysis were carried out.

MATERIALS and METHODS

Cell lines and primary brain tumor specimens- The human cell lines obtained from American Type Culture Collection (Rockville, MD) were as follows: U-87MG (Glioblastoma, HTB 14), U-138MG (Glioblastoma, HTB 16), U-373MG (Glioblastoma, HTB 17), A172 (Glioblastoma, CRL 1620), T98G (Glioblastoma, CRL 1690), Hs683 (Glioma, HTB 138), CCF-STTG1 (Glioblastoma,CRL 1718), H4 (Neuroglioma, HTB 148). Brain tumor specimens were provided by the surgical pathology division at Scripps Clinic and Research Foundation. Specimens were obtained during surgery, and were immediately frozen at -70°C.

MTA phosphorylase assay and immunoblot analysis- Enzyme activity was measured by the radiochemical method described earlier.[8] The immunoblotting procedure was performed using antibodies against the MTA phosphorylase peptides as described.[9]

DNA probes- Chromosome 9p markers used were a cDNA clone of the human interferon-αI' gene (IFNA),[10] a cDNA clone of the human interferon-β1 gene (IFNB),[11] and pCN2,[12] the 3' nontranslated region of N-*ras* oncogene. The other DNA markers, a mitochondrial RNA-processing endonuclease gene (RMRP)[13] and the galactosyltransferase gene (GALT),[14] were PCR amplified from human lymphoblastoid cell DNA and a bovine cDNA library, respectively. The PCR products were subcloned and sequenced. In slot-blot and Southern blot analyses, the J-gene segment of immunoglobulin heavy-chain (J_H) was used as a nonsyntenic control.

Purine and Pyrimidine Metabolism in Man VII, Part B
Edited by R.A. Harkness *et al.*, Plenum Press, New York, 1991

Genomic DNA extraction and slot-blot and Southern blot analyses- Extraction and Southern blot analysis of genomic DNA was performed according to standard techniques.[15] Slot-blot analysis was performed by Minifold II Slot Blot System (Schleicher & Schuell) following the manufacturer's instructions. Densitometric scanning profiles of the autoradiograms were obtained using a Bio-Rad apparatus.

Pulsed field gel analysis- DNA from cultured cells were prepared in agarose pluds. Agarose plugs were digested with a rare-cutting restriction enzyme SacII. Pulsed field gel electrophoresis was performed on CHEF-DR II (BioRad) in 0.5 x TBE. The gel was run at $14^{\circ}C$ using the switch time of 60 seconds for 12 hours, followed by the switch time of 90 seconds for 12 hours. After electrophoresis, the gel was stained in the running buffer containing 0.5µg/ml ethidium bromide. After the gel was UV irradiated, the DNA was transferred to nylon membrane.

RESULTS and DISCUSSIONS

We measured MTA phosphorylase activities in eight glioma cell lines (six glioblastomas, one glioma, and one neuroglioma). Of six glioblastoma cell lines, three cell lines (U-373MG, T98G, and CCF-STTG1) contained the enzyme activities at 0.33, 0.48, and 0.15 nmol per min per mg protein, respectively. The other 3 cell lines (U-87MG, U-138MG, A172) were enzyme-deficient. Both glioma (Hs 683) and neuroglioma (H4) cell lines lacked detectable enzyme activities. To confirm the presence or absence of the MTA phosphorylase protein, we used an anti-enzyme antibody to perform an immunoblot analysis of cell extracts from these cell lines (Figure 1A). Five cell lines lacking detectable enzyme activity had no immunoreactive enzyme protein. We then analyzed six primary brain tumors. Five of them had no immunoreactive enzyme protein (Figure 1B).

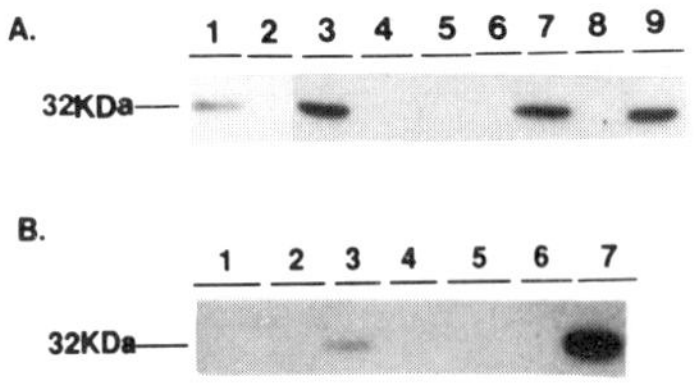

Figure 1. Immunoblot analysis of human glioma cell lines (A) and primary brain tumors (B). Cell extracts were separated by SDS-PAGE. A: 1, CCF-STTG1; 2, H4; 3, U-373MG; 4, Hs683; 5, U-138MG; 6,U-87MG; 7, T98G; 8, A172; 9, Molt4. B:1, 4, and 5, astrocytomas; 2 and 3, glioblastomas multiforme; 6, oligoastrocytoma;7, MDBK. Molt4 and MDBK were used as positive controls.

The naturally enzyme deficient cells, CEM, K-T1, NALL-1, and K562, are known to have cytogenetic abnormalities in chromosome 9p.[6] It is important to note that MTAP locus has been assigned to this same region.[5] Furthermore, these cells are hemizygous or nullizygous for the IFNA and IFNB genes.[16] Studies reported by several laboratories have shown that the loci on chromosomes 10 and 17 are frequently lost in patients with malignant astrocytomas.[17,18] A recent study using the IFNA and IFNB genes has shown that out of 19 glioma cell lines, 8 were nullizygous and 2 were hemizygous for the IFNA gene, and these ten were either nullizygous or hemizygous for the IFNB gene.[19] Therefore, we carried out genetic analysis of chromosome 9p in the glioma cell lines. At the time of these experiments, somatic DNAs from the individuals from whom these cell lines were derived were not yet available. For this reason, and because some chromosome 9p markers used are not polymorphic, we used a gene quantification method to determine the gene dosage. After hybridization with chromosome 9p markers excluding pCN2, all blots were rehybridized with a nonsyntenic J_H probe. We then compared densitometrically the gene dosage for each DNA probe.

For the IFNA gene, both Southern blot and slot-blot analyses were employed to determine genotype. As illustrated in Figure 2, Southern blots of HindIII-restricted genomic DNA with the IFNA gene probe displayed eleven bands of different size in three cell lines (U-373MG, T98G, CCF-STTG1). In A172 two bands, #1 and #4, (27kb and 8.8kb) and in H4 five bands, #1, 2, 3, 4, 5, (27, 18, 11.5, 8.8, and 7.4kb) are missing. Using slot-blot analysis to compare the signal intensity of the IFNA gene with that of the J_H gene, these five cell lines were found to be disomic for the IFNA genes (data not shown). Both Southern blot and slot-blot analyses disclosed that five of eight cell lines either completely or partially deleted members of the IFNA gene cluster. The IFNB gene, which is located telomeric to the IFNA gene cluster on chromosome 9p, was analyzed by Southern hybridization of BamHI-restricted genomic DNAs. The copy number of the IFNB gene was determined by comparing densitometer tracings of the autoradiogram of a membrane to which BamHI-restricted DNAs were transferred and then hybridized successively with the IFNB gene and J_H gene probes. Two copies of the germline J_H gene were assumed to exist in the DNA of each cell line. Two cell lines, U-87MG and U-138MG, lacked both alleles of the IFNB gene. The remainder of the cell lines tested contained two copies of the IFNB gene (Figure 3).

Figure 2

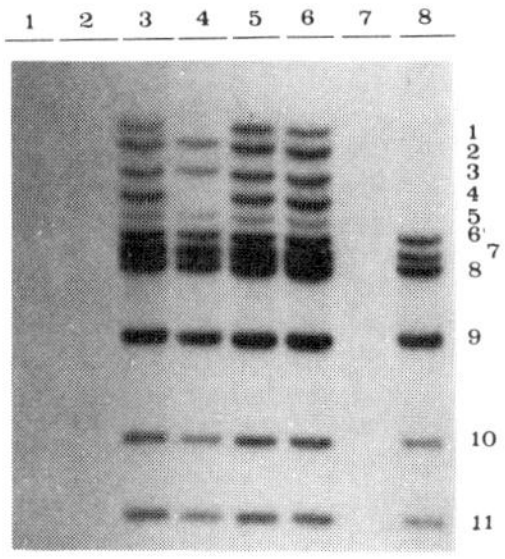

Figure 3

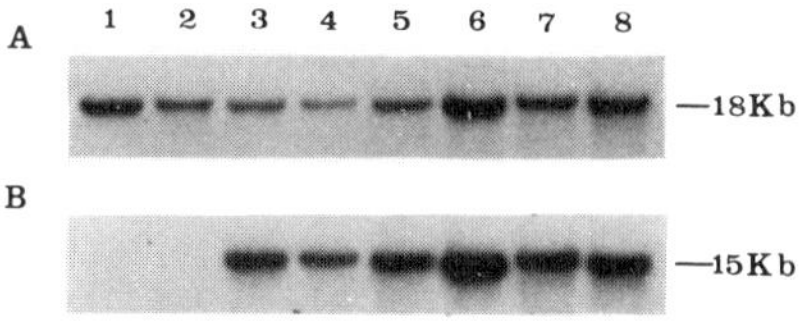

Figures 2 and 3. Absence of the IFNA genes from human glioma cell lines. In Figure 2, HindIII-restricted genomic DNAs were probed with the IFNA gene. Fragments observed were numbered and shown on the right. In Figure 3, BamHI-restricted DNAs were probed successively with the IFNB gene (panel B) and the J_H gene (panel A). Lane: 1, U-87MG; 2, U-138MG; 3, U-373MG; 4, A172; 5, T98G; 6, CCF-STTG1; 7, Hs683; 8, H4.

In addition to interferon genes, the markers, pCN2 and RMRP, have been mapped to chromosome 9p[12] and 9p21-9p12,[20] respectively. N-*ras*-like sequences detected on chromosome 9p with pCN2 have been further defined to map to the p12-cen region of chromosome 9.[21] Based on the densitometer tracings of the autoradiogram, all cell lines appeared to have two copies of N-ras-like sequences (data not shown). RMRP gene has been reported to detect a single band in BglII-restricted DNA.[20] The copy number of this gene in glioma cell lines was determined by comparing densitometer tracings of autoradiograms as described above. All glioma cell lines tested contained two copies of this gene (data not shown).

Our results are summarized in Table 1. None of 5 cell lines lacking MTA phosphorylase activity contained the complete IFNA gene cluster. Three of them were found to have two copies of the IFNB gene. Partial or complete disappearance of the IFNA gene cluster was well correlated with MTA phosphorylase deficiency. These data suggested that MTAP locus is more closely linked to the IFNA gene cluster than to the IFNB gne and other 9p markers analyzed in these studies.

Table 1. MTA phosphorylase activities and the gene copy number of the 9p genes in human glioma cell lines.

Cell line	MTAP 9pter-q12	IFNA 9p22-p13	IFNB 9p22	pCN2 9p12-cen	RMPR 9p21-p12
U-87MG	-	0	0	2	2
U-138MG	-	0	0	2	2
U-373MG	+	2	2	2	2
A172	-	2[a]	2	2	2
T98G	+	2	2	2	2
CCF-STTG1	+	2	2	2	2
Hs683	-	0	2	2	2
H4	-	2[a]	2	2	2

[a] Partial deletion of α-interferon genes.

To define the region of deletion with respect to the loci for the IFNA and IFNB genes, we performed pulsed field gel analysis of SacII restricted chromosomal DNA prepared from the cell lines A172, U-373MG, and H4 as described in Materials and Methods.

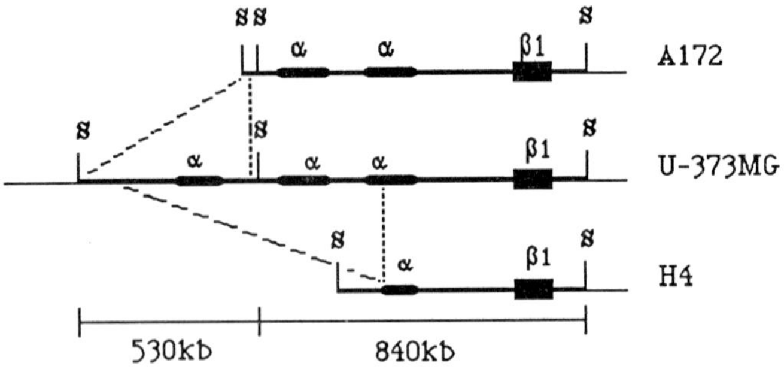

Figure 4. The schematic illustration of the deletion map in the cell lines. The size and physical map of each gene are not to scale. S, SacII; α, interferon-α gene cluster; β1, interferon-β1 gene.

The IFNA gene probe detected 840kb and 530kb SacII fragments in DNA prepared from the cell line U-373MG. However, in the cell line A172, two SacII fragments (840kb and 900kb) were detected, whereas a single 390kb SacII fragment was seen in the cell line H4. When the same blots were probed with the IFNB gene, the 840kb SacII fragment in U-373MG, both 840kb and 900kb SacII fragments in A172, and the 390kb fragment in H4 were detected. Since the IFNA gene cluster and the IFNB gene have been known to locate in the 1400kb Not I fragment,[22] the interpretation of these results is as follows (Figure 4):

The 1370kb SacII fragment has another SacII site within that fragment, which was cleaved to form two fragments of 840kb and 530kb. In A172, this internal SacII site was partially methylated or partially digested including the partial deletion of the IFNA genes, which resulted in 900kb and 840kb fragements. In H4, the deleted region including the internal SacII site was big enough to yield a 390kb fragment.

References

1. H. G. Williams-Ashman, J. Seidenfeld, and P. Galletti, Trends in the biochemical pharmacology of 5'-deoxy-5'methylthioadenosine, Biochem. Pharmacol. 31:277 (1982).
2. N. Kamatani, W. A. Nelson-Rees, and D. A. Carson, Selective killing of human malignant cell lines deficient in methylthioadenosine phosphorylase, a purine metabolic enzyme, Proc. Natl. Acad. Sci. USA 78:1219 (1981).
3. N. Kamatani, A. L. Yu, and D. A. Carson, Deficiency of methylthio-adenosine phosphorylase in human leukemic cells in vivo, Blood 60:1387 (1982)
4. J. H. Fitchen, M. K. Riscoe, B. W. Dana, H. J. Lawrence, and A. J. Ferro, Methylthioadenosine phosphorylase deficiency in human leukemias and solid tumors, Cancer Res. 46:5409 (1986).
5. C. J. Carrera, R. L. Eddy, T. B. Shows, and D. A. Carson, Assignment of the gene for methylthioadenosine phosphorylase to human chromosome 9 by mouse-human somatic cell hybridization, Proc. Natl. Acad. Sci. USA 81:2665 (1984).
6. M. O. Diaz, S. Ziemin, M. M. Le Beau, P. Pitha, S. D. Smith, R. R. Chilcote, and J. D. Rowley, Homozygous deletion of the alpha- and beta 1-interferon genes in human leukemia and derived cell lines, Proc. Natl. Acad. Sci. USA 85:5259 (1988).

7. S. H. Bigner, J. Mark, D. E. Bullard, S. M. Mahaley, Jr., and D. D. Bigner, Chromosomal evolution in malignant human gliomas starts with specific and usually numerical deviations, Cancer Genet. Cytogenet. 22:121 (1986).
8. M. Kubota, N. Kamatani, and D. A. Carson, Biochemical genetic analysis of the role of methylthioadenosine phosphorylase in a murine lymphoid cell line, J. Biol. Chem. 258:7288 (1983).
9. T. Nobori, J. G. Karras, F. Della Ragione, T. A. Waltz, P. P. Chen, and D. A. Carson, Absence of methylthioadenosine phosphorylase in human gliomas, Cancer Res. 51:3193 (1991)
10. J. Mizoguchi, P. M. Pitha, and N. B. K. Raj, Efficient Expression in *Escherichia coli* of two species of human interferon-α and their hybrid molecules, DNA 4:221 (1985).
11. N. B. K. Raj, and P. M. Pitha, Analysis of interferon mRNA in human fibroblast cells induced to produce interferon, Proc. Natl. Acad. Sci. USA 78:7426 (1981).
12. H. Middleton-Price, N. Spurr, A. Hall, and S. Malcolm, N-*ras*- like sequences on chromosomes 9, 6, and 22 with a ploymorphism at the chromosome 9 locus, Ann. Hum. Genet. 52:189 (1988).
13. H. A. Gold, J. N. Topper, D. A. Clayton, and J. Craft, The RNA processing enzyme RNase MRP is identical to the Th RNP and related to RNase P, Science 245:1377 (1989).
14. N. L. Shaper, J. H. Shaper, J. L. Meuth, J. L. Fox, H. Chang, I. R. Kirsch, and G. F. Hollis, Bovine galactosyltransferase: Identification of a clone by direct immunological screening of a cDNA expression library, Proc. Natl. Acad. Sci. USA 83:1573 (1986).
15. J. Sambrook, E. F. Fritsch, and T. Maniatis, "Molecular Cloning: a laboratory manual," Cold Spring Harbor Laboratory Press, New York (1989).
16. M. O. Diaz, C. M. Rubin, A. Harden, S. Ziemin, R. A. Larson, M. M. Le Beau, and J. D. Rowley, Deletions of interferon genes in acute lymphoblastic leukemia, New Engl. J. Med. 322:77 (1990).
17. C. D. James, E. Carlbom, J. P. Dumanski, M. Hansen, M. Nordenskjold, V. P. Collins, and W. K. Cavenee, Clonal genomic alterations in glioma malignancy stages, Cancer Res. 48:5546 (1988).
18. C. D. James, E. Carlbom, M. Nordenskjold, V. P. Collins, and W. K. Cavenee, Mitotic recombination of chromosome 17 in astrocytomas, Proc. Natl. Acad. Sci. USA 86:2858 (1989).
19. J. Miyakoshi, K. D. Dobler, J. Allalunis-Turner, J. D. S. McKean, K. Petruk, P. B. R. Allen, K. N. Aronyk, B. Weir, D. Huyser-Wierenga, D. Fulton, R. C. Urtasun, and R. S. Day, III, Absence of IFNA and IFNB genes from human malignant glioma cell lines and lack of correlation with cellular sensitivity to interferons, Cancer Res. 50:278 (1990).
20. C.-L. Hsieh, T. A. Donlon, B. T. Darras, D. D. Chang, J. N. Topper, D. A. Clayton, and U. Francke, The gene for the RNA component of the mitochondrial RNA-processing endoribonuclease is located on human chromosome 9p and on mouse chromosome 4, Genomics 6:540 (1990).
21. T. Nobori, L. E. Hexdall, and D. A. Carson, A polymorphic region defined by pCN2 (the 3' nontranslated region of N-*ras*) maps to chromosome 9cen-p12, Human Genet., in press.
22. J. D. Rowley, M. O. Diaz, R. Espinosa, III, Y. D. Patel, E. van Melle, S. Ziemin, P. Taillon-Miller, P. Lichter, G. A. Evans, J. H. Kersey, D. C. Ward, P. H. Domer, and M. M. Le Beau, Mapping chromosome band 11q23 in human acute leukemia with biotinylated probes: Identification of 11q23 translocation breakpoints with a yeast artificial chromosome, Proc. Natl. Acad. Sci. USA 87:9358 (1990).

S-ADENOSYLHOMOCYSTEINE HYDROLASE ACTIVITY IN ERYTHROCYTES FROM HIV-INFECTED PATIENTS

A. Bozzi, F. Leonardi°, M.L. De Rinaldis*, M. Ferrazzi*, and R. Strom°

Department of Biomedical and Technological Sciences University of L'Aquila; °Department of Human Biopathology, and *Institute of Infectious Diseases University "La Sapienza", Roma, Italy

INTRODUCTION

S-adenosylhomocysteine hydrolase (SAHase) catalyzes the reversible hydrolysis, to adenosine (Ado) and L-homocysteine (Hcy)[1], of S-adenosylhomocysteine (SAH), which is the product of all S-adenosylmethionine-dependent methylation reactions. The activity of this enzyme, allowing the elimination of the intracellular SAH pool, is essential to prevent SAH-dependent inhibition of most transmethylation reactions[2]. Previous work, by Cowan et al.[3] and by ourselves[4], had shown that erythrocytes from HIV-infected patients had significantly higher levels of adenosine deaminase (ADA) activity. Since a correlation appears to exist between SAHase and ADA both genetically, the structural genes of both enzymes being located on human chromosome 20[5], and functionally, both enzymes having adenosine as their substrate and/or product, it seemed of interest to investigate whether also SAHase activity was affected, in human erythrocytes, by HIV-infection. Our data indicate that there is indeed, in HIV-infected patients, a marked increase of SAHase activity, which correlates positively with the severity of the disease.

SUBJECTS AND METHODS

Human blood samples, freshly collected in EDTA-containing tubes, were obtained from HIV-infected subjects, ranging from serologically positive, disease-free individuals to overt AIDS patients. Healthy donors were used as controls. Red blood cells were then washed in isotonic buffer (PBS), pH 7.3, and filtered through an α-cellulose:microcristalline cellulose column (4 x 2 cm), to remove leukocytes and platelets[6]. Purified erythrocytes were lysed by repeated freezing and thawing, and their hemolysates immediately assayed for the enzymatic activities and for hemoglobin content[7]. ADA and pyruvate kinase (PK) activities were determined as reported by Beutler[6]. SAHase activity was assayed in the synthetic direction by measuring the amount of SAH formed, after 5 min incubation at 37°C, upon addition of 0.1 mL fresh hemolysate to a 0.4 mL of a reaction mixture containing 50 mM Hepes (pH 7.3), 2 mM D,L-homocysteine, 0.1 mM adenosine and 10 μM erythro-9-(2-hydroxy-3-nonyl)adenine (E.H.N.A.). The reaction was stopped by adding 0.1 mL of 3 M perchloric acid. The protein precipitate was removed by centrifugation, the supernatant was neutralized with 40 μL of 3 M potassium carbonate, and the precipitated salts removed by centrifugation[8]. SAH was analyzed by HPLC on an Altex ultrasphere C_{18} ODS reverse-phase column (4.6 x 15 cm) equilibrated with 0.1 M KH_2PO_4, pH 4.3, containing 0.05 M NaCl and 5.5% methanol and operating at a flow rate of 1.3 mL/min. Retention time for SAH was 7 min.

RESULTS

As shown in Table I, while in purified erythrocytes from normal healthy subjects the level of ADA activity was centered around a mean value of 1.22 ± 0.2 IU/g Hb, there was, in HIV-infected patients, a significant shift toward higher values, up to 2.33 ± 0.59 IU/g Hb. In some cases, however, severely affected patients exhibited ADA activity levels similar to those found in normal individuals. On the other hand, high ADA levels were also found in the erythrocytes of 10 HIV-negative subjects with hemolytic anemia due to hereditary spherocytosis or to congenital enzymatic deficiencies, such as glucose-6-phosphate dehydrogenase. In order to correct such"erroneous" results, the activity of another red blood cell enzyme, PK, whose levels are strictly dependent on erythrocytes age, was simultaneously determined, and the ADA/PK ratio was evaluated. As reported in Table I, the higher ADA activity values of the subjects with hemolytic anemia were indeed well corrected, thus eliminating the majority of the"false positive" samples. Also the occurrence of "false negative" cases was somewhat decreased, indicating that, in some overt AIDS patients, the relatively low ADA activity values were probably related to a less efficient erythropoyesis. These observations prompted us to verify if erythrocytes SAHase activity was also affected by HIV-infection. Table II shows indeed that the activity of this enzyme increases with the severity of the disease, with a trend similar to the one exhibited by ADA.

DISCUSSION

According to Cowan _et al._[3], the increase of ADA levels in the erythrocytes of HIV-infected patients could be due to a direct effect of the virus on the bone marrow erythroid cells. On the other hand, the elevation of ADA activity in AIDS patients and in other hemolytic syndromes[9,10] could result from a

Table I. Distribution of the erythrocytes ADA activity and of the ADA/PK ratio in HIV-infected and non-infected individuals

Subjects	No.	ADA(I.U./g Hb)	ADA/PK ratio
Normal individuals	38	1.22 ± 0.20[a]	0.117 ± 0.019
Uninfected subjects with hemolyt. anemia	10	1.75 ± 0.38	0.113 ± 0.018
LAS patients	46	1.50 ± 0.32	0.154 ± 0.022
ARC patients	20	1.78 ± 0.60	0.167 ± 0.024
AIDS patients	36	2.33 ± 0.59	0.205 ± 0.026

[a]Mean (± standard deviation)

Table II. S-adenosylhomocysteine hydrolase (SAHase) activity in erythrocytes from HIV-infected and non-infected individuals

Subjects	No.	SAHase(nmol/min/mg Hb)
Normal individuals	7	0.54 ± 0.08[a]
Serum positive clinically asymptomatic subjects	6	0.72 ± 0.12
LAS patients	8	0.77 ± 0.14
ARC patients	3	0.85 ± 0.16
AIDS patients	9	1.15 ± 0.11

[a]Mean (± standard deviation)

perturbation of ADA gene expression with aberrant maturation of the erythroid cell line. If these hypotheses are correct, the simultaneous increase of SAHase activity could be explained in terms of an accelerated catabolism of adenosine which, in turn, would "pull" the SAHase-catalyzed reaction in the hydrolytic direction. Due to the increase in the SAM/SAH ratio, a generalized acceleration of several transmethylation reactions would then occur. In mature circulating erythrocytes, as well as in HIV-infected cells, the correct balance of methylation reactions allows some crucial processes for the cell life, like protein repair-carboxymethylation or membrane phospholipids turn-over. In this context, an accelerated catabolism of adenosine, possibly triggered at the gene-expression level by HIV-infection, appears to induce a general derangement of the SAM-dependent transmethylation reactions which persists even in mature circulating erythrocytes.

AKNOWLEDGEMENTS

The financial support of the Italian Ministry of Health (Progetto AIDS 1990, Istituto Superiore di Sanità) is gratefully aknowledged.

REFERENCES

1. G. De la Haba and G. L. Cantoni, The enzymatic synthesis of S-adenosyl-L-homocysteine from adenosine and homocysteine, J. Biol. Chem. 234:603 (1959).
2. G. L. Cantoni and P. K. Chiang, The role of S-adenosylhomocysteine hydrolase in the control of biological methylation, pp. 67, in: "Natural sulfur compounds", D. Cavallini, G. E. Gaull, and V. Zappia, eds., Plenum Publ., New York (1980).
3. M. J. Cowan, R. O. Brady, and K. J. Widder, Elevated erythrocyte adenosine deaminase activity in patients with acquired immunodeficiency syndrome, Proc. Natl. Acad. Sci. USA 83:109 (1986).
4. A. Bozzi, M. Ferrazzi, F. Sorice, and R. Strom, Adenosine deaminase:pyruvate kinase ratio in erythrocytes as a useful marker of HIV infection, Clin. Chem. Enzym. Comms. 1:361(1989).
5. M. S. Hershfield and U. Francke, The human genes for S-adenosylhomocysteine hydrolase and adenosine deaminase are syntenic on chromosome 20, Science 216:739 (1982).
6. E. Beutler, "Red cell metabolism: a manual of biochemical methods", 3rd edn., Grune and Stratton Inc., N. York (1983).
7. E. J. van Kampen and W. G. Zijlstra, Standardization of hemoglobinometry. II. The hemiglobincyanide method, Clin. Chim. Acta 6:538 (1961).
8. G. De la Haba, S. Agostini, A. Bozzi, A. Merta, C. Unson, and G. L. Cantoni, S-adenosylhomocysteinase: mechanism of reversible and irreversible inactivation by ATP, cAMP, and 2'-deoxyadenosine, Biochemistry 25:8337 (1986).

9. W. N. Valentine, D. E. Paglia, A. P. Tartaglia, and F. Gilsanz, Hereditary hemolytic anemia with increased red cell adenosine deaminase (45- to 70- fold) and decreased adenosine triphosphate, Science 195:783 (1977).
10. B. E. Glader and K. Backer, Elevated red cell adenosine deaminase activity: a marker of disordered erythropoiesis in Diamond-Blackfan anaemia and other haematologic diseases, Br. J. Haematol. 68:165 (1988).

PURIFICATION AND CHARACTERIZATION OF RECOMBINANT RAT PHOSPHORIBOSYLPYROPHOSPHATE SYNTHETASE SUBUNIT I AND SUBUNIT II

Masamiti Tatibana, Sumio Ishijima, Kazuko Kita, Imtiaz Ahmad, Toshiharu Ishizuka, and Masanori Taira

Department of Biochemistry, Chiba University School of Medicine, Inohana, Chiba 280, Japan

INTRODUCTION

Phosphoribosylpyrophosphate (PRPP) synthetase catalyzes the formation of PRPP from ATP and ribose 5-phosphate. The enzyme has been purified from bacteria [1, 2] and mammalian tissues [3-5]. The rat liver enzyme exists as complex aggregates of 34-, 38-, and 40-kDa components, the 34-kDa species being the catalytic subunit [5]. The 34-kDa component is actually a mixture of two isoforms, designated as PRS I and PRS II [5, 6]. The two isoforms are composed of 317 amino acid residues, and the sequences are highly conserved, differing only by 13 residues [6]. Furthermore, the amino acid sequences of human [7, 8] and rat [6] PRS II differ only by 3 residues and those of human [8, 9] and rat [6] PRS I are completely conserved. The PRS I and PRS II are encoded by two distinct genes located on X-chromosome [10, 11]. The two genes are expressed in almost all tissues of rats but the mRNA levels differ with the tissues [12]. These observations suggest functional differences between catalytic and/or regulatory properties of PRS I and PRS II. However, separation of the two proteins from the native enzyme was impossible. Therefore, the respective rat cDNAs were expressed in Escherichia coli. The expression vector was designed to produce the unfused proteins. The recombinant isoforms (named rPRS I and rPRS II) were isolated and characterized.

METHODS

Construction of expression plasmids

The expression vectors were constructed by two modifications of pKK233-2 (Pharmacia). (a) The replication origin of pKK233-2 was replaced with that of pGEM-1 (Strategene), which ensured an increased copy number of the vector in E. coli cells. (b) 9 restriction cloning sites of pGEM-1 were inserted into the cloning sites of pKK233-2. A cDNA fragment can be conveniently inserted into these multicloning sites. The rat PRS I and PRS II cDNA fragments (1.2 and 2.1 kb, respectively) were inserted into these sites. Each fragment contained the complete coding region of 954 bp and the 3'-noncoding region. The plasmid constructs were introduced into the E. coli strain MV1304 which contains a $lacI^q$ repressor.

Purine and Pyrimidine Metabolism in Man VII, Part B
Edited by R.A. Harkness *et al.*, Plenum Press, New York, 1991

Purification of recombinant PRS I and PRS II from *E. coli* cells

The first two purification steps, polyethylene glycol precipitation and acid precipitation, were performed essentially as described previously [5]. A portion of the acid precipitates (5 mg of protein for each) was applied to a DEAE-5PW HPLC column equilibrated with 30 mM potassium phosphate buffer (pH 7.4) containing 6 mM $MgCl_2$ and 0.1 mM EDTA. Elution was performed first with 100 mM potassium phosphate (pH 6.8) containing 0.3 mM ATP, 6 mM $MgCl_2$, 0.1 mM EDTA, and 2.5 mM 2-mercaptoethanol, as enzyme-stabilizing agents [3, 4], and then with a linear gradient of 0 to 0.3 M KCl in the same buffer. The eluted enzymes were dialyzed against 50 mM potassium phosphate (pH 7.4) containing the stabilizers.

RESULTS AND DISCUSSION

Expression of rat PRS I and PRS II in *E. coli* cells

The expression plasmids consisted of a highly inducible *trc* promoter, a ribosome-binding site, the PRS I or PRS II cDNAs, and a strong *rrnB* transcription terminator. The expression of rat PRS I and PRS II was induced within 2 h after addition of isopropyl-β-D-thiogalactopyranoside (IPTG), an inducer of the *trc* promoter, and reached a maximum in 3-4 h (Fig. 1).

Purification of rPRS I and rPRS II from the transformed *E. coli* cells

The transformed *E. coli* cells were grown at 30 °C for 15 h in 10 liters of M9 medium containing 1 mM thiamine-HCl and ampicillin (20 μg/ml). After adding 1 mM IPTG, the cells were grown for an additional 3 h at 30 °C and then harvested and lysed. The rPRS I and rPRS II were purified from the soluble fractions of the extracts. The results of purification are summarized in Table 1. The purified rPRS I and rPRS II preparations showed a single band of 34 kDa, upon SDS gel electrophoresis, and had definite enzyme activity by themselves alone. Notable is that the specific activities of the rPRS I and rPRS II were 2.5- and 3.3-fold higher,

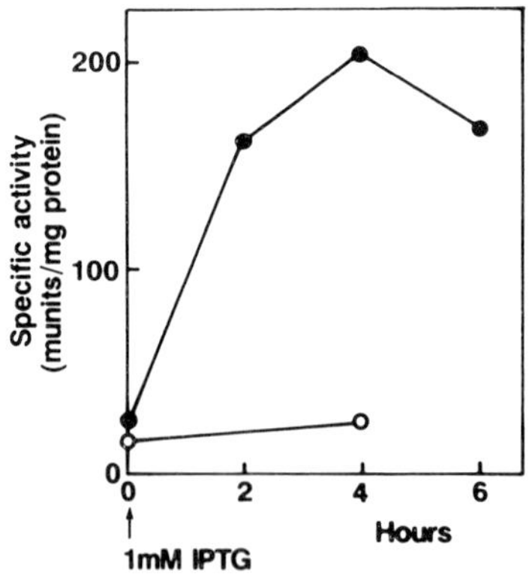

Fig. 1. Induction of the expression of rPRS II by IPTG. The *E. coli* cells transformed with the vector alone (○) or the plasmid including the rat PRS II cDNA (●) were grown at 30 °C in LB medium containing ampicillin (50 μg/ml). After adding 1 mM IPTG, the cells were harvested at the indicated times, disrupted by sonication, and centrifuged at 10,000 x g for 15 min. The supernatant was assayed for PRPP synthetase activities.

Table 1. Purification of recombinant rat PRS I and PRS II from 10-liter cultures of *E. coli* cells

Purification step	Total protein	Total activity	Specific activity
	mg	units	milliunits/mg
rPRS I			
Cell extract	1,490	1,430	960
Polyethylene glycol	188	1,820	9,730
Acid precipitation	80	1,580	19,700
DEAE-5PW	51	1,310	25,700
rPRS II			
Cell extract	942	565	421
Polyethylene glycol	70	662	9,450
Acid precipitation	27	567	21,000
DEAE-5PW	15	518	34,500

respectively, than that of the most purified rat liver enzyme, in terms of the 34-kDa components [5]. The lower activity of the rat liver enzyme is ascribable to inhibitory effects of the 38- and/or 40-kDa components within the aggregates.

Properties of rPRS I and rPRS II

The purified rPRS I and rPRS II existed as high molecular weight aggregates. Molecular masses of rPRS I and rPRS II, determined by gel filtration on a TSK G4000SW column, were 700-1,200 and 550 kDa, respectively.

The heat stabilities of rPRS I and rPRS II were quite different. The

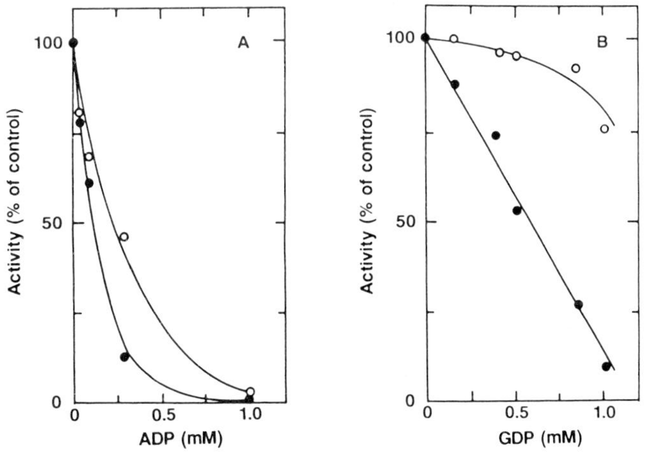

Fig. 2. Inhibition of rPRS I and rPRS II by ADP and GDP. Standard assay conditions (0.4 mM ATP, 0.2 mM ribose 5-phosphate, 4 mM $MgCl_2$, 10 mM potassium phosphate buffer, pH 7.4) were used with ADP (A) or GDP (B) added at the indicated concentrations. ●, rPRS I; ○, rPRS II.

half-life of inactivation of rPRS II at 49°C was 0.5 min, while the half-life of rPRS I was 90 min, 180-fold greater than that of rPRS II.

The apparent Km values for ATP and ribose 5-phosphate and Ka value for Pi of rPRS I were 44 µM, 40 µM, and 1.8 mM, respectively, and rPRS II showed similar values, i.e., 60 µM, 73 µM, and 2.4 mM, respectively.

Sulfate, like Pi, did activate the isoforms. rPRS I gave a maximum activity at 60 mM K_2SO_4, and the value was 43% of the activity with 50 mM Pi. In contrast, the rPRS II activity reached a maximum at 20-40 mM sulfate, and higher concentrations were inhibitory. The maximum activity was only 7% of that seen with Pi.

rPRS I and rPRS II were inhibited by nucleotides. ADP and GDP were most effective, as far as examined. rPRS I was far more sensitive than rPRS II to the inhibition (Fig. 2). The inhibition by 1 mM AMP, a reaction product, was 34% and 36% for rPRS I and II, respectively. Other nucleotides such as GTP, GMP, UTP, UDP, UMP, and CTP, only weakly inhibited both isoforms (0-19%) each at a concentration of 1 mM.

The 34-kDa component of the native enzyme is a mixture of PRS I and PRS II, and their relative amounts can vary with tissues in rats [12]. Since rPRS I and rPRS II possess different properties, as shown above, PRPP synthetase may have tissue-specific regulatory properties. In characterization of such synthetases of tissues, the different responses of the two isoforms to sulfate, ADP and GDP, and heat treatment may be helpful. Attention should be given to the existence of the two isoforms with different properties in studies of the molecular basis of PRPP synthetase disorders [13-15].

REFERENCES

1. R. L. Switzer and K. J. Gilson, Methods Enzymol. 51:3 (1978).
2. B. Hove-Jensen, K. W. Harlow, C. J. King, and R. L. Switzer, J. Biol. Chem. 261:6765 (1986).
3. I. H. Fox and W. N. Kelley, J. Biol. Chem. 246:5739 (1971).
4. D. G. Roth, E. Shelton, and T. F. Deuel, J. Biol. Chem. 249:291 (1974).
5. K. Kita, T. Otsuki, T. Ishizuka, and M. Tatibana, J. Biochem. 105:736 (1989).
6. M. Taira, S. Ishijima, K. Kita, K. Yamada, T. Iizasa, and M. Tatibana, J. Biol. Chem. 262:14867 (1987).
7. T. Iizasa, M. Taira, H. Shimada, S. Ishijima, and M. Tatibana, FEBS Lett. 244:47 (1989).
8. B. J. Roessler, G. Bell. S. Heidler, S. Seino, M. Becker, and T. D. Palella, Nucleic Acids Res. 18:193 (1990).
9. T. Sonoda, M. Taira, S. Ishijima, T. Ishizuka, T. Iizasa, and M. Tatibana, J. Biochem. 109:361 (1991).
10. M. Taira, J. Kudoh, S. Minoshima, T. Iizasa, H. Shimada, Y. Shimizu, M. Tatibana, and N. Shimizu, Somat. Cell Mol. Genet. 15:29 (1989).
11. M. A. Becker, S. A. Heidler, G. I. Bell, S. Seino, M. M. LeBeau, C. A. Westbrook, W. Neuman, L. J. Shapiro, T. K. Mohandas, B. J. Roessler, and T. D. Palella, Genomics 8:555 (1990).
12. M. Taira, T. Iizasa, K. Yamada, H. Shimada, and M. Tatibana, Biochim. Biophys. Acta 1007:203 (1989).
13. O. Sperling, P. Boer, S. Persky-Brosh, E. Kanarek, and A. de Vries, Rev. Eur. Etud. Clin. Biol. 17:703 (1972).
14. M. A. Becker, L. J. Meyer, A. W. Wood, and J. E. Seegmiller, Science 179:1123 (1973).
15. I. Akaoka, S. Fujimori, N. Kamatani, F. Takeuchi, E. Yano, Y. Nishida, A. Hashimoto, and Y. Horiuchi, J. Reumatol. 8:563 (1981).

RAPID PURIFICATION OF A BIFUNCTIONAL PROTEIN COMPLEX POSSESSING PHOSPHORIBOSYLAMINOIMIDAZOLE CARBOXYLASE (EC 4.1.1.21) AND PHOSPHORIBOSYLAMINOIMIDAZOLESUCCINOCARBOXAMIDE SYNTHETASE (EC 6.3.2.6) ACTIVITIES

Robert W. Humble, Grahame Mackenzie, Gordon Shaw *

Humberside Polytechnic, Cottingham Road, Hull
HU6 7RT. University of Bradford, Bradford BD7 1DP *.

INTRODUCTION

Previously reported partial purification protocols for the enzymes EC 4.1.1.21 and EC 6.3.2.6 have proved either unsuccessful in resolving the enzyme activities[1], or have only considered one of the enzymes[2,4,5,6]. These procedures involve numerous precipitation and lengthy chromatographic steps, with a subsequent loss in activity, especially of EC 6.3.2.6 which is highly labile. In addition, it is difficult to ascertain the purity of the enzymes obtained by the previous workers, due either to the abscence of, or only limited electrophoretic evidence reported.

Recent genetic studies, have highlighted the need for reliable and convenient assays. In particular, Parker[7] cloned the region of the E.coli pur. C gene, coding for EC 6.3.2.6 in a high copy plasmid, but was unable to purify or assay the gene product, which was found to be a protein of submit m_r 27,000. In addition, Nygaard and Saxild[8] determined the gene enzyme relationships of de novo purine biosynthesis in B. subtilis, but were unable to distinguish between mutants deficient in either the pur. C (EC 6.3.2.6) or pur. E (EC 4.1.1.21) genes due to a lack of suitable assays. They did find however, that five genes (pur. A, gua. A, gua. B, pur. F and pur. M) encode single enzymes whereas pur. B encodes a bifunctional enzyme. Tyagi et al.,[9] have reported two radiometric assays for one of the enzymes EC 6.3.2.6, both of which are lengthy, necessitating the biosynthesis of the substrate CAIR using a crude enzyme extract, with subsequent purification problems.

RESULTS AND DISCUSSION

The buffering conditions, used during the purification of the enzymes, were developed to maximise enzyme stability. In particular, it was found that during the extraction, and ammonium sulphate precipitation, the ionic strength required to maintain enzyme activity, was 0.1 M Tris HCl, at pH 7.8. At lower concentrations, (eg. 0.05 M Tris HCl) and pH 7.8 a rapid loss of activity resulted. The addition of dithiothreitol (DTT), was found to be essential in maintaining the stability of EC 6.3.2.6. In the absence of DTT, EC 6.3.2.6 activity was completely lost within two hours of extraction. In contrast, the enzyme EC 4.1.1.21 was found to be very stable in the absence of DTT, even if left at room temperature for several days. The choice of pH 7.8, was made by considering the pH activity profiles for the two enzymes. The EC 4.1.1.21 was found to be stable over the pH range 6-8, whereas, EC 6.3.2.6 was found to have a very narrow stability profile between pH 7.4-8.0, with maximum stability at pH 7.6. Since Amicon Co.,[10] comment that non-biospecific binding of protein to the Matrex Gel column can be a significant problem below pH 7.0, pH 7.8 was chosen as the maximum compatible with the enzyme's stabilities.

Binding of the enzyme pair to the Matrex Gel Blue A column required magnesium and this requirement is of particular significance, since magnesium is a cofactor for EC 6.3.2.6 but not EC 4.1.1.21. Binding of both enzymes to the group specific ligand, Cibacron Blue F3-GA might not be expected, if the enzymes were separate proteins, since the ligand functions[11] by mimicking an enzyme's requirement for a nucleotide cofactor. Since 6.3.2.6 has a requirement for ATP, it might therefore be expected, to bind whereas EC 4.1.1.21 has no such requirement and might not be expected to bind. The binding of both enzyme activities to the column therefore supports the idea first suggested by Patey and Shaw[1], working with chicken enzymes, of a bifunctional enzyme and may perhaps suggest that the binding is _via_ a Cibacron Blue-magnesium-EC 6.3.2.6-EC 4.1.1.21 complex. In order to explore this concept and elute the activities from the column several potential biospecific eluents were tried, including ATP, ethyl 5-amino-1-ß-D-ribofuranosylimidazole-4-carboxylate, aspartic acid, and AICAR. Of the eluents investigated, only ATP was slightly effective using pulse elution techniques; most of the activity was retained on the column. Effective non-biospecific elution was achieved using a salt gradient, when both activities were co-eluted in the same fractions. (Table 1).

Examination of the fractions by gradient and homogeneous SDS-gel electrophoresis showed two bands of m_r 56,000 and 40,500. The larger molecular weight sub unit is of the order, (m_r 47,000 +/- 3,000) of that obtained by Smirnov[12], working with yeast EC 4.1.1.21. Machukob[13] has determined the sequence of the Ade-1 gene (EC 6.3.2.6) in yeast, which was found to possess 306 amino acids, (m_r 34,500). Additionally, the pur.C gene product of _E.coli_, which was not assayed by Parker[7], but was indirectly identified as EC 6.3.2.6, gave a value of m_r 27,000, after two dimensional electrophoresis. Gradient gel electrophoresis of the native protein gave a value of m_r 96,000 compared to that obtained by gel filtration, m_r 420,000. This suggests that the enzyme comprises two subunits which associate in a 1:1 ratio, and are then capable of aggregating to form a larger complex. Similar behaviour has been observed with other _de novo_ pathway enzymes, including FGAR amidotransferase.[3] Isoelectric focusing, was used to determine a pI for the enzyme pair; two bands were observed at pH 7.1 and 7.2. These values correlated well with the estimated pI, determined by the method of Lampson & Tytell[14]. Both activities were co-eluted from a

TABLE 1 PURIFICATION OF EC 4.1.1.21 AND EC 6.3.2.6 FROM SHEEP LIVER.

Fraction	Vol. cm^3	Protein mg/cm^3	Sp.Activity µM/min./mg.	Yield %	Purification Factor	Sp.Activity µM/min./mg	Yield %	Purification Factor	Ratio of Sp. Activity
Crude	86	32.10	11.2	100	1	0.34	100	1	33 : 1
Super-natant	84	23.42	0.0	0	0	0.00	0	0	0
Sephadex-G25	33	23.65	19.0	48	1.7	0.61	50	1.8	31 : 1
DEAE - sepharose	32	1.78	355.1	65	32	11.06	67	33	32 : 1
Matrex Gel Blue A									
30	1.8	1.00	116)		10	3.33)		10	35 : 1
31	1.8	0.98	170)		15	4.88)		14	35 : 1
32	1.8	0.86	348)		31	10.23)		30	34 : 1
33	1.8	0.73	582)		52	16.16)		48	36 : 1
34	1.8	0.68	1176)		105	27.20)		80	43 : 1
35	1.8	0.63	1269)		113	38.88)		114	33 : 1
36	1.8	0.68	1260)	Tot.	113	38.53)	Tot.	113	33 : 1
37	1.8	0.69	1522)	37.	136	42.75)	37	126	36 : 1
38	1.8	0.78	794)		71	30.77)		91	26 : 1
39	1.8	0.86	500)		45	15.62)		46	32 : 1
40	1.8	1.00	318)		28	14.00)		41	23 : 1
41	1.8	0.97	275)		25	8.08)		24	34 : 1
42	1.8	0.98	236)		21	6.55)		19	36 : 1

column of CM-Sepharose Cl-6B with an increasing pH gradient, (0.05 M phosphate buffer pH 6.0-8.0) in a fraction of pH 6.8 (pI = elution pH + 0.4-0.6 pH units). The enzymes were retained on the column at pH 6.0. Cation exchanger gel chromatography was not exploited as a means of purifying the enzymes, because EC 6.3.2.6 is unstable at pH 6.0. Buchanan & Lukens[2,4] and Smirnov[12], working with chicken liver and yeast enzymes, repectively, found that the activities were retained on an anion exchanger column above pH 7.0, whereas, at the same pH, and in contradiction of this result, Patey & Shaw[1] reported that the chicken liver enzymes were retained on a cation exchanger column of CM-Sephadex. We found that the sheep liver enzymes would bind weakly to an anion exchanger column of DEAE-Sepharose Cl-6B Fast Flow at pH 7.8. It was found advantageous to work at this pH for reasons given and to raise the ionic strength of the buffer, so that the enzymes would pass through the column. This procedure eliminated the need for a wash step and gradient salt elution, as well as saving 24 hours chromatography time.

The behaviour of the enzyme pair throughout the purification protocol was identical; the enzymes were co-eluted from a series of chromatography columns, including, anion and cation exchangers, gel filtration, and a group specific affinity ligand. The enzymes were also irreversibly retained on a covalent chromatography gel. The ratio of specific activities (Table 1) were in good agreement throughout the purification procedure, allowing for the instability of EC 6.3.2.6. This combined evidence, supports the suggestion of an enzyme complex comprising EC 4.1.1.21 and EC 6.3.2.6. Other enzymes in *de novo* purine biosynthesis, have recently also been shown to exist as multi enzyme complexes, including a single trifunctional polypeptide, exhibiting GAR synthetase, GAR transformylase, and AIR synthetase activities[15].

REFERENCES

1. Patey, C.A. & Shaw, G. Biochem. J., 135: 543 (1973)
2. Buchanan, J.M. & Lukens, L.N. J. Biol. Chem., 234: 1799 (1959)
3. Buchanan, J.M., Frere, J.M., Schroeder, D.D., J. Biol. Chem., 246: 4727 (1971)
4. Buchanan, J.M., Lukens, L.N. & Miller, R.W., Methods in Enzymology, LI: 186 (1978)
5. Fisher, C.R. Biochem. Biophys. Acta, 178: 380 (1969)
6. Ahmad, F., Missimer, P. & Moat, A.G. Can. J. Biochem. 43: 1723 (1965)
7. Parker, J., J. of Bacteriology, 157, No. 3: 712 (1984)
8. Nygaard, P., & Saxild, H.H., Mol. Gen. Genet., 211: 160 (1988)
9. Tyagi, A.K., Cooney, D.A., Bledsoe, M. & Jayaram,H.N., J. Biochem. Biophs. Meth. 3, part 3: 123 (1980)
10. Amicon Co. 21, Hartwell Avenue, Massachusetts 02173 Publication 1-191.
11. Easterday, R.L. & Haff, L.A., in Theory and Practice in Affinity Techniques (Sundaram, P.V. & Eckstein, F., Eds.) pp 24-44 (1978)
12. Smirnov, M.N., Vestn. Leningr. Univ. Biol., 1: 92. (1982)
13. Machukob, A.H., Vestn. Leningr. Univ. Biol, 4: 92. (1986)

14. Lampson, G.P., & Tytell, A.A., Anal. Biochem., 2: 374 (1965)
15. Benkovic, S.J., Schrimsher, J.L., Schendel, F.J., Young, M., Henikoff, M., Patterson, D., Stubbe, J., & Daubner, S.C. Biochemistry 24: 7059 (1985)

HIGHEST ADA EXPRESSING MOUSE TISSUES ALSO EXHIBIT CELL-TYPE SPECIFIC COORDINATE UP-REGULATION OF PURINE DEGRADATIVE ENZYMES

Bruce Aronow, David Witte, Dan Wiginton, and John Hutton

Division of Basic Science Research, Children's Hospital
Medical Center, University of Cincinnati
Cincinnati, Ohio, USA 45229

INTRODUCTION

During previous studies of human ADA gene regulatory elements in transgenic mice (Aronow et al. 1989) we noted the profound degree of up-regulation of the endogenous mouse ADA gene in both the post-natal gastrointestinal tract (Lee 1973, Chinsky et al. 1990), and the post-implantation reproductive tract (Knudsen et al. 1988). ADA specific activity and mRNA levels in these tissues exceeds those observed in thymus, which is thought to be the principle site of biochemical pathology in ADA deficient humans. Within the proximal GI tract, there are also some striking tissue specific species differences between humans and mice. For example, humans do express ADA at high levels in the duodenum and thymus, but do not in most other tissues including tongue and esophagus (Aronow et al. 1989). In mice, ADA expression occurs in a somewhat similar pattern but very high-level expression also occurs in tongue, esophagus, forestomach. High level ADA expression has not been observed in human placenta, but the placentas of rat, cat, cow, guinea pig, and rabbit have been reported to contain very high levels of ADA at some stages of gestation (Brady and O'Donovan, 1965; Sim and Maguire, 1970). To approach the question as to why ADA is expressed at high level in some of these locations, we have attempted to observe the expression of the other purine catabolic enzymes. In this study we demonstrate that the entire series of purine catabolic pathway enzymes are expressed in identical cell types and undergo developmental co-regulation throughout the proximal GI tract. We also observe a similar co-expression of purine catabolic enzymes in a distinct population of cells of the post-implantation maternal decidua long been known to express high levels of 5'NT (Hall 1971) and more recently ADA (Knudsen et al. 1988, 1991). Like the mature GI tract, a substantial subset of cells in the reproductive tract also appear committed to the degradation of purine nucleotides.

MATERIALS AND METHODS

Mice were fed Purina 5001 chow (23.5 % protein, 6.5 % fat) ad libitum. For timed pregnancies, noon on the day of vaginal plug appearance was designated day 0.5 p.c.. Pups were born on day 19.5-20.5. Adult mice were greater than 5 weeks in age. Enzymes and purine compounds were the highest quality available from Sigma Biochemicals. Purine compounds were analyzed by means of reverse phase HPLC with the effluent monitored at 254 nm.

Tissues for histochemistry were fixed for 4 hours at 4°C in phosphate buffered saline (PBS) containing 30% sucrose and 0.5% gluteraldehyde (modified from Spencer et al. 1968, Knudsen et al. 1988), and embedded by rapid freezing in O.C.T.. The composition of each cocktail was designed to make the conversion of each substrate dependent on the presence of the enzyme(s)to be tested. Enzymes for the subsequent steps were supplemented in the mix. To test for the connectivity of all of the catabolic enzymes in situ (5'NT → XO), AMP is the substrate and no exogenous enzymes are added. The individual components (when included) are present at the following concentrations: NBT=0.8 mM, PMS=0.15 mM, AMP=1 mM, Ado=1 mM, Ino=1 mM, Hpx=1 mM, Gua=1 mM, ADA=5 ug/ml, PNP=10 ug/ml, XO=0.07 units/ml. All incubations were performed in 50mM $NaHPO_4$ buffer pH 7.4 at 37C for varying lengths of time (15 min. to 1 hour in the dark) to provided a rough quantitative index of enzymatic activity as indicated by the intensity of formazan deposition. Sections were rinsed in water and cover-slips were applied using Aqua-Poly/Mount (Polysciences, Inc., Warrington, PA). Photographs were taken within 48 hours to avoid a prominent nuclear staining artifact that occurred upon prolonged storage. For control experiments, to specifically inactivate ADA enzyme activity, adjacent sections were incubated in 1 uM dCF for 30 minutes at room temperature. To specifically inactivate XO activity, sections were preincubated with 1 mM oxypurinol for 30 min. To observe the inhibitory effect of oxypurinol, PMS had to be omitted.

RESULTS

Cell-type Specific ADA Regulation.

The cell types responsible for generating the high level of ADA mRNA present in total RNA from the proximal GI and reproductive tracts were identified by in situ hybridization (not shown). Intense accumulations of ADA mRNA occurred in a highly localized fashion within a diverse set of maturing epithelial structures of the proximal GI tract. In the tongue, esophagus, and forestomach, ADA mRNA accumulation was confined to the superficial layers of the squamous epithelium with little or no signal present in the more immature basal layers. The duodenal mucosa showed a gradient of signal that was highest in the villous tips, lower around the villous base, and virtually absent in the crypts. Because villous epithelial cells are derived from mitoses that occur in the crypts, and squamous epithelial cells are derived by maturation from basal cell mitoses (for review see Potten and Hendry 1983, Gordon 1989), ADA gene expression to form ADA mRNA must be subject to stringent regulation during the post-mitotic maturation of these cells. From duodenum to terminal ileum, each of the small bowel segments that we have examined exhibited similar crypt to villous gradients of ADA mRNA. However, the distal small bowel exhibited less total signal. In situ hybridization of lung, uterus, cervix, colon, skin, brain, pancreas, lymph nodes, and spleen failed to reveal any restricted cell types that accumulate a high level of enzyme or mRNA. Thus, since ADA expression does not occur at high level in epithelial or placental cell types that are morphologically quite similar to the expressing cell-types, there appears to be strong regional specificity to ADA's expression in the GI tract and placenta.

Histochemical Detection of Purine Catabolic Enzymes

To evaluate the relative cell-type specific expression of other enzymes in the purine catabolic pathway, we modified standard histochemical techniques (Spencer et al. 1968, Knudsen et al. 1988) to allow the detection of each of the enzymes of the purine catabolic pathway. We further sought to demonstrate that the product of one enzymatic step could efficiently provide substrate for the next, in situ. The potential of this approach is that an entire pathway, with a series of metabolically connected individual enzymes, could be demonstrated in situ. The results of these studies indicated that each of the catabolic enzymes (5'NT, ADA, PNP, XO, and GUAase) is present within each of the cell types of the proximal GI tract. Table 1 indicates only the relative intensities of the staining reactions, and should not be construed as strict quantitative measures of the amount of enzyme present. For each tissue examined there was complete concordance with results that were obtained by in situ hybridization for ADA mRNA (not shown). Specifically, the great majority of the differentiated mucosal epithelial cells of the tongue, esophagus, forestomach, and duodenum exhibited intense expression of each of the 5 enzyme activities that were tested. A strong signal was also detected when the complete purine catabolic pathway was tested indicating that the enzymes were also able to function in a linked fashion in situ. In all tissue sections that we have analyzed, for each site of high level ADA expression, the omission of purine substrate from the cocktail fully prevented the appearance of

Table I. Summary of histochemical staining reactions performed for each of the purine catabolic enzyme activities. Visual assessment of each enzyme reaction was made and recorded as trace when just visible, + light staining, and ++ when dark staining was present.

Tissue		Location of Enzyme Activity	5' NT, ADA, PNP, XO assayed individually	GUAase	AMP→ XO pathway
adult mouse GI	tongue	very strong signals in	++	++	++
	esophagus	mature squamous muco-	++	++	++
	forestomach	sal epithelial cells	++	++	++
	duodenum	villi enterocytes -brush border & basalar	++	++	++
3-day old pup	tongue	weak signals- surface epithelial	+	+	+
	duodenum	weak signals- villous epithelial	+	+	+
1-day old pup	tongue	trace signals-surface epithelial	tr.	tr.	tr.
	duodenum	trace signals-villous epithelial	tr.	tr.	tr.
term-fetal pup	tongue	no signals	-	-	-
	duodenum	no signals	-	-	-
reproductive tract	day e7	peri-embryonic decidual cells	++	-	++
	day e9	antimesometrial decidual cells	++	-	++
	day e16	decidua basalis cells and spongiocyto-trophoblasts	++	-	++

formazan precipitate. When the complete pathway was assayed using AMP + NBT, the preincubation of the tissue with dCF, a potent inhibitor of ADA, or with oxypurinol, the active metabolite of the clinically potent XO inhibitor allopurinol , there was complete inhibition of formazan appearance.

Because ADA activity has been noted to increase progressively in the proximal mouse GI tract over the first 1-3 weeks following birth (Lee 1973; Chinsky 1990) and immediately following implantation in the reproductive tract (Knudsen et al 1991), we evaluated histologically whether the expression of each of the other purine catabolic enzymes increased in parallel. These results are also summarized in Table 1. In the GI tract, term fetal duodenum (~day 19.5 p.c.) failed to express any of the purine catabolic enzymes. However, even 1 day following birth, the histochemical staining technique demonstrated faintly detectable 5'NT, ADA, and PNP in the duodenum. By post-natal day 7, each of the purine catabolic enzymes was present at low yet easily detectable levels within the same cell types of tongue and duodenum. We conclude that each of the purine catabolic pathway enzymes is subject to postnatal developmental activation in a coordinated manner.

Each of the purine catabolic enzymes also was expressed intensely by a series of distinct cell types of the uterus and placenta during the post-implantation maternal decidual reaction (days 6.5 - 9) and as well in the maturing placenta (days 11- 18). The only exception was that GUAase activity was not detectable at any of these stages. These data are consistent with the observation of Van Kveel (1982) that uric acid and guanine accumulate in rodent amniotic fluid. At an early post-implantation stage (approximately day 6.5 p.c.), the purine catabolic pathway enzymes co-localized to a subset of maternal decidual cells immediately surrounding the embryo (Table I). In contrast, decidual cells further from the embryo failed to express the purine catabolic enzymes at a detectable level. At the next stage that we have studied in detail, day 9 p.c., there is a very high level of expression of the purine catabolic enzymes within the maternal antimesometrial decidual cells (decidua capsularis), a population of cells that has been characterized to be in a state of degeneration and regression (Welsh and Enders 1985, Katz and Abrahamson 1987). Nearer to term (day 16 p.c.), there was high level expression of the catabolic enzymes at the myometrial base of the placenta in fetal spongiotrophoblasts, maternal decidual basalis cells, and maternal derived metrial gland cells (Stewart and Peel, 1978). There may be some differences among these cell-types with respect to their quantitative expression of each purine catabolic enzyme, but the histochemical method is not sufficiently quantitative or of sufficient resolution to be sure. What is clear however, is that each enzyme is expressed by the predominant cell type of each zone at very much higher levels than surrounding placental cell-types and at a level similar to that observed in the proximal GI.

IMPLICATIONS

The general implication of these results is that there appears to be a commitment by the above described cell types to the digestion and degradation of purine nucleotides. It is likely that there are common regulatory mechanisms that coordinately control the expression of these enzymes, but this remains to be determined. It also certainly remains to be determined what the biological significance is of this elaborately organized expression pattern, but a variety of evidence from several sources suggests several considerations.

Since there is little evidence for abnormal GI tract function in humans with three different gene defects affecting purine catabolism (ADA, PNP, XO), it is unlikely that low concentrations of free adenosine *must* be maintained locally for the basic functioning (e.g. peristalsis or circulatory control) of the GI tract. Thus, whatever physiological role is played by GI tract expression of the complete purine catabolic pathway, a failure to do so is not a cause of marked GI tract dysfunction per se. Since the GI tract is a frequent target site in the life cycles of a variety of purine auxotrophic parasites, such as Toxoplasma, Entamoeba, Giardia, and Helminthic organisms, it is possible that the in situ degradation of purines has served as an evolutionarily significant line of defense against parasitism.

Decidual cells have been long considered to play a role in the nutrition of the early postimplantation embryo, but given the high level of XO expression by the peri-embryonic decidual cells, but it is unlikely that decidual cell degeneration leads to the production of preformed and non-toxic purines for the rapidly proliferating cells of the embryo. More likely, the purine catabolic enzymes are expressed to cope with a considerable amount of purine nucleotides that are generated by a high level of cell death that accompanies remodelling of the maternal decidua (Welsh and Enders 1985, Katz and Abrahamson 1987). Thus, a segment of the cell-death program may be directed towards the expression of catabolic enzymes. It is possible that a failure to express this or any of several catabolic pathways during implantation, placental, and embryonic development could lead to infertility, intrauterine growth retardation, birth defects, or fetal death.

REFERENCES

Aronow, B., Lattier, D, Silbiger, R., Dusing, M, Hutton, J, Stock, J., Jones, G., McNeish, J., Potter, S., Witte, D., and D. Wiginton (1989) Evidence for a complex regulatory array in the first intron of the human adenosine deaminase gene. Genes Dev. **3**:1384-1400.

Chinsky, J.M., Ramamurthy, V., Fanslow, W.C., Ingolia, D.E., Blackburn, M.R., Shaffer, K.T., Higley, H.R., Trentin, J.J., Rudolph, F.B., Knudsen, T.B., and Kellems, R.E. (1990) Developmental expression of adenosine deaminase in the upper alimentary tract of mice. Differentiation **42**: 172-183.

Hall, K. (1971) 5-nucleotidase, acid phosphatase and phosphorylase during normal, delayed and induced implantation of blastocysts in mice: a histochemical study. Endocrinology **51**: 291-301.

Katz, S., and Abrahamsohn, P.A. (1987) Involution of the antimesometrial decidua in the mouse. An ultrastructural study. Anat. Embryol. (Berl.) **176**: 251-258.

Knudsen, T. B., Blackburn, M.R., Chinsky, J. M., Airhart, M. J., and R. D. Kellems. 1991. Ontogeny of adenosine deaminase in the mouse decidua and placenta: Immunolocalization and Embryo transfer studies. Biology of Reproduction. **44**:171-184.

Knudsen, T.B., Gray, M.K., Church, J.K., Blackburn, M.R., Airhart, M.J., Kellems, R.E., and Skalko, R.G. (1989) Early postimplantation embryolethality in mice following in utero inhibition of adenosine deaminase with 2'-deoxycoformycin. Teratology **40**: 615-625.

Lee, P. C.. 1973. Developmental changes of adenosine deaminase, xanthine oxidase, and uricase in mouse tissues. Dev. Biol. **31**:227-233.

Spencer, N., Hopkinson, D. H., and H. Harris. (1968) Adenosine deaminase polymorphism in man. Ann. Hum. Gen. **32**: 9-15.

Stewart, I., and Peel, S. (1978) The differentiation of the decidua and the distribution of metrial gland cells in the pregnant mouse uterus. Cell Tissue Res. **187**: 167-179.

Van Kveel, B. K. (1982) Placental transfer and metabolism of purines and nucleosides in the pregnant guinea pig. Placenta **3**:127-136.

Welsh, A. O., and A.C. Enders (1985) Light and electron microscopic examination of the mature decidual cells of the rat with emphasis on the anti-mesometrial decidua and its degeneration. Am J Anat. **172**:1-29.

CYTIDINE DEAMINASE: A RAPID METHOD OF PURIFICATION AND SOME PROPERTIES OF THE ENZYME FROM HUMAN PLACENTA

Alberto Vita, Silvia Vincenzetti, Adolfo Amici, Ersilia Ferretti and Giulio Magni*

Dipartimento di Biologia C.M.A., Università di Camerino
*Istituto di Biochimica , Università di Ancona (Italy)

INTRODUCTION

Cytidine deaminase serves a dual role in nucleoside pyrimidine metabolism. In microorganisms the physiological function of the enzyme consists in the reutilization of exogenous and/or endogenous cytidine and deoxycytidine for the nucleotides synthesis through a series of enzymatic reaction indicated as "salvage pathway". Furthermore the enzyme enables the microorganisms to utilize cytidine as the sole carbon source (Hammar-Jaspersen, 1983). The enzyme has been previously purified and characterized in our laboratory from Escherichia coli (Vita et al., 1985); studies in vivo have shown that the enzyme is inducible and it is involved in catabolic events (Hammar-Jaspersen, 1983; Magni et al., 1985; Albrechtsen and Ahmad, 1980). In animal cells its role is less well defined. In humans, cytidine deaminase causes deactivation of several cytidine-based drugs including antineoplastic and antiviral agents. Therefore in order to synthesize powerful cytidine deaminase inhibitors able to improve the chemoterapic strategy, we have studied some structural requirements for the specific binding enzyme-ligand. For this purpose pure human placental citidine deaminase was used, obtained by a rapid and efficient purification procedure.

EXPERIMENTAL PROCEDURES

Enzyme assay. Cytidine deaminase activity was determined by a direct spectrophotometric assay based on the decrease of absorbance at the appropriate wavelength when cytidine or the cytidine analog is deaminated to the corresponding deamino-nucleoside. The most suitable analytical wavelenght and the molar absorptivity changes for each substrate were determined by following the spectral changes which occurred after prolonged incubation of a known amount of the substrate with an excess of cytidine deaminase. The standard reaction mixture consisted of 0.167 mM cytidine, 100 mM Tris-HCl, pH 7.5, in a final volume of 1.0 ml. The reaction was initiated by the addition of 0.02-0.04 enzyme units. In pH and kinetic studies the substrate was deoxycytidine in a range 0.018-0.085mM; 0.1M KCl was present to quench the different ionic strength in the assay mixtures. On enzyme unit is defined as μm of substrate deaminated per minute at 37°C.

Purine and Pyrimidine Metabolism in Man VII, Part B
Edited by R.A. Harkness *et al.*, Plenum Press, New York, 1991

RESULTS AND DISCUSSION

Enzyme purification. Two human placentas from normal full-term pregnancies were immersed, immediately after parturition, in a solution containing 0.129M sodium citrate pH 7.5 at 4°C to reduce the blood coagulation. The tissue was trimmed free of umbilical cord and adherental membranes, cut into small pieces, and washed with 0.9% NaCl to remove blood clots. The tissue, suspended in 3 volumes of 20mM buffer A (Tris-HCl pH 7.5, 1mM EDTA, 1mM DTT)containing 0.25M sucrose and 0.01mM PMSF was homogenized in a Waring Blendor and centrifuged at 11,000xg for 30 min. The supernatant was collected (crude extract) and then slowly transferred to a sintered glass funnel containing 30ml of affinity resin (Ashley and Bartlett, 1984) previously equilibrated with 50mM buffer A. After the resin was washed with 50mM buffer A containing 0.5M KCl, the enzyme was eluted adding 2.5mM cytidine to the washing buffer. The active pool, concentrated to 1ml and dialyzed against 20mM buffer A by ultrafiltration on an Amicon PM-10 membrane, was applied onto a Mono Q HR 5/5 column (Pharmacia-LKB), previously equilibrated with 20mM buffer A (solvent A). The column, connected to an FPLC system, was eluted with a discontinuous gradient formed between solvent A and solvent B (20mM buffer A, 0.5M KCl). The gradient program was: isocratic step at 0%B for 15 min, isocratic step at 30%B for 10 min, a linear step from 30%B to 70%B in 65 min and a isocratic step at 100%B for 10 min. The flow rate was 0.5 ml/min. The cytidine deaminase activity eluted in a range 35-45%B. The results of a typical enzyme purification, starting from 600g of placental tissue, are summarized in Table 1. The key step in this purification procedure is the affinity chromatography, which make possible the enzyme purification within two days. The enzyme was eluted from the affinity resin with cytidine 2.5mM at pH 7.5 instead of 0.5M borate at pH 9.0 as previously reported for the *E.coli* enzyme purification (Ashley and Bartlett, 1984). This suggests a specific binding of the enzyme's catalytic site with the affinity resin, containing a cytidine-analog covalently bound to a crosslinked agarose matrix.

Table 1. Purification of human placental cytidine deaminase

Step	Total Protein (mg)	Total Activity (units)	Specific Activity (units/mg)	Purification (-fold)	Yield (%)
Crude extract	$2.90x10^{3}$	25.0	$8.6x10^{-4}$	-	100
Affinity column	1.04	25.0	24.0	27907	100
Mono Q	0.19	14.0	73.7	85676	56

Purity and stability. When the final enzyme preparation was submitted to 15% SDS-PAGE a single sharp band was detected by silver staining. On 7.5% PAGE the gel showed one diffuse band (Rm=0.5) correlated to the cytidine deaminase activity. The final enzyme preparation retained activity for up to three weeks at 4°C in 20mM buffer A, pH 7.5. After two months the enzyme lost about 20% of its original activity. However, at -20°C, 20% glycerol appeared to be necessary to prevent the inactivation occurring after freezing and thawing. The maintainement of the enzyme activity is strictly dependent on the presence of reducing agents.

Molecular weight and subunit structure. The molecular weight of the native enzyme, extimated by gel filtration on FPLC system, was about 52 KDa. The final enzyme preparation when run on denaturating acrylamide gels consisted of a single protein band of 13.5 KDa. These results suggest that the enzyme is an oligomer composed of four apparently identical subunits. We have previously reported tetrameric structure for cytidine deaminase from chicken liver (Vita et al., 1987) and human spleen (Vita et al., 1989a). In human reticulocyte a tetrameric structure was also hypothesized by Teng (1975) based upon an electrophoretic multi-banded pattern obtained by using a crude enzyme preparation. Recently four identical subunits were also reported for the Bacillus subtilis cytidine deaminase (Song and Neuhart, 1989). The identical SDS polyacrylamide gel electrophoresis pattern, obtained by incubating the enzyme either in the presence or in the absence of 2-mercaptoethanol, with and without heat treatment at 100°C for 5 min, revealed the absence of interchain disulfide bridges.

pH studies. Plots of pKm and log Vmax vs. pH, indicate four stages of ionization at pKs: 4.5, 5.8, 7.5 and 9.5, each correlable, as proposed by Tripton and Dixon (1979), to ionizable groups of the substrate and/or aminoacid residues on the catalytic site. Since the pKa value at 4.5 closely corresponds to that of the conjugate acid of deoxycytidine (pka_1=4.30) and since the 3-methylcytidine doesn't act either as substrate or inhibitor, it seems plausible that the decrease in Vmax at the low pH edge might result from the protonation of the substrate in the N-3 position. The inflection in Vmax at pH 9.5 appears to reflect the ionization of the free enzyme, being the pka_2 of deoxycytidine about 13. The pka value of 9.5 appears consistent with the ionization of an ε-amino group of a lysine residue. The inflection points of the wave, in the plot of pKm vs. pH, giving pKs of 5.8 and 7.5 can be correlated with the ionization of an imidazole group of an histidine residue and a -SH group of a cysteine residue, respectively. The assignement of a group for any pK value on the basis of kinetic studies can only be tentative because of the many factors that might influence the ionization constant. However, evidence of an involvement of an histidine residue on the catalytic site of the enzyme has been suggested by preliminary data of photoinactivation pH-dependent of the enzyme, when incubated with methylene blue as described by Weil (1965). The presence of an essential sulphidryl group on the catalytic site was further evidenced by the complete reversal of the p-chloromercuribenzoate inhibition by the substrate and by the strong competitive inhibition exerted by the active site-directed mercurial 5-chloromercuricytidine (Ki=1,2 μm)

Kinetic analysis. All the kinetic constants values are in good agreement with the values previously reported for the enzyme purified from human spleen (Vita et al.,1989a). Instead, there are marked differences in the values relative to cytidine deaminase obtained from other sources as canine liver (Fanucchi et al., 1986), E.coli (Vita et al., 1985) and chicken liver (Vita et al., 1989). These results imply a due caution in the use of non human source in studies of the pharmacological effect of cytidine-analogs. In an attempt to improve the chemotherapeutic approach by using these antimetabolites in combination with cytidine deaminase inhibitors, we have tested a variety of cytidine-analogs for their capacity to inactivate the deaminase activity. Interesting is the inhibition exerted by 1-n-hexylcytosine and 1-(3-hydroxypropyl)cytosine, in which the ribose moiety of cytidine is replaced by aliphatic chains. This result correlates with the inhibition of adenosine deaminase by 9-alkyl-adenines (Schaeffer and Schwender, 1974). Modification of the pyrimidine ring led to 3-deaza derivative, 4-hydroxy-1-B-D-ribofuranosyl pyperidin-2-one, whose inhibitory activity is three orders of magnitude lower than that of the corresponding tetrahydrouridine derivative. This result indicates that the nitrogen in position 3 of the pyrimidine ring is important for the interaction of the inhibitor with the active site of the enzyme, as discussed above under "pH studies".

REFERENCES

Albrechtsen, H., and Ahmad, S.I., 1980, Regulation of the synthesis of nucleoside catabolic enzymes in Escherichia coli: further analysis of a deo O^c mutant strain, Molec.gen.Genet.,179:457

Ashley, G.W., and Bartlett, P.A., 1984, Purification and properties of cytidine deaminase from Escherichia coli, J.Biol.Chem. ,259:13615

Fanucchi, M.P., Watanabe, K.A., Fox, J.J. and Chao, T.C., 1986, Kinetics and substrate specificity of human and canine cytidine deaminase, Biochem. Pharmacol.,35:1199

Hammar-Jaspersen, K., 1983, Nucleoside catabolism, In: "Metabolism of nucleotides, nucleosides and nucleobases in microorganisms", Munch-Petersen, A. (Ed.) Academic Press, London

Magni, G., Vita, A. and Amici, A.,1985, Pyrimidine nucleoside-catabolizing enzymes in Escherichia coli B,in: "Current Topics in Cellular Regulation" Horecker, B.L. and Stadtman, E.R. (Eds), vol.26, Academic Press, New York

Schaeffer, H.J. and Schwender, C.F.,1974, Enzyme inhibitors. 26. Bridging hydrophobic and hydrophilic regions on adenosine deaminase with some 9-(2-Hydroxy-3-alkyl)adenines, J.Med.Chem. ,17:6

Song, B-H., and Neuhard, J.,1989,Chromosomal location, cloning and nucleotide sequence of the Bacillus subtilis cdd gene encoding cytidine/deoxycytidine deaminase, Mol Gen Genet,216:462

Teng,Y.S., Anderson,J.E. and Giblett, E.R.,1975 Am.J.Hum.Genet.,27:492

Tipton, K.F. and Dixon, H.B.F., 1979, Effects of pH on enzymes,in: Methods in enzymology, vol. 63, Academic Press, New York

Vita, A., Amici, A., Cacciamani, T., Lanciotti, M. and Magni, G.,1985, Citidine deaminase from E. coli B. Purification and enzymatic and molecular properties, Biochemistry, 24:6020

Vita, A., Cacciamani, T., Natalini, P., Ruggieri, S., Santarelli, A. and Magni, G.,1987, Purification of cytidine deaminase from chicken liver Ital.J.Biochem.,36:275

Vita, A., Cacciamani, T., Natalini, P., Ruggieri, S. and Magni, G.,1989a, Citidine deaminase from human spleen, in: "Purine and pyrimidine metabolism in man VI, part B", Mikanagi, K., Nishioka, K. and Kelley, W.N.(eds),Plenum Publishing Corporation, New York

Vita, A., Cacciamani, T., Natalini, P., Ruggieri, S., Raffaelli, N. and Magni, G.,1989b, A comparative study of some properties of cytidine deaminase from E. coli and chicken liver, Comp.Biochem.Physiol. 93B:591

Weil,L.,1965, On the mechanisms of the photo-oxidation of amino acids sensitized by methylene blue, Arch.Bioch.Biophys.,110:57

EXPRESSION AND SUBSTRATE SPECIFICITIES OF HUMAN THYMIDINE KINASE 1, THYMIDINE KINASE 2 AND DEOXYCYTIDINE KINASE

Staffan Eriksson, Birgitte Munch-Petersen,
Borys Kierdaszuk and Elias Arnér

Medical Nobel Institute, Department of Biochemistry I
Karolinska Institute, Stockholm, Sweden

INTRODUCTION

Deoxynucleoside kinases catalyze 5′-phosphorylation of deoxynucleosides and deoxynucleoside analogs used in chemotherapy. Cytosolic thymidine kinase (TK1), mitochondrial thymidine kinase (TK2) and cytosolic deoxycytidine kinase (dCK) are key enzymes in this metabolism. They have been isolated from many sources and diverging results have been reported on their biochemical properties (reviewed in 1,2).

TK1, TK2 and dCK have recently been purified from human tissues and cells (3-6), which have enabled a characterization of their subunit structure and enzyme kinetics. The genes for TK1 and dCK are cloned and sequenced (7,8) and the mechanism of cell cycle regulation of TK1, which is found only in S-phase cells, is now the focus of intensive molecular genetic studies (1,9). However, it has been less clear in wich tissues dCK and TK2 are expressed and much of the confusion is due the fact that these enzymes show overlapping substrate specificities.

We have recently prepared pure TK1, TK2 and dCK from human leukemic spleen (4,5) and in this communication results on the kinetic properties, substrate specificities and relative levels of these three enzymes in extracts from different cells and tissues are presented.

MATERIALS AND METHODS

Enzyme and extract preparations: TK1 was purified approximately 20 000-fold from extracts of mitogen activated pheripheral human blood leukocytes or from leukemic spleen as described (5). Its subunit molecular weight on SDS PAGE was approximately 25 kD and the active form was a tetramer as determined by gel filtration chromatography in the presence of ATP. TK2 was purified 20 000-fold from extracts of leukemic spleen and it was active as a monomer of 29 kD (5). dCK was purifed 7000-fold from leukemic spleen and it was active as a dimer of 30 kD polypetides (4).

Purine and Pyrimidine Metabolism in Man VII, Part B
Edited by R.A. Harkness *et al.*, Plenum Press, New York, 1991

Cell extracts were prepared by freeze-thawing cells three times in a 50 mM Tris-HCl buffer, ph 7.6, containing 2 mM dithiotreitol, 20% glycerol, 5 mM benzamidine, 0.5 mM phenylmethylsulfonyl fluoride and 0.5% of the detergent NP-40. Tissues were suspended in the same buffer and homogenized in a Waring type blender. Extracts were centrifuged at 10 000 x g and protein (determined with the Bradford method,10) and enzyme activites measured in the supernatants. In some cases extracted proteins were subjected to DEAE chromatography to separate TK1, TK2 and dCK activities as described (5).

Enzyme assays. Enzyme activities in crude extracts were determined with 10 uM ^{3}H-Thd or dCyd as decribed (4,5). Substrate kinetic constants were determined from saturation curves covering broad concentration intervals using the Michaelis-Menten or Hill equations and non linear regression analysis with the Enzfitter program from Elsevier-Biosoft. Values are from one experiment which have been repeated at least twice with very similar results.

Table I. Activities of deoxynucleoside kinases in extracts from human tissues and cells.

Extracts from	TK1	TK2 (pmole/min/mg)[a]	dCK
CEM T-lymphoblasts	250	10	200
Resting leukocytes	<1	5	60
Activated leukocytes	200	20	110
Adherent Monocytes	<1	20	80
Macrophages (5 weeks)	<1	20	50
Spleen	100	15	80
Colon	10	10	20
Liver	<1	2	<1
Heart	<1	3	<1
Brain	<1	10	<1

a) Values are from several different experiments and the standard deviation was +/- 20%.

RESULTS

Levels of TK1, TK2 and dCK in extracts from human cells and tissues. Table 1 shows the specific activity of TK1 , TK2 and dCK in extracts from various cells i.e. cultured lymphoblastoid cells, normal human blood leukocytes (resting, PHA activated for 72h, adhered to plastic culture flasks or the same type of cells cultivated for 5 weeks so that the monocytes were transformed into macrophages) and from normal human tissues obtained after surgery. To distinguish between TK1, TK2 and dCK activities we had to use DEAE chromatography (5). In some cases the relative activity of these two enzymes were determined by using the dCyd kinase assay performed in the presence and absence of 1 mM Thd. This addition block TK2 but not dCK activity.

TK1 activity was found only in extracts from S-phase cells, while TK2 appeared to be present in all extracts but at levels about 10-20 fold lower than TK1. Total TK2 activity varied even less than its specific activity so that all cells appeared to have approximately the same level of TK2 within a factor of three. dCK activity was found in resting and activated lymphocytes, monocytes, macrophages and colon but not in extracts from heart, liver and brain.

Kinetic constants and substrate specificities of TK1, TK2 and dCK. Table II shows a summary of K_m, V_{max} and Hill coefficients for the preferred natural substrates of the three enzymes, determined as described earlier (4,5). In case of TK2 and dCK where we observe negative cooperativity, the lowest K_m and highest V_{max} values that are shown. Hill coefficients for dCK with dCyd, AraC and dAdo were below 1, but varying values were obtained with different enzyme preparations. The purity of the enzymes and their subunit compositions are described in the Method section. The examples of modifications (ie different halogeno substitutions) indicated are based on a large series of γ-^{32}P-ATP phosphate transfer experiments performed as described earlier (11). The relative activities of the analogs denotes the phosphorylation relative to Thd (for TK1 and 2) and dCyd (for dCK), using 0.1 mM ATP as phosphate donor and 0.1 mM nucleoside analog.

Table II. Kinetic parameters and accepted substrate analogs for TK1, TK2 and dCK.

Enzymes		Preferred substrates			Accepted substitutions	
		K_m (μM)	V_{max}[a]	Hill (n)	Relative activity	
TK1	Thd	0.5	9500	1.0	5-halogeno	1.0
	dUrd	9	12000	1.0	5-etyl	0.8
					3′-halogeno	0.5
					3′-azido	0.5
TK2	Thd	0.3	560	0.5	5-halogeno	0.8
	dUrd	6	690	1.0	5-propyl	0.4
					5-Brvinyl	0.2
					arabinoside	0.2
	dCyd	36	900	1.0	5-halogeno	0.2
dCK	dCyd	0.8	200	0.5-0.9	5-halogeno	0.2
					5-butyl	0.2
					2′or 3′-halogeno	0.5-1.0
					2′-3′-dideoxy	0.3
					arabinoside	0.8
					acyclic	0.2
	dAdo	60	1200	0.6	2-halogeno	2.0
					arabinoside	0.5
	dGuo	120	800	-		
					arabinoside	0.1

a) n moles/min/mg

DISCUSSION

Different tissues and cell types show great variation in their capacity to salvage d oxynucleoside analogs due to both cell cycle specific as w ll as tissue specific expression of the deoxynucleoside kinases. Our studies like those of several others (1,9), demonstrate that TK1 activity is found only in S phase cells or in tissues containing a substantial proportion of proliferating cells, indicating that TK1 serves as a source for the synthesis of thymidine nucleotides in situations of high demand for DNA precursors.

TK2 on the other hand is found in extracts from all cells but at much lower levels than TK1. This could be due to inefficient extraction of this enzyme from its mitochondrial compartment, although our studies indicate that the enzyme in fact is easily extractable from the mitochondria. The tissue distribution of TK2 support its presumed role in the synthesis of pyrimidine deoxynucleotides for mitochondrial DNA replication.

dCK activity was found both in resting and proliferating leukocytes but not in tissues like liver, heart and brain. The reason for this differential expression is not known but the underlying mechanism can now be investigated using molecular genetic techniques.

TK1 showed the most restricted substrate specificity phosphorylating only Thd and dUrd and a few 5 and 3′ analogs, all with Michaelis-Menten kinetics. TK2 showed a broader specificity including dCyd, dUrd and Thd with substantial modifications at the 5 position. TK2 could phosphorylate arabinonucleosides but showed very low activity with 3′ analogs. It also showed clearcut negative cooperativity with Thd analogs (5).

dCK showed the broadest specficity of these enzymes, phosphorylating both cytosine containing nucleosides with very different sugar modifications as well as dAdo and dGuo. The K_m and V_{max} values for the purine and pyrimidine substrates were very different and this enzyme apparently have a very high adaptability to accommodate the various nucleoside acceptors in its active site. Furthermore, a Hill coefficient lower than 1 was usually found but it differed depending on the conditions for preparation of enzyme prior to assay (12). Therefore, we do not at present know if the negative cooperativity of dCK is due to a heterogenous population of modified enzyme molecules or if it is an important regulatory mechanism.

A substantial body of knowledge exist regarding conserved domains of amino acid sequences found in viral induced deoxynucleoside kinases (13). This information has been used together with the known 3D structure of adenylate kinase to model the active site of herpes (HSV1-TK) and human TK1 (14). Suggestions have been presented for the most likely interactions between certain conserved amino acid (ie asp162 and arg163 in HSV1-TK) and deoxynucleosides in order to explain the substrate specificity of the enzyme (14).

These predictions will most likely soon be verified by chrystallographic data, which thereby will provide a framework for future rational drug design of improved antiviral and cytostatic nucleoside analogs.

ACKNOWLEDGMENTS

This work was supported by grants from the Swedish Medical Research Council, the Swedish Cancer Society, The Swedish Board for Thechnical Development and the Medical-Faculty of the Karolinska Institute.

REFERENCES

1. Kit, S. (1985) Microbiol. Sci 2, 369-375.
2. Bohman, C. & Eriksson, S. (1990) Biochem. (Life Sci. Adv.) 9, 11-35.
3. Sherley, J. L. & Kelly, T. J. (1988) J. Biol. Chem. 263, 375-382.
4. Bohmann, C. & Eriksson, S. (1988) Biochemistry 27, 4258-4265.
5. Munch-Petersen, B., Cloos, L., Tyrstedt, G. & Eriksson, S. (1991) J. Biol. Chem. 266, 9032-9038.
6. Datta, N.S., Shewach, D.S., Mitchell, B.S: & Fox, I. H. (1989) J. Biol. Chem. 264, 9359-9364.
7. Bradshaw Jr. H. D. & Deninger, P. L. (1984) Mol. Cell Biol. 4, 2903-2909.
8. Chottiner, E. G., Shewach, D. S., Datta, N. S., Ashcraft, E., Gribbin, D., Ginsburg, D., Fox, I. H. & Mitchell, B. S. (1991) Proc. Natl. Acad. Sci. U S A, 88, 1531-1535.
9. Bradely, D. W., Dou, Q-P., Fridovich-Keil, J. L. & Pardee, A. (1990) Proc. Natl. Acad. Sci. U S A 87, 9310-9314.
10. Bradford, M. (1976) Anal. Biochem. 72, 248-254.
11. Eriksson, S., Kierdaszuk, B., Munch-Petersen, B., Öberg, B. & Johansson, N.-G. (1991) Biochem. Biophys. Res. Commun. 176, 586-59.
12. Kierdaszuk, B. & Eriksson, S. (1990) Biochemstry 29, 4109-4114.
13. Balasubramaniam, N. K., Veerisetty, V. & Gentry. G.A. (1990) J. Gen. Virol. 71, 2979-2987.
14. Zimmermann, N., Beck-Sickinger, A. G., Folkers, G., Krickl, S., Müller, I. & Jung, G. (1991) Eur. J. Biochem. in press.

PYRIMIDINE 5'-NUCLEOTIDASE (S) OF HUMAN ERYTHROCYTES:

ENZYMATIC AND MOLECULAR CHARACTERIZATION

A. Amici , E. Ferretti , S. Vincenzetti , A. Vita , and G. Magni

Istituto di Biochimica, Facoltà di Medicina e Chirurgia, Università di Ancona, 60100 Ancona, Italy
Dipartimento di Biologia Cellulare MCA, Università di Camerino, 62032 Camerino, Italy

INTRODUCTION

Human erythrocytes have been shown to possess specific pyrimidine nucleotidase activities in their soluble fraction[1,2]. Two enzymatic forms, called P5N-I and P5N-II, were able to hydrolyze pyrimidine nucleotides in vitro [3]. The deficiency of one of the two forms (P5N-I) is associated to a nonspherocytic haemolitic anemia, which is characterized by marked basophilic stippling and abnormous accumulation of pyrimidine nucleotides. Nothing is known about the mechanism inducing haemolysis and the role of pyrimidine 5'-nucleotidase in the preservation of red cell integrity. Therefore in order to get insight into such a mechanism a number of studies on the major properties of pyrimidine 5'-nucleotidases have been conducted in erythrocytes, including purification and characterization of the two isoenzymes.

METHODS

Enzyme Assay

The standard assay mixture contains 0.05 M Tris-HCl pH 7.5, 1 mM substrate, 1 mM $MgCl_2$, 0.5 mM DTT and 0.001 unit/ml of enzymatic activity. One unit is defined as the amount of enzyme which hydrolyzes 1 μmol of substrate per minute. Incubation were carried out at 37°C for 30', stopped by the addition of $HClO_4$ to a final concentration of 0.4 M and centrifuged. The supernatant was then brought to pH 6.0 by the addition of 1 M K_2CO_3 and centrifuged. The determination of the amount of nucleoside produced was performed by analyzing an aliquot of the neutralized sample in a C-18 reverse phase column (4.6 x 20 mm), equilibrated with 0.1 M potassium phosphate buffer, pH 6.0 at a flow rate of 2.0 ml/min. The separation and quantitation were performed by following the absorbance of samples and standards at 254 nm[4,5].

Isoenzyme Purification

P5N-I and P5N-II were purified to apparent homogeneity by classical chromatographic steps, including: ion exchange chromatography on DE-52 (Watman), hydrophobic interaction chromatography on CL-4B phenyl-sepharose (Pharmacia), hydroxylapatite chromatography (Biorad), ion exchange in FPLC using a TSK DEAE-5PW column (Pharmacia) and hydrophobic interaction FPLC on a TSK phenyl-5PW column (Pharmacia). The key step in the separation of the two isoenzymes is the hydroxylapatite chromatography [6]. All steps in the procedure were carried out at 4°, and all buffers used contained 0.5mM DTT, 0.5 mM EDTA and 1 mM $MgCl_2$ unless otherwise specified.

Molecular and kinetic properties determination

The native molecular weights of P5N-I and P5N-II were estimated by gel filtration on a Superose 12 FPLC column (10 x 300 mm). Purity and subunit molecular weights were estimated by SDS-PAGE. Isoelectrofocusing was conducted on a 110-ml LKB 8100 ampholine column containing 1% of ampholines (pH 4.0-6.0) and stabilized by a glycerol gradient from 0 to 65% (w/v). pHs optima were determined by using 0.05 M tris and 0.05 M bis-tris adjusted to pH with HCl. Apparent K_m were derived from Lineweawer-Burk plot, using a substrate concentration ranged from 0.01 to 10 mM.

TABLE I
PURIFICATION OF PYRIMIDINE 5'-NUCLEOTIDASE ISOZYME I

Step	prot. mg	act. U	S.A. U/mg	rec. %
Hemolysate	214800	34.8	0.00016	100
DE-52	2291	29.1	0.0129	85
Phenyl-sepharose	522	22.0	0.0425	63
Hydroxylapatite	12	11.9	0.960	34
TSK-DEAE 5PW	0.7	5.6	8.00	16
TSK-phenyl 5PW	0.118	3.3	28.2	6.6

TABLE II
PURIFICATION OF PYRIMIDINE 5'-NUCLEOTIDASE ISOZYME II

Step	prot. mg	act. U	S.A. U/mg	rec. %
Hemolysate	137300	107	0.00078	100
DE-52	3600	95	0.026	85
Phenyl-sepharose	640	88	0.138	63
Hydroxylapatite	5.4	58	10.7	34
TSK-DEAE 5PW	0.27	23	85	16

RESULTS AND DISCUSSION

The preparation obtained from the last step of the purification procedure of P5N-I (table I) submitted to gel filtration gave a single protein peak corresponding to the activity and migrating as a 34,000 daltons molecular weight macromolecule. The SDS-PAGE gave a protein stained pattern with a broad

TABLE III

MOLECULAR AND kINETIC PROPERTIES OF P5N ISOENZYMES

	Isozyme I	Isozyme II
Molecular weight:		
native	36,000	45,000
denaturated	34,000	23,000
Isoelectric point	5.1	5.4
Optimum pH	7.5-8.0	6.5
Km	30 μM (C5'MP)	250 μM (U3'MP)
	400 μM (dU5'MP)	412 μM (dU5'MP)
		192 μM (dU3'MP)

TABLE IV

SPECIFICITY OF PYRIMIDINE 5'-NUCLEOTIDASE ISOZYMES

SUBSTRATE (1mM)	P5N I RESIDUAL %	P5N II RESIDUAL %	SUBSTRATE (1mM)	P5N I RESIDUAL %	P5N II RESIDUAL %
UMP	100	2	U3'MP	2	64
CMP	95	0	C3'MP	0	0
GMP	0	0	G3'MP	0	30
AMP	0	0	A3'MP	0	0
IMP	4	2			
dUMP	14	13	dU3'MP	3	72
dCMP	52	0	dC3'MP	0	0
dTMP	43	7	dT3'MP	0	100
dGMP	0	0	dG3'MP	0	4
dAMP	0	0	dA3'MP	0	0
dIMP	0	20			
cU2'3'MP	0	0	cU3'5'MP	0	0
cC2'3'MP	0	0	cC3'5'MP	0	0
cG2'3'MP	0	0	cG3'5'MP	0	0
cA2'3'MP	0	0	cA3'5'MP	0	0
			cT3'5'MP	0	0
U2'MP	4	36	G2'MP	0	1
C2'MP	3	0	A2'MP	0	0

band corresponding to a molecular weight of about 36,000 daltons. An identical SDS-PAGE pattern was observed in the sample obtained from gel filtration seggesting a monomeric structure of apparently omogeneus enzyme.

The final preparation of P5N-II (table II) was apparently omogeneous when analyzed in SDS-PAGE, showing a molecular weight of 23,000 daltons. The same sample submetted to gel filtration gave a single protein peak which superimpsed to the activity and migrated with a molecular weight of 45,000, suggesting a dimeric structure for P5N-II with identical subunits.

Both purified isoenzymes were stable at 4°C when stored in high ionic strenght buffers (1 M KCl), on the contrary they were inactivated by freezing and thawing. They were also sensitive to oxydation of -SH groups, which was prevented by the addition of 0.5 mM DTT. EDTA was needed to protect enzymatic ativities from metal ions inactivation.

Electrofocusing gave completely separated activities with a mayor peak at pH 5.1 and a shoulder at pH 5.0 for P5N-I and a single peak at pH 5.4 for P5N-II.

The pH optima for P5N-I and P5N-II enzyme assay are in the range of 7.5-8.0 and 6.5 respectively.

Substrate specificity studies, shown on table IV, indicate that P5N-I is mainly involved in pyrimidine 5'monophosphate degradation, whereas P5N-II preferentially catalyzes 3'monophosphate hydrolysis with the exclusion of all cytosine and adenine monophosphate derivatives.

FOOTNOTES

This work has been supported by C.N.R. Target project "Biotecnology and Biostrumentation" and by Regine Marche "Ricerca Finalizzata" Program.

REFERENCES

1. W. N. Valentine, K. Fink, D. E. Paglia, S. R. Harris, and W. S. Adams, Hereditary hemolytic anemia with human erythrocyte pyrimidine 5'-nucleotidase deficiency, J.Clin. Invest., 54:866 (1974).
2. D. E. Paglia, and W. N. Valentine, Characteristics of a pyrimidine-specific 5'-nucleotidase in human erythrocytes, J. Biol. Chem., 250:7973 (1975).
3. A. Hirono, H. Fujii, H. Natori, I. Kurokawa, and S. Miwa, Chromatographic analysis of human erythrocyte pyrimidine 5'-nucleotidase from five patients with pyrimidine 5'-nucleotidase deficiency, Brith. J. Haem., 65: 35 (1987).
4. A. Amici, P. Natalini, S. Ruggieri, A. Vita and G. Magni, A spectrophotometric method for the assay of pyrimidine 5'-nucleotidase in human erythrocytes, Brit.J.Haem. 73: 392 (1989).
5. G. Magni, A. Amici, S. Ruggieri, A. Vita and P. Natalini, Erythrocyte pyrimidine 5'-nucleotidase: HPLC assay, observation on the enzyme behavior, in "Advances in experimental medicine and biology" vol.253B:165-171, (1989).
6. A. Amici, E. Ferretti and G. Magni, Nucleotide metabolism in human erythrocytes: purification and properties of specific pyrimidine nucleotidase, Coll.Czec.Chem.Comm., 55:153-156, (1990).

GTP ACTIVATES TWO ENZYMES OF PYRIMIDINE SALVAGE FROM THE HUMAN INTESTINAL PARASITE *GIARDIA INTESTINALIS*

William J. O'Sullivan, Barbara M. Jiminez, Yan Ping Dai and Choy Soong Lee

School of Biochemistry, University of New South Wales
P.O. Box 1, Kensington, NSW, Australia, 2033

INTRODUCTION

Giardia intestinalis (also referred to as *Giardia lamblia*) is an anaerobic flagellated parasitic protozoan which infects the human small intestine, leading to a spectrum of disorders, including severe diarrhoea, abdominal pain, anorexia and stunted growth in children[1]. Giardiasis is listed by the World Health Organisation as one of the ten leading parasitic diseases of humans. About 200 million people are affected world wide each year in both developed and developing countries. Although rarely life threatening on its own, it represents a major public health problem, particularly in children and contributes significantly to malnutrition.

The current treatment for Giardiasis, which relies on metronidazole and related compounds, is less than satisfactory, with a considerable incidence of side effects. As an approach to the development of new anti-giardial agents, we have been exploring basic aspects of metabolism in the parasite, an approach made possible by the development of procedures for the cultivation of the trophozoite stage *in vitro*[2].

Any organism that is undergoing rapid development and reproduction requires an ample supply of nucleic acid precursors. Based on incorporation and enzyme studies it has been established that *G. intestinalis* is dependent upon salvage of preformed bases and nucleosides for its supply of both purines and pyrimidines[3-5].

It now appears certain that the synthesis of the pyrimidine ribonucleotides in *G. intestinalis* is channelled via uracil phosphoribosyltransferase (UPRTase)[5], any uridine taken up by the parasite being first converted to uracil by a phosphorylase[6]. The phosphorylase has been purified to homogeneity and shown to be also capable of acting on deoxyuridine and thymidine but not cytidine[6].

While cytidine is rapidly incorporated into the parasite, its main fate appears to be deamination to uridine. Aldritt et al[4] reported a cytosine phosphoribosyltransferase activity but we have not been able to confirm this observation. On the contrary, we have recently demonstrated the presence of CTP synthetase.

This article reports the partial purification of UPRTase and CTP synthetase from *G. intestinalis*. In particular, it deals with observations on the activation of both enzymes by the purine nucleotide, GTP.

EXPERIMENTAL

G. intestinalis trophozoites (Portland 1 Stock ATCC 30888) were grown and crude supernatants prepared as described previously[6]. The assay used for UPRTase has been described[5]. The standard assay for CTP synthetase was based on that of Kang *et al*[7] and included GTP (usually 2mM) in 40mM Hepes-KOH (pH 7.0) with mercaptoethanol (35 mM), ATP, (6 mM), GTP, (2 mM), ^{3}H-UTP (1.6 mM), L-glutamine, (5 mM) Mg Cl_2, (5 mM) and enzyme in a total volume of 100 μl. The reaction was carried out at 37°C for 60 min and substrate and product separated by thin layer chromatography on a PEI-cellulose plate with 1M KH_2PO_4 as the mobile phase.

Partial purification of both UPRTase and CTP synthetase was achieved by successive passage through a Mono Q HR 5/5 anion exchange column and a Mono P 5/5 chromatofocusing column in a fast protein liquid chromatography (FPLC) system, followed by chromatography on Superose 12.

RESULTS AND DISCUSSION

A 3100 fold purification of UPRTase was achieved. The enzyme appeared to be substantially homogeneous with only minor contaminants demonstrated by SDS- polyacrylamide electrophoresis (PAGE). Final specific activities of up to 1030 nmole/min/mg protein were achieved. The enzyme had a molecular weight of 76,000 and appeared to be a dimer. Preliminary characterisation yielded the following information : pH optimum, 6.0, K_m (uracil), 0.13 mM, K_m (PRPP), 0.29 mM. A number of metal ions were effective, the observed order being $Mn^{2+}>Mg^{2+}>Co^{2+}>Ni^{2+}>Cd^{2+}>Ca^{2+}>Zn^{2+}$.

A similar approach yielded a 1,600-fold purification, with a 3.3% yield, for CTP synthetase. PAGE demonstrated three major bands and one minor; further purification is proceeding. Properties of the partially purified enzyme included a K_m for UTP of approximately 0.33 mM with saturating ATP, and a K_m for ATP of 0.11 mM. The enzyme could not utilise NH_4^+ as a replacement for glutamine and appeared to have a requirement for mercaptoethanol, indicating that it contained sulphydryl groups essential for activity.

GTP was found to modify the activities of both UPRTase and CTP synthetase, with enhanced activation being observed in both crude supernatants and with the purified enzymes. An example of the effects seen as a function of GTP concentration for the purified preparation of UPRTase, is shown in Fig. 1. dGTP was almost as effective as GTP and a small degree of activation was also observed with GDP.

The effect of GTP on CTP synthetase activity was even more striking, the requirement being absolute. This is illustrated by the results in Fig. 2. from which a K_d value for GTP of 0.4 mM was calculated.

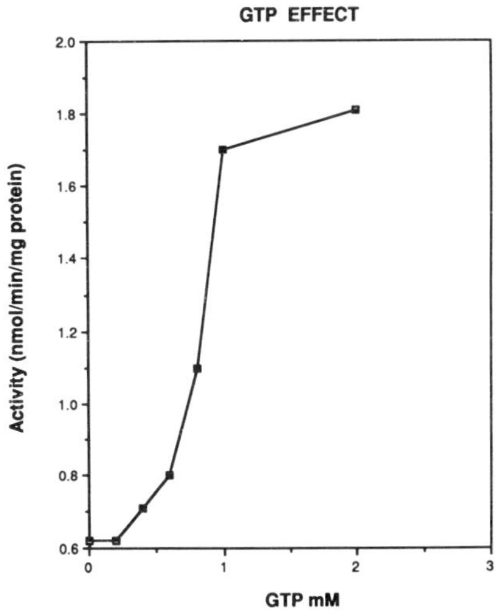

Fig. 1. Effect of GTP on the activity of UPRTase. Conditions were: uracil, 0.43 mM; PRPP, 2.5 mM; $MnCl_2$, 10 mM; MES-KOH, 50mM, pH 6.0.

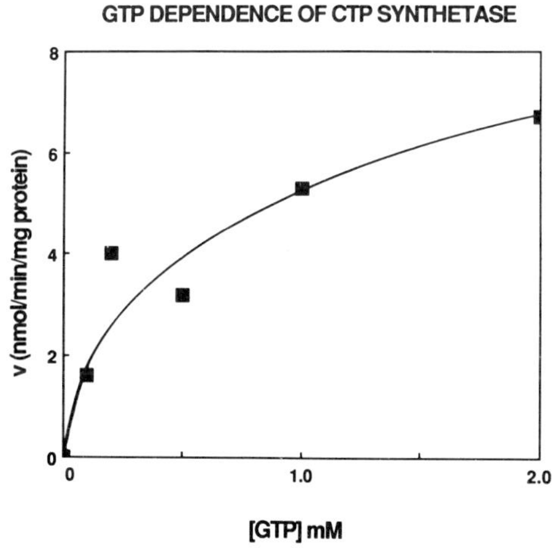

Fig. 2. Effect of GTP on activity of CTP synthetase. conditions were: ATP, 5 mM; $MgCl_2$, 5 mM; L-glutamine, 5 mM; ^{3}H-UTP, 1.6 mM; Hepes-KOH, 40 mM pH 7.4; mercaptoethanol, 35 mM.

Activation of UPRTase by GTP has been observed for the enzyme from *E.coli*[8] but not for that from bakers yeast[9] or *Crithidia luciliaie*[10]. (The enzyme is not found in human tissues). By comparison, GTP appears to be required for significant activity of CTP synthetase from a number of sources, including *E.coli*[11] and bovine liver[12], through the effect appears to be much more dramatic with *G. intestinalis*.

The activation of both UPRTase and CTP synthetase by GTP implies a tight linkage between purine and pyrimidine biosynthesis for the parasite, at least for ribonucleotides. The nature of the activation has not yet been definitively established. Under nearly saturating condition for CTP synthetase, the activation appears to be hyperbolical. However, from our preliminary experiments it appears that the activation of UPRTase as a function of GTP concentration follows an allosteric type relationship; this has to be confirmed.

There is evidence from sequence studies of 16S-like ribosomal RNAs that *Giardia* is a very primitive eukaryote, representative of the earliest surviving branch of the eukaryotic evolutionary tree[13]. The role of GTP for the parasite enzymes, which is remarkably similar to that observed for those from *E. coli* would be consistent with this hypothesis.

REFERENCES

1. E.A. Meyer and E.L. Jarroll. Giardiasis. Am. J. Epidemol. 111:1 (1980).
2. L.S. Diamond, D.R. Harlow and C.C. Cumich. A new medium for the axenic cultivation of *Entamoeba histolytica* and other *Entamoeba*. Trans. R. Soc. Trop. Med. Hyg. 72:431 (1978).
3. D.G. Lindmark and E.J. Jarroll. Pyrimidine metabolism in *Giardia lamblia* trophozoites. Mol. Biochem. Parasitol. 5:291 (1982)
4. S.M. Aldritt, P. Tien, and C.C. Wang. Pyrimidine salvage in *Giardia lamblia*. J. Exp. Med. 161:437 (1985)
5. G.F. Vitti, W.J. O'Sullivan and A.M. Gero. The biosynthesis of uridine 5'-monophosphate in *Giardia lamblia*. Int. J. Parasitol. 17:805 (1987).
6. C.S. Lee, B.M. Jimenez and W.J. O'Sullivan. Purification and characterization of uridine (thymidine) phosphorylase from *Giardia lamblia*. Mol. Biochem. Parasitol. 30:271 (1988).
7. G.J. Kang, D.A. Cooney, J.D. Moyer, J.A. Kelly, H.Y. Kim, V.E. Marquez and D.G. Johns. Cyclopententylcytosine triphosphate, formation and inhibition of CTP synthetase. J.Biol.Chem. 264:713 (1988).
8. R. Fast, and O. Sköld. Biochemical mechanism of uracil uptake regulation in *Escherichia coli*. J. Biol. Chem., 252:7620 (1977).
9. P. Natalini, S. Ruggieri, I. Santarolli, A. Vita and G. Magni. Bakers yeast UMP: pyrophosphate phosphoribosyltransferase. Purification, enzymatic and kinetic properties. J. Biol Chem., 254:1558 (1979).
10. T. Asai, C.S. Lee, A. Chandler and W.J. O'Sullivan. Purification and characterization of uracil phosphoribosyltransferase from *Crithidia luciliae*. Comp. Biochem. Physiol. 95:159 (1990).
11. D.E. Koshland, Jr., and A. Levitzki. CTP synthetase and related enzymes. The Enzymes, 10:539 (1974).
12. H. Weinfeld, C.R. Savage, Jr. and R.P McPartland. CTP synthetase of bovine calf liver. Methods Enzymol. 51:84 (1978).
13. M.L. Sogin, J.H. Gunderson, H.J. Elwood, R.A. Alonso and D.A. Peattie. Phylogenetic meaning of the kingdom concept: an unusual ribosomal RNA from *Giardia lamblia*. Science, 243:75 (1989).

POSSIBLE METABOLIC BASIS FOR GTP DEPLETION IN RED CELLS OF PATIENTS WITH PRPP SYNTHETASE SUPERACTIVITY

A. Giacomello and C. Salerno

Institute of Medical Physiopathology, University of Chieti, and Department of Human Biopathology, University of Rome La Sapienza, Italy

Three different inborn enzyme abnormalities have been identified as causing excessive purine production in man [1]: deficiency of hypoxanthine guanine phosphoribosyltransferase (HGPRT; EC 2.4.2.8), superactivity of phosphoribosylpyrophosphate synthetase (PRPP synthetase; EC 2.7.6.1), and deficiency of purine nucleoside phosphorylase (PNP; EC 2.4.2.1).

There is controversy regarding the mechanisms leading to purine overproduction in these disorders. One view suggests that increased PRPP availability can account for the enhanced rate of cellular purine synthesis. The proposal is in line with the observation that (i) cells from patients with HGPRT deficiency [1,2] or PRPP synthetase superactivity [1,3] and erythrocytes of subjects with PNP deficiency [4] elicit increased intracellular PRPP concentrations; (ii) PRPP acts both as rate limiting substrate and allosteric activator of amidophosphoribosyltransferase (amido PRT; EC 2.4.2.14) [5]; (iii) many studies have documented coordinate changes in PRPP concentrations and rates of purine synthesis *in vitro* or *in vivo* [3]. However, many circumstances in which PRPP availability and rates of purine synthesis *de novo* are dissociated have been identified [3]. It has been found [2] that PNP deficient fibroblasts and limphoblasts have PRPP contents characteristic of normal cells. Moreover, during incubation in a purine-free medium, normal cells are capable of synthesizing purine nucleotides *de novo* at rates approaching those of cells deficient in HGPRT or in PNP despite a severalfold higher PRPP concentration in the mutant cells [2].

The last observation has redirected attention to the contribution of purine reutilization to purine nucleotide homeostasis. Since purine nucleotides inhibit PRPP synthetase and amido PRT [3,5], not only increased PRPP availability but also purine nucleotide depletion could lead to purine overproduction. Evidence for purine nucleotide depletion in inherited purine disorders associated with purine overproduction comes from the observation that (i) low erythrocyte GTP levels have been reported in HGPRT and PNP deficiency as well as in PRPP synthetase superactivity [6]; (ii) in HGPRT deficiency, guanosine plasma levels are undetectable following fructose infusion [7].

Human red blood cells depend almost exclusively on guanine salvage for the maintenance of their guanine ribonucleotide pool [8]. The HGPRT-catalyzed reaction between guanine and PRPP to form GMP and PP_i is usually considered the main if not the only path involved in this metabolism. The mechanism of the reaction has been investigated by initial velocity, product inhibition, and isotope exchange studies [9-11]. The minimum kinetic model, which still fits the experimental data, is that substrates must bind to HGPRT in an ordered sequence, first the dimagnesium complex of PRPP and then the purine base, while products (the monomagnesium complexes of PP_i and of the nucleotide) can dissociate from enzyme in a random fashion. The equilibrium constant is far toward the ribonucleotide: using hypoxanthine as substrate a value between 200 and 500 can be estimated from the Haldane's relationship.

A strong competition should exist between guanase (EC 3.5.4.3) and HGPRT for their common substrate, guanine. Guanase has a Km value for the shared substrate (0.26 to 0.56 µM) [12] one order of magnitude lower than that of HGPRT (4.0 µM) [9]. Thus, the high ratio of guanase to HGPRT specific activities in brain, liver, gut, and kidney (Table I) should be operative. An important part of the production of urate seems to occur via the guanine-xanthine conversion [14].

Assuming that the HGPRT-catalyzed reaction between guanine and PRPP occurs *in vivo*, it is simple to explain guanine nucleotide depletion observed in HGPRT and PNP deficiency. In the first case the reaction cannot occur owing the deficiency of HGPRT. In PNP deficiency, guanine is not formed. However, in PRPP synthetase superactivity, the increased intracellular levels of PRPP should lead to an increase in guanine reutilization and not to a depletion of guanine nucleotides.

We suggest that guanine nucleotide depletion in PRPP synthetase superactivity could be explained taking into account the HGPRT-catalyzed IMP-GMP exchange, a reaction that has been studied in depth in our laboratories [15-17]. When IMP reacts with guanine in the presence of PP_i, magnesium ions, and HGPRT, equimolar amounts of GMP and hypoxanthine are formed. The kinetic model for the HGPRT-catalyzed IMP-GMP exchange is shown in Fig. 1. A Theorell-Chance mechanism for guanine binding and hypoxanthine release has been postulated to justify the deviation from parallel of the reciprocal plots obtained studying guanine activation of IMP pyrophosphorolysis with PP_i and IMP

Table I. Specific activity of HGPRT (purine substrate: guanine) and guanase in some human tissues.

Tissue	HGPRT activity	Guanase activity	Reference
	(nmol/mg protein/hour)		
Liver	30-66	453-2372	12,13
Gut	27	169	12,13
Kidney	48	108-151	13
Basal ganglia	413-1137	2662-5027	13

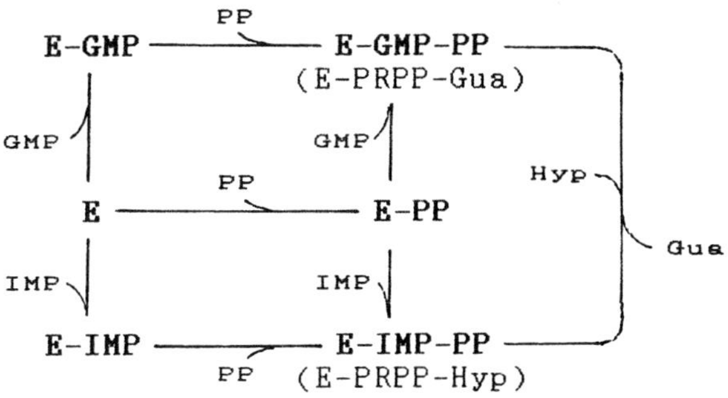

Fig. 1. Basic interconversion figure for the HGPRT-catalyzed IMP-GMP exchange.

as variable substrates [15]. According to this hypothesis, guanine should directly utilize the phosphoribosyl moiety of IMP for GMP synthesis without PRPP liberation from the enzymic active site. This mechanism is in agreement with the observation that (i) no appreciable amount of PRPP is formed during the IMP-GMP exchange; (ii) the maximum increase of the initial rate of IMP pyrophosphorolysis in the presence of guanine is achieved at concentrations of the purine base lower than the Km value for guanine obtained by studying the reaction between guanine and PRPP [9,15]. The last result confirms that guanine binds to different enzyme forms in the phosphoribosylation of the purine base by PRPP and in the IMP-GMP exchange. The apparent equilibrium constant of the IMP-GMP exchange is ranging from 0.35 to 0.45 [16]. Since the ratio between IMP and GMP concentrations is higher than 1 in most human cells (Table II) and, at least in biological fluids, hypoxanthine level is higher than guanine level, IMP-GMP exchange could occur *in vivo*. Evidence has been obtained that the reaction is working in intact human erythrocytes [17].

According to the reaction mechanism in Fig. 1, the E-PRPP complex should be not formed during the IMP-GMP exchange and PRPP should behave as an inhibitor of the reaction by forming a dead-end complex with the free enzyme. If so, guanine nucleotide depletion in PRPP synthetase superactivity can be simply explained by relating the impairment of guanine salvage with the increased level of PRPP (Fig. 2).

Table II. Concentrations of IMP and GMP in human cells.

	IMP	GMP	IMP/GMP	References
Liver biopsies	0.12-0.30⁺	0.04-0.06⁺	2.0-7.5	18
Kidney biopsies	0.05-0.11⁺	0.02-0.03⁺	1.7-5.5	18
Erythrocytes	5-85 §	n.d.	1.7-25¶	19,20,21

⁺ μmol/g wet weight; § nmol/ml cells; ¶ GMP has been calculated from GTP and GDT assuming close equilibria for purine nucleoside monophosphate and adenylate kinase.

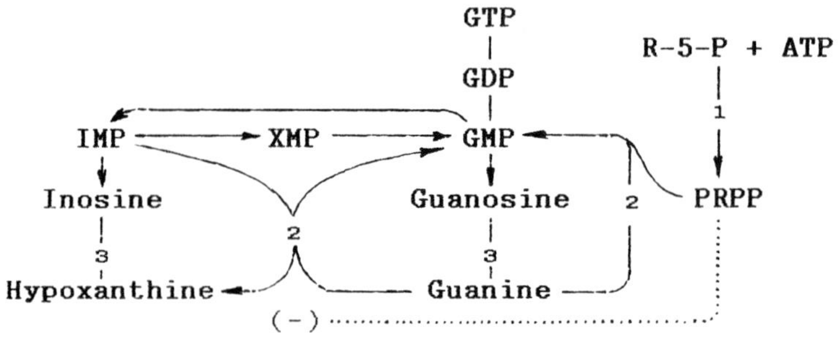

Fig. 2. Schematic representation of the pathways of guanine salvage and proposed inhibition by PRPP of the HGPRT-catalyzed IMP-GMP exchange (dotted line). Numbers 1 to 3 indicate the enzymes whose inherited defects lead to purine overproduction (1: PRPP synthetase; 2: HGPRT; 3: PNP).

REFERENCES

1. I.H.Fox and W.N.Kelley, Adv. Exp. Med. Biol. 41B:471-477 (1974)
2. M.S.Hershfield and J.E.Seegmiller, J. Biol. Chem. 252:6002-6010 (1977)
3. M.A.Becker and M.Kim, J. Biol. Chem. 262:14531-14537 (1987)
4. A.Cohen, D.Doyle, D.W.Martin and A.J.Ammann, N. Engl. J. Med. 295:1449-1454 (1976)
5. E.W.Holmes, J.B.Wyngaarden and W.N.Kelley, J. Biol. Chem. 248:6035-6040 (1973)
6. H.A.Simmonds, L.D.Fairbanks, G.S.Morris, D.R.Webster and E. H.Harley, Clin. Chim. Acta 171:197-210 (1988)
7. M.L.Jimenez, T.Ramos, F.A.Mateos, J.G.Puig, I.H.Fox, Klin. Wochenschr. 65(X):11-12 (1987)
8. Y.Sidi, I.Gelvan, S.Brosh, J.Pinkhas and O.Sperling, Adv. Exp. Med. Biol. 253A:67-71 (1989)
9. J.F.Henderson, L.W.Brox, W.N.Kelley, F.M.Rosenbloom and J. E.Seegmiller, J. Biol. Chem. 243:2514-2522 (1968)
10. A.Giacomello and C.Salerno, J. Biol. Chem. 253:6038-6044 (1978)
11. C.Salerno and A.Giacomello, J. Biol. Chem. 256:3671-3673 (1981)
12. R. Kuzmits, H. Stemberger and M.M. Muller, Adv. Exp. Med. Biol. 122B:183-188 (1980)
13. F.M.Rosenbloom, W.N.Kelley, J.Miller, J.F.Henderson and J. E.Seegmiller, J. Am. Med. Ass. 202:175-177
14. G.Van Waeg, F.Niklasson and C.H.De Verdier, Adv. Exp. Med. Biol. 195A:425-430 (1986)
15. C.Salerno and A.Giacomello, J. Biol. Chem. 254:10232-10236 (1979)
17. C.Salerno and A.Giacomello, Biochemistry 24:1306-1309 (1985)
16. C.Salerno and A.Giacomello, Experientia 38:1196-1197 (1982)
18. G.Van den Berghe and J.Jaeken, Adv. Exp. Med. Biol. 195A: 27-33 (1986)
19. V.Stocchi, L.Cucchiarini, F.Canestrari, M.P.Piacentini and G.Fornaini, Anal. Biochem. 167:181-190 (1987)
20. B.M.Dean, D.Perret and M.Sensi, Biochim. Biophys. Res. Comm. 80:147-154 (1978)
21. A.Ericson, F.Niklasson and C.H.De Verdier, Clin. Chim. Acta 127:47-59 (1983)

GUANINE METABOLISM IN PRIMARY RAT NEURONAL CELLS

Sara Brosh, Oded Sperling, Esther Dantziger and Yechezkel Sidi

Departments of Clinical Biochemistry and Med. D Sackler Faculty of Medicine, Tel-Aviv University, Address correspondence to Dr Y. Sidi, Department of Medicine D, Beilinson Medical Center, Petah-Tikva 49100, Israel

INTRODUCTION

Guanine ribonucleotides (GuRN) play a crucial role in many functions related to normal cellular and neuronal activities. Depletion of intracellular GuRN pools was reported in red blood cells from patients affected with the hereditary Lesch-Nyhan syndrome (1,2) and suggested as the possible metabolic basis for the neurological manifestations associated with this syndrome. Despite the apparent connection between guanine and GuRN metabolism and neuronal function, knowledge regarding the specific metabolism of guanine and GuRN in neuronal cells is rather limited.

The aim of this study was to characterize the pathways of guanine and GuRN metabolism in neurons. Primary rat neuronal cultures served as a model for this purpose.

MATERIALS AND METHODS

Primary neuronal cultures were prepared according to Yavin and Yavin (3).

For measuring the metabolic fate of (8-^{14}C)guanine, the culture medium was replaced with 1 ml of fresh, serum-free medium containing (8-^{14}C)guanine. After incubation for the time period specified, the medium was transferred to prechilled tubes containing 0.2 ml of 3 N PCA. The cells were placed on ice and washed with ice cold saline. 250 µl of 0.5N PCA were added and cell extracts were collected.

For the estimation of the metabolic fate of GuRN, cells were incubated for 60 min with (8-^{14}C)guanine, as described above. Following incubation, the medium was discarded and replaced by 1 ml of fresh, prewarmed guanine-free medium. Incubation was resumed for the time period specified.

Nucleotides, nucleosides and bases, were separated by bidimensional thin layer chromatography on microcrystalline cellulose as described before (4).

RESULTS

Incubation of mature (14 days old) cultured neurons with (8-^{14}C)guanine revealed two main pathways for guanine metabolism (Table 1), incorporation into nucleotides and

Purine and Pyrimidine Metabolism in Man VII, Part B
Edited by R.A. Harkness *et al.*, Plenum Press, New York, 1991

further on to nucleic acids (approximately 18 % of the total cellular nucleotides were incorporated into nucleic acids after 1 h), and degradation to xanthine. At all guanine concentrations studied, the rate of guanine deamination to xanthine exceeded that of guanine incorporation into nucleotides by 7.6-11.7 folds. 99 % of the label in the nucleotides was in the form of guanine nucleotides. Only negligible amounts of the guanine label (<2 %) were found in guanosine, inosine, hypoxanthine and adenosine. A lower ratio (2.7) between guanine deamination and incorporation into nucleotides was found in immature cultures (24-h-old), which at 6 μM guanine deaminated 12.7 % of the guanine into xanthine and incorporated 4.7 % of the guanine into nucleotides.

Table 1 Metabolic fate of labeled guanine in mature neurons

Labeled guanine (μM)	Distribution of radioactivity in purines in the acid soluble fraction (pmol/mg of protein/h)				
	Medium		Intracellular		
	Guanine	Xanthine	Guanine	Xanthine	Total nucleotides
1.5	106	918	2.5	2.8	107
3.0	771	1830	13.6	8.4	241
6.0	1823	3376	23.6	18.5	407
12.0	3446	6490	80.1	40.9	556

In order to clarify the metabolic fate of GuRN, the medium containing the labeled guanine was replaced after 1 h of incubation with medium without guanine and incubation was continued for various time periods, up to 22 h. The decrease in the radioactivity of GuRN was fast, from 560 pmol of labeled guanine/mg protein to 115 pmol/mg of protein, within 22 h. The radioactivity lost from the GuRN pool was found in nucleic acids (insoluble in PCA; 32.8 %), adenine nucleotides (7.9 %), xanthine (49.3 %), guanine (4.3 %), hypoxanthine (4.1 %), guanosine (1.1 %) and inosine (0.5 %). The distribution of label originating from GuRN among the acid soluble purine derivatives, at various time periods, is depicted in Fig 1. Addition to the culture of 0.4 mM 8-aminoguanosine (8-AGuo), did not affect the initial labeling of the various intracellular and extracellular purines, but during the post labeling degradation period, it markedly decreased the production of labeled xanthine by more than 3 folds, associated with a marked increase in the label in guanosine, guanine and inosine, by 12, 3 and 10 folds, respectively. The accumulation of radioactivity in guanosine exceeded that in inosine by 2.3 folds. The addition of 8-AGuo did not affect the shift of the label in the acid soluble fraction from guanine nucleotides to adenine nucleotides.

The flux from IMP to GMP was assessed by tracing the flux of (8-^{14}C)hypoxanthine to GuRN. Incubation of the neurons at 10 μM (8-^{14}C)hypoxanthine for 1 h, resulted in incorporation of 2100 pmol/mg of protein. 6.2 % of the label incorporated into nucleotides resided in GuRN, 2.3 % in IMP and 91.5% in adenine nucleotides. Addition of mycophenolic acid (MA; 0.75 μM), resulted in 99 % inhibition of the flux of (8-^{14}C)hypoxanthine through IMP to GuRN. Addition of MA to the cultures after the removal of the (8-^{14}C)guanine-containing medium, resulted (after 22 h) in a 5 folds increase in the label in inosine and 1.3 folds increase in the label in hypoxanthine (Fig 1). No change in the labeling pattern of nucleotides was observed.

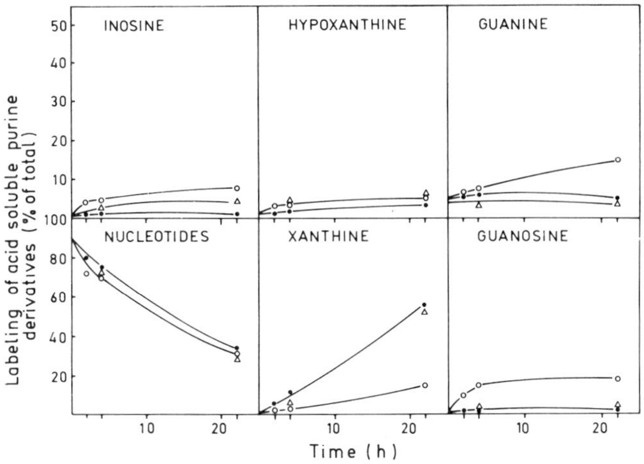

Fig 1 Distribution of label in acid soluble purine derivatives at various time intervals, following prelabeling of the neurons with (8-^{14}C)guanine. o, control; ●, with 8-AGuo (0.4 mM); △ , with MA (0.75 µM). Values are means of three experiments, each in triplicates.

DISCUSSION

The results demonstrate clearly that the degradative deamination of guanine exceeds markedly the anabolic incorporation of guanine into nucleotides. The preferential deamination of guanine in the neurons is greater in mature (7.6-11.7 folds), as compared to immature cells (2.7 folds). These results are compatible with a previous observation in our laboratory that the Vmax ratio of guanase to HGPRT in cultured neurons is higher in mature (7.3 folds), compared to immature neurons (2.7 folds)(5). Despite the relatively much higher activity of guanase, significant incorporation of labeled guanine into nucleotides, mainly GuRN, could be measured, at a rate of 8.5-13.1% of that of deamination. The incorporation of hypoxanthine into nucleotides was found to be about 3 folds that of guanine. The difference in the rate of incorporation can be explained by the difference in the availability of the two substrates, caused by the high activity of guanase and the absence of xanthine oxidase.

The fast transfer of radioactivity from GuRN to nucleic acids and to purine degradation products, resulted in a fast rate of turnover of the GuRN pool. Preliminary results in our laboratory suggest that the turnover rate of GuRN is at least double that of adenine nucleotides. Since the neurons do not contain xanthine oxidase activity, the accumulation of label from GuRN mainly in xanthine, can be taken to indicate that the main pathway of GuRN catabolism is that through dephosphorylation of GMP to guanosine, its phosphorolysis by PNP to guanine, and subsequently deamination of the guanine by guanase to xanthine. Nevertheless, the finding, in presence of 8-AGuo, of accumulation of label also in inosine, indicates that GMP is degraded also through IMP to inosine (and subsequently to hypoxanthine). Part of the IMP was also converted to AMP, as reflected by the marked shift of radioactivity from GuRN to adenine nucleotides.

The replenishment of the GuRN pool, to compensate for the marked loss due to nucleic acid synthesis and degradation, can occur, through IMP by de novo synthesis, or by salvage from hypoxanthine, and directly to GMP, by salvage from guanine. The results indicate that the fluxes from IMP and from guanine to GMP are relatively inefficient. Apparently, under physiological conditions, these fluxes suffice to compensate for the loss of GuRN. However, a congenital absence of either of the pathways for GMP generation may render the neurons vulnerable to GuRN depletion. Such depletion may actually occur in HGPRT and PNP deficiency.

The results furnish information concerning the activity of the guanine nucleotide cycle (GMP--->IMP --->XMP --->GMP), which is analogous to the previously described purine nucleotide cycle (6). We demonstrated that the two arms of this putative cycle are present in the neuron: reduction of GMP to IMP (incorporation of labeled guanine through GMP to IMP, inosine and hypoxanthine), and incorporation of label from (8-^{14}C)hypoxanthine to GuRN. The regulation of the activity of the pathway from GMP to IMP differs in several cell types. It is a major catabolic pathway in red blood cells, but in other mammalian cell types this pathway is inactive (7,8). The finding that the addition of MA to the neuronal cells, incubated with labeled guanine, resulted in increased labeling in inosine and hypoxanthine, may be taken to reflect the actual metabolite flow through the guanine cycle. Nevertheless, it may also reflect increased flux of unlabeled precursors through the purine de novo biosynthesis pathway towards IMP, as a response to GuRN depletion, induced by MA. It remains to be seen whether the guanine nucleotide cycle has any role in the homeostasis of the GuRN pool size in neuronal cells and if it has any other role in the metabolism of the neuron.

Acknowledgment: This work was supported in part by USA - Israel Binational Science Foundation grant 85-00325.

References

1. Sidi Y. and Mitchell B.S. (1985) J. Clin Invest. 76, 2496-2419.
2. Simmonds H.A., Fairbanks F.S., Morris G., Morgan A., Watson K., Timms P. and Singh B. (1987) Arch. Dis. Child. 62, 385-391.
3. Yavin Z. and Yavin E. (1977) Exp. Brain. Res. 29, 137-147.
4. Zoref-Shani E. Kessler-Icekson G. and Sperling O. (1988) J. Mol. Cell. Cardiol. 20, 23-33.
5. Brosh S., Sperling O., Bromberg Y. and Sidi Y. (1990) J. Neurochemistry 45, 1776-1781.
6. Shultz N. and Lowenstein J.M. (1976) J. Biol. Chem. 251, 485-492.
7. Hershko A., Razin A., Shoshani T. and Mager J. (1967) Biochim. Biophys. Acta. 149, 59-73.
8. Henderson J.F., Zombor G. and Burridge P.W. (1978) Can. J. Biochem. 56, 474-479.

URINARY PTERINS IN LESCH-NYHAN SYNDROME

Ivan Sebesta, Jakub Krijt, Stanislav Kmoch, and Josef Hyanek

Centre for Inherited Metabolic Disorders,
Department of Clinical Biochemistry, 1st Medical Faculty,
Charles University, Prague, Czechoslovakia

INTRODUCTION

The Lesch-Nyhan syndrome results from a complete deficiency of the purine salvage enzyme hypoxanthine guanine phosphoribosyltransferase (HPRT E.C.2.4.2.8). Clinical manifestations include muscle hypotonia, torsion dystonia, delayed motor development, compulsive self-mutilative behaviour, and mental handicap, as well as increased quantities of uric acid in body fluids, which may lead to nephropathy, renal calculi, tophi and gouty arthritis. The latter manifestations are effectively controlled using allopurinol and the pathophysiological mechanism of overproduction of uric acid is known. However, there is no effective treatment for the neurologic and behavioural features and the processes by which a lack of HPRT brings about the neurological dysfunction remain unclear (1).

The inappropriate development of nigrostriatal dopaminergic neurons may be important. A relationship between the degree of neurological involvement and the low guanosine triphosphate (GTP) levels in the erythrocytes in the HPRT deficiency was found (2). GTP is the substrate for the enzyme which catalyzes the first step in the pterin de novo synthesis pathway (Figure 1).

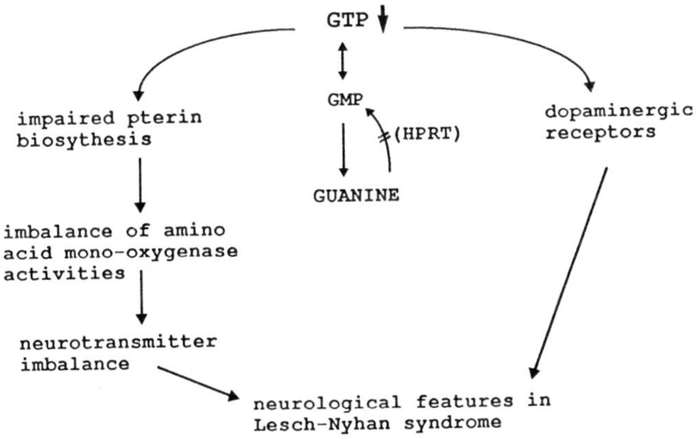

Fig. 1. Schematic description of suggested role for GTP in the pathogenesis of neurological symptoms in the Lesch-Nyhan syndrome.

Purine and Pyrimidine Metabolism in Man VII, Part B
Edited by R.A. Harkness *et al.*, Plenum Press, New York, 1991

It has been suggested that the deficiency of HPRT constrains the metabolic flux along the pterin de novo synthesis pathway and therefore compromises the production of neurotransmitters (3).

The present study was carried out to determine whether the enzyme defect in the Lesch-Nyhan syndrome affects the urinary pterin excretion.

PATIENTS

Case 1: Patient L.K. born 09/08/88. Diagnosis of the Lesch-Nyhan syndrome was made at 2 years of age.

Case 2: Patient 3.K born 03/09/82. Lesch-Nyhan syndrome was confirmed at 3 years.

The pterin concentrations in both complete HPRT deficient patients were determined from three separate urine samples taken during an 8 month period. All subjects, including control healthy children (2-6 years) showed a creatinine clearance above 80ml/min/1,73m2.

METHODS

Total biopterin and total neopterin were determined after oxidation of reduced pterins (dihydrobiopterin, tetrahydrobiopterin, dihydroneopterin) with magnesium dioxide using methods described previously (4). All chemicals were of the highest analytical grade available.

Table 1. Urinary excretion of pterins in patients and controls expressed in mmoles per mole creatinine.

	Total biopterin (B)	Total neopterin (N)	% B**	BNCR**
Patient 1	0.7	0.9	44	31
	0.6	0.7	46	28
	0.6	0.6	50	30
Patient 2	1.1	1.4	44	48
	1.3	1.4	50	65
	1.1	0.6	65	72
Normal boys* (n=14)	1.1 (0.4)	1.0 (0.7)	56 (14)	64 (24)
range	(0.6-1.8)	(0.2-2.6)	(31-79)	(23-94)

* Values are means (SD)

** Percentage of total biopterin: 100.B / (B+N)

*** Biopterin-Neopterin-Creatinine ratio : $\frac{B \cdot B}{Cr. (B+N)} \cdot 10^5$

RESULTS

Table 1. compares the results of the biochemical investigation of the levels of total biopterin and total neopterin in urine in our patients and controls. Although the pterins concentrations in patient 1 were persistently lower than means for controls, they fell within the control range. Patient 2 showed values comparable with the normals.

DISCUSSION

The specific relationship between the neurologic symptoms associated with the Lesch-Nyhan syndrome and HPRT deficiency remains unclear. The most troublesome symptom, self-mutilation, is still poorly understood as well. Biochemical and histologic studies underline the importance of inappropriate development of nigrostriatal dopaminergic neurons (1).

Findings of high levels of HPRT in the area of the basal ganglia of normal postmortem human brain tissue and reduction of dopamine function in the basal ganglia of Lesch-Nyhan patients implicate cells of this brain area in pathogenesis. Athetosis and choreiform movements are also seen with lesions of the putamen and caudate (1).

GTP may play an important role in the pathogenesis of the neurological symptoms in Lesch-Nyhan patients. The brain like the erythrocyte is largely dependent on salvage to sustain its GTP levels. Simmonds (2) found low GTP levels in the erythrocytes of Lesch-Nyhan patients which were in striking contrast to the normal GTP levels in patients only partially deficient in HPRT who lacked neurological problems. The same differences were found also in the other two purine disorders with neurological problems. The hypothesis that slight depletion of nucleoside triphosphate is responsible of intermittent cessation of neural function in very active areas in Lesch-Nyhan brain has also been presented (5).

GTP provides a link between purine and pterin metabolites. GTP is the substrate for GTP-cyclohydrolase which catalyses the first step in the pterin de novo synthetic pathway. Reduction of GTP supply due to HPRT deficiency would be expected to constrain pterin de novo synthesis and the production of the tetrahydrobipterin cofactor for amino acid monooxygenase activities and therefore compromises the production of neurotransmitters (Fig. 1).

However we did not find any alteration in the urinary excretion of pterins in Lesch-Nyhan patients. Suprisingly little has been written on pterins levels in body fluids of Lesch-Nyhan syndrome. Singh (6) found even higher excretion of urinary pterins in a case of Lesch-Nyhan syndrome with delayed onset of self-mutilation. Despite the fact that the normal basal ganglia have been reported to have a higher concentration of tetrahydrobiopterin than other parts of the brain, Manzke (7) found that neither tetrahydrobiopterin treatment (BH4) alone nor a combination of BH4 with 5-hydroxytryptophan and levodopa + carbidopa significantly influenced the neurological symptoms in this syndrome. It is a reasonable presumption that the link between GTP and neurological symptoms may be directly through the modulating effect of GTP on the activity state of dopaminergic receptors (1).

Our results suggest that the link between pterin and purine metabolism in this syndrome remains uncertain and requires further study. However inability to sustain GTP concentrations in the CNS at levels compatible with normal physiological functions may play an important role in the pathogenesis of neurological and behavioural symptoms of Lesch-Nyhan syndrome.

REFERENCES

1. J. T. Stout, and C. T. Caskey, Hypoxanthine phosphoribosyltransferase deficiency: The Lesch-Nyhan syndrome and gouty arthritis, in: "The metabolic basis of inherited disease". 6th ed. C.R. Scriver, A.L. Beaudet, W.S. Sly, D. Valle, ed., McGraw-Hill, New York (1989).

2. H. A. Simmonds, L. D. Fairbanks, J. A. Duley, and V. Micheli, Importance of the human erythrocyte in the diagnosis of inherited purine and pyrimidine disorders, Biomed Biochim Acta. 49:259 (1990).

3. R. W. E. Watts, Defects of Tetrahydrobiopterin synthesis and their possible relationship to a disorder of purine metabolism (Lesch-Nyhan syndrome), Adv Enzyme Regul. 23:25 (1984).

4. A. Niederwieser, W. Staudenmann, and E. Wetzel, High-performance liquid chromatography with column switching for the analysis of biogenic amine metabolites and pterins, J Chromatogr. 290:237 (1984).

5. R. A. Harkness, G. M. McCreanor and R. W. E. Watts, Lesch-Nyhan syndrome and its pathogenesis: purine concentrations in plasma and urine with metabolite profiles in CSF, J Inher Metab Dis. 11:239 (1988).

6. S. Singh, J. Willers, K. Ullrich, H. Gustmann, A. Niederwieser, and H. W. Goedde, A case of Lesch-Nyhan syndrome with delayed onset of self-mutilation: Search for abnormal biochemical, immunological and cell growth characteristic in fibroblasts and neurotransmitters in urine, Adv Exp Med Biol. 195A:20 (1986).

7. H. Manzke, H. Gustmann, H. G. Koke, and W. L. Nyhan, Hypoxanthine and tetrahydrobiopterin treatment of a patient with features of the Lesch-Nyhan syndrome, Adv Exp Med Biol. 195A:197 (1986).

BASIS FOR THE CHONDRO-OSSEOUS DYSPLASIA ASSOCIATED WITH ADENOSINE DEAMINASE DEFICIENCY: SELECTIVE TOXICITY TO IMMATURE CHONDROCYTES

R.L. Wortmann, K.K. Tekkanat, J.A. Veum, R.A. Meyer, Jr., J.O. Hood, and W.A. Horton

Medical College of Wisconsin, VA Medical Center, and Marquette University School of Dentistry, Milwaukee, Wisconsin and University of Texas Health Science Center, Houston, Texas, U.S.A.

An inherited deficiency of adenosine deaminase (ADA) causes severe combined immunodeficiency disease and changes in cartilage and bone termed chondro-osseous dysplasia (1). The chondro-osseous dysplasia develops after birth, resolves with successful enzyme replacement therapy, and manifest characteristic radiographic features and distinctive histopathology (2,4). The histopathologic changes include absence of transition from proliferative to hypertrophic cells, lack of organized columns of hypertrophic cells, and uninterrupted calcified cartilage formation (3). The metabolic abnormalities that underlie the chondro-osseous dysplasia of ADA deficiency are poorly understood.

In vitro and in vivo studies were performed to further elucidate the effects of ADA deficiency on chondrocytes and to gain insight into the pathogenesis of the chondro-osseous dysplasia. In vitro experiments examined the effects of a model of ADA deficiency, the combination of deoxyadensoine and ADA inhibitors on chondrocytes cultured in agarose. Previous studies of chondrocytes have been limited because these cells are difficult to obtain in adequate numbers and rapidly change phenotypic expression with traditional culture techniques. The agarose culture method allows study of a larger number of cells than are usually available from one source and examination of chondrocytes in different stages of differentiation (5).

Cartilage was obtained from stifle joints of adult pigs and the chondrocytes were released from matrix by sequential enzymatic digestion with trypsin and collagenase. The released chondrocytes were then "dedifferentiated" by culturing in monolayer and passaged to generate a sufficient number of cells from the single source. Third generation cells were then allowed to "redifferentiate" to phenotypic chondrocytes and proliferate by placing them in agarose culture (5,6). Tissue culture plates were coated with a thin layer of 1% high Tm agarose. A mixture containing 2% low Tm agarose prepared in double-distilled water and an equal volume of 2 x high glucose DMEM containing 10% fetal bovine serum and cells was added to the plates providing an initial cell density of 4.5×10^5 cells per 35 mm diameter plate. The plates were incubated at 37°C for 5 minutes followed by gelation at 4°C for 15 minutes. Cultures were supplemented with high glucose DMEM containing 50 µg/ml ascorbate, 10% fetal bovine serum, and antibiotics. The phenotypic changes were monitored by determining glycosaminoglycan staining with alcian blue (6) and collagen typing using SDS-polyocrylamide electrophoresis (7). By the fourth week in culture these chondrocytes produced abundant amounts of glycosaminoglycan and only type II collagen, similar to chondrocytes immediately after release from cartilage matrix.

To determine the effects of ADA inhibition in this system media was supplemented with deoxyadenosine (0-200µM) and erythro-9[2-hydroxy-3-nonyl] adenine and (EHNA) (0-

5μM). At the time of analysis, the contents of the plates were washed, removed into microfuge tubes, and incubated at 40°C for 1 hour with agarase (240 U in 1 ml of 40 mM Tris-acetate, pH 6.9). After incubating another hour with collagenase (2 mg/ml) and hyaluronidase (0.5 mg/ml) added to the tubes, the cells were filtered through a wire mesh and rinsed before counting and assessing viability using trypan blue exclusion.

In one set of experiments the combination of deoxyadenosine and EHNA was added to the agarose cultures at week one and continued to be added throughout the culture period. Cell viability was determined at weeks 2, 3, 4, and 5 to determine the effect of continuous ADA inhibition. The results (Table 1) demonstrate a dose-dependent toxicity of deoxyadenosine to redifferentiating chondrocytes in agarose culture.

Table 1. Continuous Incubation with Deoxyadenosine Viability

[dAdo]uM	2wk[a]	3wk[a]	4wk[b]	5wk[b]
0	74.4 ± 7.3	73.4 ± 7.8	70.7 ± 3.8	65.1 ± 0.2
50	63.2 ± 4.9	65.0 ± 2.6	50.7 ± 7.2	34.2 ± 5.1
100	47.7 ± 5.1	46.2 ± 1.1	98.9 ± 4.1	34.0 ± 4.4
150	46.0 ± 5.4	43.4 ± 3.3	44.0 ± 1.4	27.4 ± 4.1
200	40.6 ± 6.2	31.5 ± 2.1	39.0 ± 4.2	14.1 ± 1.9

[a] 8 samples for each condition
[b] 4 samples for each condition

In a second series of experiments the effects of adding deoxyadenosine for only one week were assessed. These results (Table 2) data demonstrate that the toxic effect of ADA inhibition is selective to less differentiated cells and that chondrocytes that function similar to cells freshly released from matrix are not affected by the model.

Table 2. One-Week Incubations with Deoxyadenosine Viability

[dAdo]uM	2wk[a]	3wk[a]	4wk[a]	5wk[b]
0	74.4 ± 7.3	70.2 ± 4.8	67.7 ± 3.1	64.0 ± 3.6
50	63.2 ± 4.9	65.1 ± 2.3	65.5 ± 7.1	65.9 ± 5.5
100	51.7 ± 5.1	55.8 ± 8.5	66.3 ± 4.7	64.0 ± 3.8
150	46.0 ± 5.4	58.9 ± 1.3	67.2 ± 5.7	65.1 ± 2.6
200	40.6 ± 6.2	59.5 ± 8.7	67.6 ± 4.0	70.1 ± 0.1

[a] 8 samples for each condition
[b] 4 samples for each condition

An in vivo model of ADA deficiency was established by injecting immature mice with the ADA inhibitor, 2' deoxycoformycin (DCF) (8). Doses of DCF (2 or 5 μg/gram body weight) were administered by intraperitoneal injection to C57BL/6J mice. All experiments included littermate controls injected with normal saline. At the conclusion of the experiments the mice were sacrificed and the activities of ADA and a control enzyme purine nucleoside phosphorylase were determined for erythrocytes, thymocytes, and spleen cells to determine the degree of ADA inhibition achieved. In addition the knees were dissected, fixed, and sectioned for histology. Growth plate size was determined by histomorphometric measurements.

Administration of DCF in the first two weeks of life proved lethal. However DCF administered twice weekly beginning at age 14 or 16 days caused an inhibition of thymocyte ADA activity and decreased growth plate size in proportion to the dosage and duration of treatment (Table 3). Purine nucleoside phosphorylase activity was not effected.

Table 3. Mice Treated with 2'-Deoxycoformycin

Treatment (day begun	sacrifice)	DCF dosage (ug/g wt)	Doses (No)	Thymocyte ADA (% control)	Growth Plate Width (% control)	N
14	21	2.4	2	62	76[a]	12
14	21	5.0	2	36	62[a]	12
16	28	2.0	4	11	68[a]	4
16	28	5.0	4	3	45[a]	4

[a] = $p < 0.05$

Histology of the growth plates of mice treated with DCF were essentially normal except for the smaller size and occasional areas of decreased organization in the hypertrophic zones. Thus the treatment of immature mice with the ADA inhibitor causes significant changes in immature cartilage, but does not recapitulate the changes observed in human disease (3).

Because successful enzyme replacement therapy leads to resolution of the chondro-osseous dysplasia in ADA-deficient people (4), experiments were performed in which DCF was administered for a limited time, then discontinued, and the animal allowed to live. Additional experiments in which DCF was not initiated until age 41 days were performed to determine if the toxicity was selective for immature mice. These results (Table 4) indicate that ADA inhibition is selectively toxic to cartilage of immature mice, and that the effects of this model may be overcome or "outgrown" with restoration of enzyme activity.

Table 4. Mice Treated with 2'-Deoxycoformycin

Treatment (day begun	sacrifice)	DCF dosage (ug/g wt)	Doses (No)	Thymocyte ADA (% control)	Growth Plate Width (% control)	N
14	55	5.0	2	96	115[a]	5
41	48	5.0	4	2	100[a]	7

[a] = no significant difference

In summary, the results of these in vitro and in vivo experiments indicate that inhibition of ADA activity is selectively toxic to immature chondrocytes. The reasons for the sensitivity of immature cells and the resistance of more mature cells is unresolved. Previous biochemical studies in chondrocytes suggests ATP depletion may be the critical factor (9). Observations from studies using lymphocytes have provided evidence that the toxicity in these cells is caused by a block in DNA synthesis resulting from accumulation of deoxy-ATP, inhibition of ribonucleotide reductase, and depletion of deoxy CTP (10); inhibition of S-adenosylhomocysteine hydrolase by the accumulation of adenosine and deoxyadenosine (11); or by NAD depletion triggered by increased poly (ADP-ribose) synthetase activity (12).

REFERENCES

1. H. J. Meuwissen, B. Pollara, and R. J. Pickering, Combined immunodeficiency disease associated with adenosine deaminase deficiency, Pediatrics 86:169-181 (1975).
2. J. J. Wolfson and V. F. Cross, The radiographic findings in forty-nine patients with combined immunodeficiency, Combined Immunodeficiency Disease and Adenosine Deaminase Deficiency: A Molecular Defect, Academic Press, New York (1975).

3. S. D. Cederbaum, I. Kaitila, D. L. Rimoin, and E. R. Stiehm, The chondro-osseous dysplasia of adenosine deaminase deficiency with severe combined immunodeficiency, Pediatrics 89:737-742 (1976).
4. B. S. Yullish, R. C. Stern, and S. H. Polmar, Partial resolution of bone lesions: a child with severe combined immunodeficiency disease and adenosine deaminase deficiency after enzyme-replacement therapy, Am. J. Dis. Child 134:61-63 (1980).
5. P. D. Benya and J. D. Shaffer, Dedifferentiated chondrocytes reexpress the differentiated collagen phenotype when cultured in agarose gels, Cell 30:215-224 (1982).
6. K. A. Skantze, C. E. Brinckerhoff, and J. P. Collier, Use of agarose culture to measure the effect of transforming growth factor B and epidermal growth factor on rabbit articular chondrocytes, Cancer 45:4416-4421 (1985).
7. W. A. Horton, O. J. Hood, A. S. Ahmed, and E. S. Griftey, Abnormal ossifications in thanatophoric dysplasia, Bone, 9:53-61 (1988).
8. H. Ratech, R. Hirschhorn, and G. J. Thorbecke, Effects of deoxycoformycin in mice III. A murine model reproducing multi-system pathology of human adenosine deaminase deficiency, A.J.P. 119:65 (1985).
9. R. L. Wortmann, J. A. Veum, L. M. Ryan, and H. S. Cheung, Differential deoxyadenosine toxicity to immature rabbit cartilage in vitro: A model for the chondro-osseous dysplasia of adenosine deaminase deficiency, Arthritis and Rheumatism Vol. 32, No. 8 (1989).
10. B. Ullman, L. J. Gudas, A. Cohen, D. W. Martin Jr., Deoxyadenosine metabolism and cytotoxicity in cultured mouse T lymphoma cells: a model for immunodeficiency disease, Cell 14:365-375, 1978.
11. T. D. Palella, R.A. Schatz, T. E. Wilens, and I. F. Fox, S-adenosylhomocysteine accumulation and selective cytotoxicity to cultured T- and B-lymphoblasts, J. Lab. Clin. Med. 100:269-278 (1982).
12. S. Seto, C. J. Carrera, M. Kibota, D. B. Wasson, D. A. Carson, Mechanism of deoxyadenosine and 2-chlorodeoxyadenosine toxicity to nondividing human lymphocytes, J. Clin. Invest. 75:377-383 (1985).

ANALYSIS OF FOREBRAIN DOPAMINERGIC PATHWAYS IN HPRT - MICE

DJ Williamson,[1] J Sharkey,[2] AR Clarke,[1] A Jamieson,[3] GW Arbuthnott,[3] PAT Kelly,[2] DW Melton[4] and ML Hooper[1]

Departments of Pathology,[1] Clinical Neurosciences,[2] Pharmacology,[3] and Molecular Biology[4] University of Edinburgh, Teviot Place, Edinburgh, EH8 9AG

INTRODUCTION

Lesch-Nyhan Syndrome (LNS) is an X-linked disease caused by deficiency of hypoxanthine-guanine phosphoribosyl transferase (HPRT; EC 2.4.2.8). Predictably this results in excessive de novo synthesis of purines which in turn leads to hyperuricaemia, gout, renal failure and its consequences. The syndrome is also characterised by extrapyramidal movement disorders, self-mutilatory behaviour (SMB) and mental retardation, the pathogenesis of which is uncertain. However, HPRT activity is high in normal brain tissue, particularly the basal ganglia, which suggests dependence of these areas on purine salvage pathways.

An $HPRT^-$ strain of mice was derived using embryonal stem cells (ES cells) from preimplantation strain 129/Ola embryos.[1] Analysis of the disrupted HPRT gene ($\underline{hprt}^{b-m3}$) revealed a deletion extending from the coding region into upstream regulatory sequences so that no functional enzyme could be produced[1,2] and no mRNA was detectable (S.Thompson, DWM personal observations).

MATERIALS AND METHODS

Male inbred 129/Ola mice hemizygous for the $\underline{hprt}^{b-m3}$ allele and age-matched controls were examined. In some experiments young mice were treated for 2d with allopurinol (250 or 400 µg/ml drinking water which reduced allantoin excretion by 40% or 95% respectively). They were killed by cervical dislocation and the brains were rapidly dissected. The olfactory bulbs were removed and the forebrains were separated by a coronal section just anterior to the optic chiasm. Tissues were snap frozen in liquid nitrogen and stored at -70°C before analysis. Samples were thawed in 0.1M perchloric acid, weighed, homogenised and filtered (Spin-X; Costar). HPLC analysis with electrochemical detection was performed as previously described.[3]

Purine and Pyrimidine Metabolism in Man VII, Part B
Edited by R.A. Harkness *et al.*, Plenum Press, New York, 1991

Table 1. Forebrain amine metabolite concentrations in HPRT⁻ mice treated with allopurinol

Group*	Drug (μg/ml)	Mean Concentration (μg/g)					
		DA	DOPAC	DOPAC/DA	HVA	5-HT	5-HIAA
$HPRT^-$	-	2.93	0.55	0.19	0.28	1.05	0.20
$HPRT^+$	-	5.96	0.58	0.10	0.32	1.07	0.18
$HPRT^-$	250	2.10	0.37	0.18	0.27	0.92	0.17
$HPRT^+$	250	3.15	0.36	0.12	0.23	0.98	0.13
$HPRT^-$	400	2.43	0.48	0.20	0.29	1.08	0.26
Mutation Effect		$P<0.001$	NS	$P<0.001$	NS	NS	$P<0.05$
Drug Effect		$P<0.01$	$P<0.001$	NS	NS	NS	NS

* Data from 2 experiments (5-6 mice group) were combined using regression analysis and compared by computing F statistics.

Dissected brain tissues from 3 mice per group were pooled for assay of [^{3}H]-SCH 23390 and [^{3}H]-Spiperone (Amersham) binding to D_1-type and D_2-type receptors respectively. Bound ligand was separated by filtration but the protocol was otherwise as previously described.[4] Receptors in the caudate nucleus and substantia nigra were also localised and quantified by receptor autoradiography[5,6] following binding of 1nM [^{3}H] SCH 23390 or 0.4nM [^{3}H] Spiperone to serial coronal sections. Image analysis and quantitation were performed using a Quantimet-970 (Cambridge Instruments). Non-specific binding accounted for 10% of total [^{3}H]-SCH 23390 binding and 25% of [^{3}H]-Spiperone binding.

RESULTS AND DISCUSSION

Unlike boys with LNS, young $HPRT^-$ mice showed no evidence of neurological disease, in particular no spontaneous extrapyramidal movement disorders.[7,8] Depletion of dopamine (DA) and its metabolite homovanillic acid (HVA) has been noted in the CSF of some,[9] but not all,[10] patients and an autopsy study found 70-90% depletion of the former in the basal ganglia.[11] The present study confirmed previous reports of forebrain DA depletion in $HPRT^-$ mice[8,12] (Table 1). However, the extent of forebrain DA depletion was much less than in humans with LNS. The normal levels of dihydroxyphenylacetic acid (DOPAC) and HVA were also in accord with published results.[8,12] Together these data suggest an increase in DA turnover as reflected by an increase in the DOPAC/DA ratio.[13]

Small increases in basal ganglia 5-hydroxytryptamine (5-HT) and 5-hydroxyindoleacetic acid (5-HIAA) have been noted in LNS patients.[11] The significance of the apparent increase in 5-HIAA

Table 2. Autoradiographic receptor binding data from the caudate nucleus and S.nigra of young (3 mo) and old (18-24 mo) $HPRT^-$ mice.

Mice	Ligand Binding (fmol/mg)			
	D_1 Receptor		D_2 Receptor	
	Caudate	S.Nigra	Caudate	S.Nigra
Young	421	332	138	9
Old #1	435	366	134	32
Old #2	422	327	134	15
Old #3	420	324	134	5

in the $HPRT^-$ mice is uncertain given no effect on its precursor 5-HT. Normal levels of striatal 5-HT were found in studies of inbred mice carrying the $\underline{hprt}^{b-m2}$ allele although detailed regional analysis revealed slight falls in the prefrontal cortex and substantia nigra.[8,12] No differences in 5-HIAA were found in the latter study. Selective lesions of nigro-striatal dopaminergic neurons can be made by intracisternal injection of 6-hydroxydopamine (6-OH DA) into neonatal rats. Subsequent striatal hyperinnervation by serotonergic neurons results in an increase in 5-HT and 5-HIAA.[14,16] In this model SMB may be induced by administration of L-DOPA, a response which is dependent on D_1-receptors.[14]

There was no difference in the density or affinity of D_1-type or D_2-type receptors in the basal ganglia and substantia nigra between young $HPRT^-$ and $HPRT^+$ mice (Figure 1). Thus there was no evidence of receptor up-regulation in response to

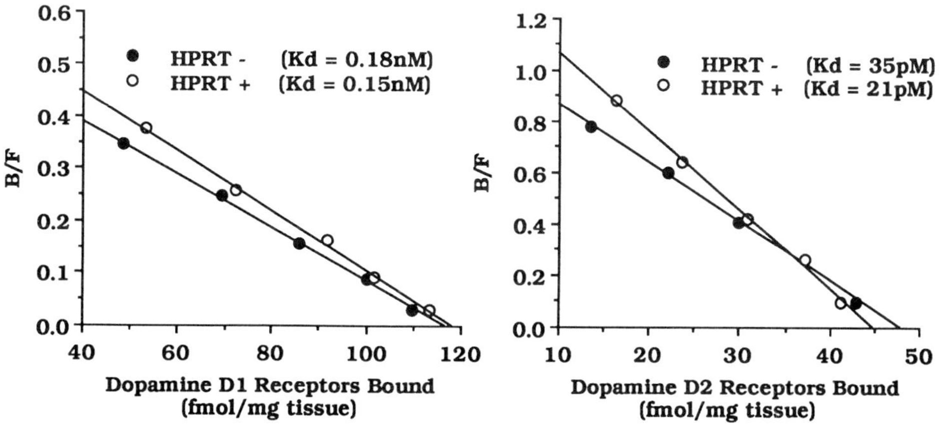

Fig.1. Scatchard plots of $[^3H]$-SCH 23390 and $[^3H]$-Spiperone binding

the reduction in DA level. Nor was there a consistent difference in binding to these receptors between young and old HPRT⁻ mice (Table 2). Apomorphine (a DA agonist) failed to induce extrapyramidal movements in HPRT⁻ mice,[8] in contrast to its effect on rats with neonatal 6-OH DA lesions.[14] Although there was no evidence of receptor supersensitivity, HPRT⁻ mice were more sensitive than normal mice to amphetamine-induced extrapyramidal movements.[17]

To examine the link between purine and brain amine metabolism, allopurinol was administered. No overt behavioural changes were induced but it did reduce forebrain DA concentrations in both normal and HPRT⁻ mice (Table 1). The effect of allopurinol could be distinguished from that of the mutation in that the former also resulted in a reduction in DOPAC concentration so that the DOPAC/DA ratio remained unchanged. These findings suggest that differences in the purine degradation pathway between mouse and man, such as the presence of uricase, do not play a role after maturity in mediating the behavioural consequences of HPRT-deficiency.

REFERENCES

1. M. Hooper, K. Hardy, A. Handyside, S. Hunter and M. Monk. Nature, 326:292 (1987)
2. S. Thompson, A.R. Clarke, A.M. Pow, M.L. Hooper and D.W. Melton. Cell, 56:313 (1989)
3. S.P. Butcher, I.S. Fairbrother, J.S. Kelly and G.W. Arbuthnott. J. Neurochem., 50:346 (1988)
4. P. Seeman, N.H. Bzowej, H.C. Guan, C. Bergeron, G.P. Reynolds, E.D. Bird, P. Riederer, K. Jellinger and W.W.Tourtellotte. Neuropsychopharmacology., 1:5 (1987)
5. T.M. Dawson, D.R. Gehlert, H.I. Yamamura, A. Barnett and J.K. Wamsley. Eur. J. Pharmacol., 108:323 (1985)
6. E.B. De Souza. Endocrinology, 119:1534 (1986)
7. M.L. Hooper. Discussions in Neurosciences, 5:124 (1988)
8. S. Finger, R.P. Heavens, D.J.S. Sirinathsinghji, M.R. Kuehn and S.B. Dunnett. J. Neurol. Sci., 86:203 (1988)
9. F.S. Silverstein, M.V. Johnston, R.J. Hutchinson and N.L. Edwards. Neurology, 45:907 (1985)
10. R.A. Harkness. Adv. Exp. Med. Biol., 253A:159 (1990)
11. K.G. Lloyd, O. Hornykiewicz, L. Davidson, K. Shannak, I. Farley, M. Goldstein, M. Shibuya, W.N. Kelley and I.H. Fox. N. Engl. J. Med., 305:1106 (1981)
12. S.B. Dunnett, D.J.S. Sirinathsinghji, R. Heavens, D.C. Rogers and M.R. Kuehn. Brain Res., 59:401 (1989)
13. M.J. Zigmond, A.L. Acheson, M.K. Stachowiak and E.M. Stricker. Arch. Neurol., 41:856 (1984)
14. G.R. Breese, H.E. Criswell, G.E. Duncan and R.A. Mueller. Brain Res. Bull., 25:477 (1990)
15. M.K. Stachowiak, J.P. Bruno, A.M. Snyder, A.M. Stricker and M.J. Zigmond. Brain Res., 291:164 (1984)
16. J. Luthman, B. Bolioli, T. Tsutsumi, A. Verhofstad and G. Jonsson. Brain Res. Bull., 19:269 (1987)
17. H.A. Jinnah, F.H. Cage & T. Friedmann J. Cell Biochem. (Suppl.) 14A:369 (1990)

DJW is supported by Action Research and the S.Davidson Fund.

ABSENCE OF dGTP ACCUMULATION AND COMPENSATORY LOSS OF DEOXYGUANOSINE KINASE IN PURINE NUCLEOSIDE PHOSPHORYLASE DEFICIENT MICE

J.P. Jenuth, J.E. Dilay, E. Fung, E.R. Mably, and F.F. Snyder

Departments of Paediatrics and Medical Biochemistry University of Calgary, Calgary Alberta Canada T2N 4N1

INTRODUCTION

The first reported cases of purine nucleoside phosphorylase (PNP) deficiency in man were characterized by selective cellular immune dysfunction[1]. Patients have a pronounced decrease in T-cell numbers with normal or exaggerated B-cell function[2-4]. They excrete PNP substrates and have elevated plasma levels of these metabolites. Intracellular nucleotide pools of erythrocytes show an increase in deoxyguanosine triphosphate (dGTP), nicotinamide adenine dinucleotide (NAD) and a decrease in guanosine triphosphate (GTP)[4-7]. The metabolite thought to be responsible for the T-cell dysfunction is dGTP[5].

Two independent mutations at the PNP locus have been recovered in the progeny of C57BL/6J male mice treated with ethylnitrosourea and mated to C3H/HeHa female mice[8]. The normal C57BL/6J PNP allele has been designated NP-1D[9] and the mutant alleles are assigned NP-1E and NP-1F. We describe here our findings for cellular nucleotide pools in PNP deficient mice which led us to uncover a secondary enzyme deficiency of deoxyguanosine kinase.

MATERIALS AND METHODS

The recovery of mice with two independent mutations at the PNP locus has been described[8]. Mutant mice used in this study have been backcrossed to C57BL/6J (B6) for six generations and are designated B6-NPE and B6-NPF.

Nucleotides were examined by extraction of cell pellets with 2.5 volumes of 12% trichloroacetic acid followed by neutralization with 0.5 M tri-N-octylamine in 1, 1, 2-trichlorotrifluoroethane. Nucleotides were separated using a Partisphere-5-SAX column (Whatman). Elution buffers were: A, sodium acetate, 0.5 M; potassium phosphate, 0.25 M; phosphoric acid, 0.25 M; pH 4.7; and B, 1/100 dilution of A, pH 3.5 with phosphoric acid. Samples were eluted at 100% B for 5 minutes followed by a linear gradient to 100% A, 0% B for 15 minutes and held at 100% A for an additional 20 minutes. Peaks were identified by retention time and absorbance ratios.

Deoxycytidine kinase and deoxyguanosine kinase were assayed using [5-H^3]-deoxycytidine and [8-^{3}H]-deoxyguanosine respectively as described[10].

RESULTS AND DISCUSSION

We have previously described some of the biochemical and metabolic features of two mutations at the PNP locus in the mouse[8]. The mutations were examined here at the sixth generation backcross to C57BL/6J. The B6-NPE mutant has significant residual PNP activity as shown by tissue survey: erythrocytes, 5.4%; kidney, 11.7%; liver, 5.4%; spleen leucocytes, 28%; and thymocytes, 3.5%. The B6-NPF mutant strain is more severely deficient in PNP activity: erythrocytes, 2.0%; kidney, 0.2%; liver, 0.1%; spleen leucocytes, 3.6%; and thymocytes, 1.4%. Thus the NP-1F mutation is severely deficient, particularly in organs such as kidney and liver. Both the NP-1E and NP-1F mutations exhibited normal Michaelis constants for inosine and phosphate, indicating that the residual activity functions with normal affinity for substrates. These mutants also excrete purine nucleoside substrates in proportion to the severity of the enzyme deficiency[8]. Thus NP-1E mutants excrete inosine and guanosine at greater than 10-fold the normal level but they do not excrete detectable amounts of deoxyribonucleosides. The NP-1F mutant excretes inosine, guanosine, deoxyinosine and deoxyguanosine, together at greater than 100-fold control levels.

Changes in cellular nucleotide pools are observed in human PNP deficiency[4-7]. Analysis of erythrocyte nucleotides showed no significant difference in NAD or ATP for either of the two PNP mutations as compared to controls. The ratio of mutant to control GTP pools showed a significant ($p = 0.05$) increase: 1.45 for B6-NPE and 1.31 for B6-NPF. Thus the increase in erythrocyte and NAD pools and the decrease in GTP characteristic of human PNP deficiency are not observed in the mouse model. In contrast there is for erythrocytes an apparent increase in GTP in the PNP deficient mice and no change in NAD.

The accumulation of dGTP is presumed to be of significance in the metabolic mechanism causing T cell deficiency in the human disease. We examined dGTP levels in mutant mice. We found no detectable level of dGTP in either erythrocytes, < 0.01 pmole/10^6 cells, or spleen lymphocytes or thymocytes, < 1 pmole/10^6 cells, of PNP deficient mice. Thus within the limits of HPLC detection we have not observed an accumulation of dGTP in the mouse model.

The absence of dGTP accumulation in the PNP deficient mouse which shows evidence of *in vivo* deoxyguanosine accumulation, B6-NPF, suggests some differences between human and mouse metabolism. The most likely sites which would affect dGTP production from deoxyguanosine are deoxyguanosine kinase, dGMP kinase or dGMP nucleotidase.

Assay of deoxyguanosine kinase activity in erythrocytes revealed a marked deficiency in the mutants as compared to control: 8.1% for B6-NPE and 3.6% for B6-NPF. Control B6 deoxyguanosine kinase activity was 0.13 ± 0.02 nmole/min/mg protein. Analysis of deoxyguanosine kinase activity in heterozygotes has given equivocal results with a possible partial reduction in the NP-1DF heterozygote. The reduction of deoxyguanosine kinase activity was also apparent in other tissues where the mutant had the following residual activity in liver: 15.7%; B6-NPE; and 7.5%, B6-NPF, and thymocytes: 75.8%, B6-NPE; and 59%, B6-NPF. The decrease in deoxyguanosine kinase does not appear to be due to simple metabolite inhibition as dialysis of cell lysates does not restore activity. In addition this does not appear to be deoxycytidine kinase activity as there is no appreciable activity in mouse erythrocytes and this activity was not significantly reduced in mutant thymocytes.

We have previously described a kinetic approach to modelling metabolism at a metabolic branch point[11]. We have applied this procedure to the metabolism of deoxyguanosine via PNP or deoxyguanosine kinase by use of experimentally determined Michaelis constants and maximal velocities. The ratio of phosphorylation/phosphorolysis can be predicted as a function of deoxyguanosine concentration. For control B6 mice, this ratio in erythrocytes varies from 0.03 to 0.1. Thus phosphorylation is at most $1/10^{th}$ of phosphorolysis. In the heterozygote, NP-1DF, the ratio varies from 0.07 to 0.25, indicating that even with half normal PNP levels, phosphorylation of deoxyguanosine is unfavourable. In the homozygote mutant, NP-1F, the ratio varies from 1.2 to 5, assuming that deoxyguanosine kinase activity is unchanged. Thus at all substrate concentrations deoxyguanosine phosphorylation would be equal or greater than phosphorolysis. Correcting the model for the reduced levels of deoxyguanosine kinase actually observed in the NP-1F mutant, the ratio varies from 0.2 to 0.7, again becoming unfavourable for the accumulation of dGTP. The observed reduction in deoxyguanosine kinase activity and the kinetic modelling studies indicate that the reduced levels of the kinase in the mutant appear to be responsible for the absence of significant dGTP accumulation in the PNP deficient mouse.

In summary we have recovered two mutants at the PNP locus in the mouse which are partially and severely deficient in PNP activity. They excrete PNP substrates at greater than 10- and 100-fold the level of normal mice. Examination of nucleotide levels in PNP deficient mice showed differences from human PNP deficiency. There was no evidence of increased NAD, and GTP pools were not decreased but showed a marginal increase in the PNP deficient mice. In addition, within the limits of HPLC methodology, we found no evidence for the accumulation of dGTP in the PNP deficient mouse cells. Examination of deoxyguanosine kinase activity revealed a reduction in PNP deficient mouse tissues. This reduction in activity does not appear to be due to metabolite inhibition and does not appear to be deoxycytidine kinase activity. Models of deoxyguanosine metabolism suggest that PNP deficient mice would preferentially phosphorylate deoxyguanosine whereas control mice would not. Because of the decrease in deoxyguanosine kinase activity in the PNP mutants, phosphorylation is predicted to be less than phosphorolysis at all substrate concentrations. Thus the decrease in deoxyguanosine kinase activity appears to be a compensatory mechanism which prevents the accumulation of dGTP in the PNP deficient mouse model.

ACKNOWLEDGMENTS

This work was supported by the Medical Research Council of Canada grant MT-6376.

REFERENCES

1. Giblett ER, Ammann AJ, Sandman R, Wara DW, Diamond LK. Nucleoside phosphorylase deficiency in a child with severely deficient T-cell immunity and normal B-cell immunity. Lancet, 3:1010-1013 (1975).
2. Stoop JW, Zegers BJM, Hendricks GFM, Siegenbeek van Heukelom LH, Staal GEJ, De Bree PK, Wadman SK, Ballieux RE. Purine nucleoside phosphorylase deficiency associated with selective cellular immunodeficiency. New England J Med, 296:651-655 (1977).
3. Gelfand EW, Lee JJ, Dosch H. Selective toxicity of purine deoxynucleosides for human lymphocyte growth and function. Proc Natl Acad Sci USA 76(4):1998-2002 (1979).
4. Rijksen G, Kuis W, Wadman SK, Spaapen LJM, Duran M, Voorbrood BS, Staal GEJ, Stoop JW, Zegers BJM. A new case of purine nucleoside phosphorylase deficiency: Enzymologic, clinical, and immunologic characteristics. Pediat Res 21(2):137-141 (1987).
5. Cohen A, Gudas LJ, Amman AJ, Staal GEJ, Martin DW Jr. Deoxyguanosine triphosphate as a possible toxic metabolite in immunodeficiency associated with purine nucleoside phosphorylase deficiency. J Clin Invest 61:1405-1409 (1978).
6. Simmonds HA, Watson AR, Webster DR, Sahota A, Perrett D. GTP depletion and other erythrocyte abnormalities in inherited PNP deficiency. Biochem Pharmacol 31:941-946 (1982).
7. Simmonds HA, Fairbanks LD, Morris GS, Morgan G, Watson AR, Timms P, Singh B. Central nervous system dysfunction and erythrocyte guanosine triphosphate depletion in purine nucleoside phosphorylase deficiency. Arch Dis Child 62:385-391 (1987).
8. Mably ER, Fung E, Snyder FF. Genetic deficiency of purine nucleoside phosphorylase in the mouse. Characterization of partially and severely enzyme deficient mutants. Genome 32:1026-1032 (1989).
9. Mably ER, Carter-Edwards T, Biddle FG, and Snyder FF. Purine nucleoside phosphorylase heterogeneity in the mouse as analyzed by isoelectric focusing and specific activity. Comp Biochem Physiol 89B:427-431 (1988)
10. Lukey T and Snyder FF. Purine ribonucleoside and deoxyribonucleoside kinase activities in thymocytes. Specificity and optimal assay conditions for phosphorylation. Can J Biochem 9:677-682 (1980)
11. Snyder FF and Lukey T. Kinetic considerations for the regulation of adenosine and deoxyadenosine metabolism in mouse and human tissues based on a thymocyte model. Biochim Biophys Acta 696:299-307 (1982).

ADENYLOSUCCINASE ACTIVITY AND SUCCINYLPURINE PRODUCTION IN FIBROBLASTS OF ADENYLOSUCCINASE-DEFICIENT CHILDREN

Françoise Van den Bergh, M. Françoise Vincent, Jaak Jaeken* and Georges Van den Berghe

Laboratory of Physiological Chemistry, International Institute of Cellular and Molecular Pathology, UCL 7539, B-1200 Brussels, Belgium, and *Department of Pediatrics, University of Leuven, B-3000 Leuven, Belgium

INTRODUCTION

Adenylosuccinase (adenylosuccinate lyase, EC 4.3.2.2, ASase) catalyzes two steps in the biosynthesis of purine nucleotides : the conversion of succinylaminoimidazole carboxamide ribotide (SAICAR) into AICAR along the *de novo* pathway, and the formation of AMP from adenylosuccinate (S-AMP) in the conversion of IMP into adenine nucleotides. Both reactions involve the cleavage of a succinyl group, yielding fumarate. ASase deficiency is the first enzyme deficiency reported in man (1) along the *de novo* pathway of purine synthesis. The defect is transmitted as an autosomal recessive trait and results in the accumulation in cerebrospinal fluid, plasma and urine, of two normally undetectable compounds, SAICAriboside and succinyladenosine (S-Ado). These are the products of the dephosphorylation, by cytosolic 5'-nucleotidase (2), of the two substrates of ASase. From a clinical and biochemical study of 8 children with ASase deficiency (3) two main subtypes of the defect can be distinguished : the first one, identified in 7 patients and hereafter referred to as type I, is characterized by very profound psychomotor retardation and by S-Ado/SAICAriboside ratios between 1 and 2. In type II, diagnosed in one girl, mental retardation is slight, and S-Ado/SAICAriboside ratios are about 4.

The aim of this work was to characterize fibroblast ASase in the two subtypes, and to evaluate the consequences of the defect on purine metabolism in intact cells.

METHODS

Skin fibroblasts were cultured by standardized techniques. ASase was assayed radiochemically (4) in cell sonicates. For measurements of the incorporation of [^{14}C]hypoxanthine, cells were washed 3 times, resuspended in Krebs-Ringer bicarbonate buffer at the concentration of ~ 4.10^{6} cells/ml, and incubated with 20 μM [^{14}C]hypoxanthine for 20 min. Incubations were stopped by the addition of PCA, followed by neutralization. Radioactivity in nucleotides and nucleosides was determined by TLC and by HPLC, coupled with radioactivity detection. For measurements of *de novo* synthesis, cells were collected and inoculated as monolayers (2-3.10^5 cells/35 mm dish), in medium containing dialyzed fetal calf serum, for 24 h. Incorporation was started by replacing the medium by the same medium containing 0.2 mM [^{14}C]formate. After 7 h, cells were trypsinized, centrifuged, extracted with PCA, and radioactivities determined as above.

RESULTS

Activities of ASase in control and in patient fibroblasts

In the fibroblasts of type I patients, the activities of ASase with S-AMP and with SAICAR were decreased in parallel, to ~ 30 % of the respective control values (Table I). In contrast, in the type II patient, the activity with S-AMP was reduced to ~ 3 % of normal; that with SAICAR was ~ 30 % of control. The much more pronounced loss of activity with S-AMP as compared to SAICAR, provides an explanation for the higher S-Ado/SAICAriboside ratio in the body fluids of the latter patient.

Table 1 . Adenylosuccinase activities in control and in patient fibroblasts. Control values represent means ± SEM for three cell lines. Patient numbers correspond to those given in ref. 3. -, non measurable.

		Vmax (nmol/min/mg protein)		*Km* (µM)	
		S-AMP	SAICAR	S-AMP	SAICAR
Controls		1.44 ± 0.1	0.92 ± 0.20	10 ± 2	7 ± 2
Patients					
Type I	# 1	0.44	0.22	10	~ 30
	# 2	0.46	0.31	11	~ 30
	# 3	0.43	0.22	9	~ 20
	# 7	0.47	0.26	11	~ 15
Type II	# 8	0.04	0.31	-	~ 25

Characteristics of residual ASase activities

The apparent *Km* for S-AMP of fibroblast ASase from type I patients was not modified (Table 1). Similar results were obtained with ASase from cultured lymphoblasts from these patients by Barshop et al. (5). Owing to the very low residual activity with S-AMP in the cells from the type II patient, *Km* for S-AMP could not be measured. In both mutant cell types, the *Km* for SAICAR was increased 2- to 4-fold as compared to control values, although low activities hampered precise measurements.

In control cells, *Vmax* and *Km* of ASase, measured both with S-AMP and SAICAR, were not influenced by the addition of 100 mM KCl. In cells of type I patients, the addition of 100 mM KCl barely influenced *Vmax* but decreased *Km* for S-AMP and for SAICAR 2- to 3-fold. In contrast, the activity of ASase with SAICAR was markedly inhibited by KCl in type II fibroblasts. Further studies showed that this inhibition was due to the anion, competitive (*Km* increased to 270 μM in the presence of 90 mM KCl), and also exerted by other anions which inhibited in the following order : KH_2PO_4 > K_2SO_4 > KI > KBr > KCl > KF > KCOOH.

In cells from controls and from the two types of patients the activity of ASase, measured with SAICAR, was inhibited by AMP and by AICAR, products of the reaction. In cells from controls and from type I patients the activity with SAICAR was not inhibited by purine and pyrimidine nucleoside triphosphates. In contrast, the activity of ASase in cells of the type II patient was markedly (60 - 90 %) inhibited by 2.5 mM ATP, GTP, ITP, ZTP, UTP and CTP.

Studies in intact fibroblasts

Fibroblasts from type I patients, as well as those of the type II patient, displayed growth curves that were similar to those of control cells in medium containing undialyzed fetal calf serum. In medium prepared with dialyzed serum, in which purine bases such as hypoxanthine have been removed, growth of control fibroblasts was distinctly reduced. However, cells from ASase-deficient patients growed more rapidly than control cells, particularly after 12 days of culture, suggesting an adaptation of the cells to the enzyme defect.

Total incorporation of labelled hypoxanthine into the purine nucleotide pool of fibroblasts of ASase-deficient patients was comparable to that measured in controls. In both control fibroblasts and in cells from type I patients incubated with labelled hypoxanthine, 5 % of radioactivity was recovered in IMP, 7 % in guanine nucleotides and 88 % in adenine nucleotides. Neither S-AMP nor S-Ado were detected. In contrast, in the cells of the type II patient, 9 % of radioactivity was recovered in IMP, 7 % in guanine nucleotides, only 25 % in adenine nucleotides, but 56 % in S-AMP, and 3 % in S-Ado. This indicates that only the pronounced deficiency of ASase activity with S-AMP hampers metabolic flux through the enzyme step, but nevertheless still allows conversion of S-AMP into adenine nucleotides to proceed.

Total incorporation of labelled formate into purine nucleotides was about 40 % higher in cells of type I patients, and 60 % higher in cells of the type II patient, as compared to control fibroblasts. In accordance with previous work (6), this indicates a compensatory increase of the *de novo* synthesis of purine nucleotides in the ASase-deficient cells, which may explain their higher growth rates in dialyzed fetal calf serum. In both control fibroblasts and in cells from type I patients, SAICAR and S-AMP could not be detected. In cells from the type II patient, however, 6 % of radioactivity was recovered in SAICAR and 5 % in S-AMP. This indicates that even the partial defect of ASase activity with SAICAR influences conversion of SAICAR into AICAR. This decreased flux might be related to the inhibition exerted by KCl on the activity of ASase with SAICAR in fibroblasts of the type II patient.

CONCLUDING REMARKS

This study confirms the genetic heterogeneity of ASase deficiency. That two main subtypes of the defect exist was already apparent from its clinical picture and from the determination of S-Ado/SAICAriboside ratios in body fluids. This variation is now further unfolded by the markedly different kinetic characteristics of the residual enzyme activities in cultured skin fibroblasts of both types of patients. That isoforms of ASase exist had already been revealed by the non-uniform effect of starvation on the activity of the enzyme in various rat tissues (7). This was further documented by the observation that ASase-deficient patients display a marked decrease of the enzyme activity in liver and kidney, a partial defect in lymphocytes, fibroblasts and occasionally muscle, and normal enzyme activities in erythrocytes and granulocytes (1,3). Rat muscle ASase is composed of 4 subunits with MW of 52,000 (8). The present study suggests that the native enzyme in some tissues may be a heteropolymer formed from subunits encoded by different genes. The mutation in type I patients might have affected one species of subunit, whereas that in the type II patient might have modified another. Alternatively, in the latter, the mutation may have affected differently the binding of S-AMP and of SAICAR to the enzyme.

ACKNOWLEDGEMENTS

This work was supported by grant 3.4539.87 of the Fund for Medical Scientific Research (Belgium) and by the Belgian State - Prime Minister's Office for Science Policy Programming. G. Van den Berghe is Director of Research of the Belgian National Fund for Scientific Research.

REFERENCES

1. Jaeken J, Van den Berghe G. An infantile austistic syndrome characterised by the presence of succinylpurines in body fluids. *Lancet* **2** : 1058-61, 1984
2. Van den Berghe G, Jaeken J. Adenylosuccinase deficiency. *Adv Exp Med Biol* **195A** : 27-33, 1986
3. Jaeken J, Wadman SK, Duran M, van Sprang FJ, Beemer FA, Holl RA, Theunissen PM, De Cock P, Van den Bergh F, Vincent MF, Van den Berghe G. Adenylosuccinase deficiency : an inborn error of purine nucleotide synthesis. *Eur J Pediatr* **148** : 126-31, 1988
4. Van den Bergh F, Vincent MF, Jaeken J, Van den Berghe G. Radiochemical assay of adenylosuccinase : demonstration of parallel loss of activity toward both adenylosuccinate and succinylaminoimidazole carboxamide ribotide in liver of patients with the enzyme defect. *Analyt Biochem* **193** : 287-91, 1991
5. Barshop, BA, Alberts AS, Gruber HE. Kinetic studies of mutant human adenylosuccinase. *Biochim Biophys Acta* **999** : 19-23, 1989
6. Laikind PK, Gruber HE, Jansen I, Miller L, Hoffer M, Seegmiller JE, Willis RC, Jaeken J, Van den Berghe G. Purine biosynthesis in chinese hamster cell mutants and human fibroblasts partially deficient in adenylosuccinate lyase. *Adv Exp Med Biol* **195B** : 363-9, 1986
7. Brand LM, Lowenstein JM. Effect of diet on adenylosuccinase activity in various organs of rat and chicken. *J Biol Chem* **253** : 6872-8, 1978
8. Casey PJ, Lowenstein JM. Purification of adenylosuccinate lyase from rat skeletal muscle by a novel affinity column. Stabilization of the enzyme, and effects of anions and fluoro analogues of the substrate. *Biochem J* **246** : 263-9, 1987

PURINE NUCLEOTIDE CYCLE, MOLECULAR DEFECTS AND THERAPY

G. Van den Berghe, F. Bontemps and M. F. Vincent

Laboratory of Physiological Chemistry, International
Institute of Cellular and Molecular Pathology
UCL 7539, B-1200 Brussels, Belgium

In the late 1920's, Parnas and coworkers showed that muscle contraction is accompanied by the production of ammonia and that this production is equivalent to the conversion of adenine nucleotides into IMP. Moreover, they hypothesized that amino acids are used for the regeneration of AMP (reviewed in 1). In 1971, Lowenstein and Tornheim (2) formulated the concept that three enzymes of purine metabolism, adenylosuccinate synthetase, adenylosuccinate lyase, and AMP deaminase (Fig. 1), formed together a functional unit, which they termed the "purine nucleotide cycle" (PNC). In this lecture, the main properties of the enzymes of the PNC, the connections of the PNC with other pathways, its functions, and its molecular defects in man will be reviewed.

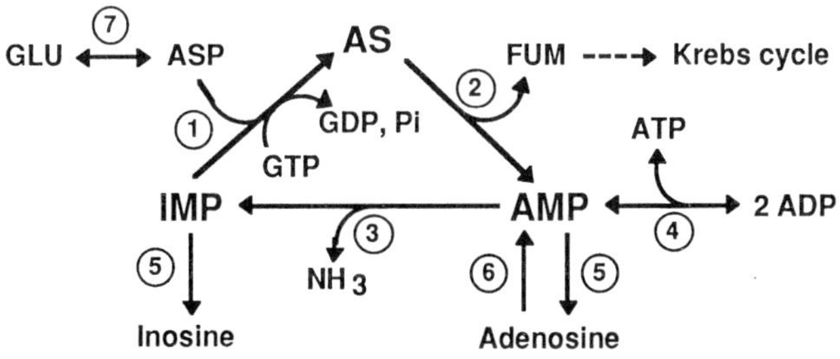

Figure 1. The purine nucleotide cycle. (1) Adenylosuccinate synthetase; (2) adenylosuccinate lyase; (3) AMP deaminase. Also shown are some adjacent reactions : (4) adenylate kinase; (5) cytosolic 5'-nucleotidase; (6) adenosine kinase; (7) glutamate oxaloacetate transaminase. AS, adenylosuccinate.

Purine and Pyrimidine Metabolism in Man VII, Part B
Edited by R.A. Harkness *et al*., Plenum Press, New York, 1991

ENZYMES OF THE PURINE NUCLEOTIDE CYCLE

Adenylosuccinate synthetase (reviewed in 3) catalyzes the condensation of IMP with aspartate in a reaction which requires GTP. It has a M.W. of ~ 100,000 and is composed of 2 subunits. Its activity is inhibited by its products, adenylosuccinate, GDP and Pi, and by AMP, GMP, and fructose-1,6-P_2. The enzyme exists under two isoforms : type M, a basic, and type L, an acidic protein. Skeletal and heart muscle contain only type M, liver contains both type M and L in near equal amounts, and brain and kidney contain predominantly type L. Adenylosuccinate synthetase is absent in human erythrocytes.

Adenylosuccinate lyase (also called adenylosuccinase, reviewed in 4) cleaves the C-N bond of adenylosuccinate, yielding AMP and fumarate. It also catalyzes the cleavage of SAICAR into AICAR and fumarate, the 8th step of the *de novo* pathway of purine synthesis. The rat muscle enzyme has a M.W. of ~ 200,000 and is composed of 4 subunits (5). Adenylosuccinate lyase is inhibited by AMP and AICAR, products of the reaction. That isoforms of adenylosuccinate lyase exist is apparent from the study of patients in whom the enzyme is deficient (see below), but these isozymes have not yet been characterized.

AMP deaminase (reviewed in 6) converts AMP into IMP and NH_3. It has a M.W. of ~ 320,000, is composed of 4 subunits, and displays complex, allosteric kinetic properties. In the absence of effectors, its substrate kinetics is sigmoid. ATP and ADP, potent stimulators of AMP deaminase, abolish sigmoidicity, whereas GTP and Pi, its main inhibitors, accentuate sigmoidicity. Several isoforms of AMP deaminase have been described, termed A (or M) in muscle, B (or L) in liver and kidney, C in heart, E_1 and E_2 in erythrocytes (7). Ongoing work is bringing insight in this complex pattern (8,9). Two different genes for AMP deaminase have been identified : AMPD 1, which is only expressed at high levels in adult skeletal muscle and encodes isoform M; AMPD 2, which encodes isoform L and is also expressed in embryonic muscle. The other isozymes might be mixtures of products of both genes and/or of additional genes (Bausch-Jurken *et al.*, this symposium).

CONNECTIONS OF THE PURINE NUCLEOTIDE CYCLE

The PNC is connected with a series of enzymes of purine metabolism. IMP is also formed by the *de novo* pathway, can be converted into guanine nucleotides (not shown in Fig. 1), and degraded into inosine by cytosolic 5'-nucleotidase. AMP is maintained in equilibrium with ATP and ADP by adenylate kinase, and can be dephosphorylated into adenosine. The necessity to maintain intracellular ATP requires a strict control of the catabolism of AMP. This is achieved in part by the low activity of cytosolic 5'-nucleotidase(s) toward AMP (10,11), and/or by recycling of adenosine by adenosine kinase (12). However, owing to the much higher activity of cytosolic 5'-nucleotidase toward IMP, the preservation of AMP requires in addition, either a profound inhibition of AMP deaminase, or a high rate of conversion of IMP into AMP.

The PNC is linked to amino acid metabolism by glutamate-oxaloacetate transaminase which converts glutamate into aspartate. It is also connected with the Krebs cycle, since fumarate is an intermediate of the latter.

FUNCTIONS OF THE PURINE NUCLEOTIDE CYCLE

Three main functions have been proposed for the PNC : [1] removal of AMP, in order to pull the adenylate kinase reaction in the direction of formation of ATP; [2] generation of NH_3; [3] production of Krebs cycle intermediates. From the observation that the isoforms of adenylosuccinate synthetase and of AMP deaminase vary in their regulatory properties, it can be inferred that the role of the PNC will not be the same in all tissues.

Muscle contraction requires ATP to drive actomyosin ATPase. As long as exercice has not depleted the muscle energy store, creatine-P, ATP can be regenerated by myofibrillar creatine kinase. However, when the intensity of exercice has exhausted creatine-P, regeneration of ATP depends on other processes. Operation of the PNC can, in theory, regenerate ATP by three mechanisms : [1] formation of IMP, removing AMP and pulling adenylate kinase in the direction 2 ADP –> AMP + ATP; [2] liberation of NH_3, stimulating muscle glycolysis at the level of phosphofructokinase, thereby enhancing the regeneration of ATP; [3] release of fumarate, producing Krebs cycle intermediates and consequently increasing the capacity of the latter to regenerate ATP. The trigger which stimulates AMP deaminase during sustained contraction is probably an increase in free ADP. That the PNC functions during intense muscle contraction has been shown by numerous *in vitro* and *in vivo* experiments. Among the latter, the observation that the following components of the PNC : NH_3, IMP, adenylosuccinate, and fumarate together with its derivative malate, increase more or less simultaneously in muscle *in situ* during exercice (13,14), indicates that the cycle functions as a whole during contraction rather than, as claimed by some authors, in two phases : deamination during contraction, followed by reamination during recovery. The importance of the PNC in muscle contraction is further evidenced by its defects (see below). However, the observation that AMP deaminase-deficient patients produce lactate normally upon muscle contraction renders the role of NH_3 in the stimulation of glycolysis less likely.

Liver forms urea, the end product of protein catabolism in mammals. Urea is classically stated to arise from NH_3 generated by mitochondrial glutamate dehydrogenase. This scheme has been questioned on the basis that the *in vitro* activity of glutamate dehydrogenase was too low to account for the *in vivo* rates of urea synthesis, and it has been proposed that NH_3 was produced by the PNC (15). Subsequent experiments have, however, invalidated this theory. Hadacidin, an inhibitor of adenylosuccinate synthetase, was found not to affect ureogenesis (16). Studies of the incorporation of [^{15}N]alanine into the 6-NH_2 group of hepatic adenine nucleotides have shown that its initial rate reached only ~ 5 % of that in urea (17). This is in agreement with kinetic studies of liver AMP deaminase which have shown that it is 95 % inhibited under physiological conditions by the concentrations of GTP and Pi prevailing in the liver cell (18).

Kidney plays an important role in acid-base homeostasis by producing NH_3. One of the enzymes involved in this process is Pi-dependent glutaminase. Whether the resulting glutamate is in turn deaminated by glutamate dehydrogenase, or enters the PNC via aspartate, has also been the subject of numerous investigations. Present evidence indicates that in normal acid-base status, the turnover of the adenine nucleotides, and thus the activity of the

PNC, can account for the rate of release of NH_3 from glutamate. However, under acidotic conditions, the increased formation of NH_3 would to a major extent be accounted for by an increased activity of glutamate dehydrogenase (19).

MOLECULAR DEFECTS AND THERAPY

Muscle AMP deaminase deficiency, discovered in 1978 (20), is characterized clinically by muscular weakness, fatigue, and cramps or muscle pain following various degrees of exercise (reviewed in 21). Typically, the several-fold increase in venous plasma NH_3, recorded in normal subjects upon vigorous exercise, is absent. A "semi-ischemic" forearm exercise test is therefore an easy diagnostic procedure. The symptoms are most likely caused by deficient "extraction of energy" from the adenine nucleotide pool, and by depletion of Krebs cycle intermediates. An impairment of muscle glycolysis caused by the deficient release of NH_3 is unlikely owing to the normal increase in lactate in exercising patients. Measurements of the activity of AMP deaminase in several large series of diagnostic muscle biopsies show that it is deficient in ~ 2 % of all specimens. In about half of these, AMP deaminase deficiency seems secundary to a neuromuscular disease, such as amyotrophic lateral sclerosis or Werdnig-Hoffmann disease. In the other half the deficiency is apparently a primary genetic defect, most likely transmitted as an autosomal recessive trait. The frequency of the defect in the general population is unknown, although recent investigations (Gross *et al.*, this symposium) indicate that the mutation might be very frequent in Caucasians.

Patients with primary AMP deaminase deficiency should be advised to exercise with caution to prevent rhabdomyolysis. Administration of ribose (2 to 60 g per day orally in divided doses) has been reported very beneficial in some patients but ineffective in others. The effect of ribose has been ascribed to its role as a glycolytic substrate and/or as a precursor of phosphoribosyl pyrophosphate (Wagner *et al.*, this symposium).

Erythrocyte AMP deaminase deficiency, identified in 1984 (22), is clinically completely asymptomatic. Erythrocyte concentrations of ATP are about 150 % of normal (23), which can be explained by impairment of the degradation of AMP by way of AMP deaminase. Inheritance is autosomal recessive. The frequency of homozygotes is ~ 1/5000 in Japan, Korea and Taiwan, and the gene frequency is estimated at ~ 1/30.

Regulatory mutations of liver AMP deaminase which would render the enzyme less sensitive to its physiological inhibitors have been proposed as a cause of primary gout with overproduction of uric acid (24). In one gouty patient, a decreased sensitivity of hepatic AMP deaminase to GTP could be demonstrated (25).

Adenylosuccinate lyase deficiency, discovered in 1984 (26, reviewed in 27) is characterized by the accumulation in cerebrospinal fluid, plasma and urine of SAICAriboside and succinyladenosine (S-Ado). These derive from dephosphorylation of the two substrates of the enzyme by cytosolic 5'-nucleotidase. Adenylosuccinase deficiency is autosomal recessive and displays marked genetic heterogeneity (28). From a clinical and biochemical viewpoint, two main subtypes of the defect can be distinguished : the first one, identified hitherto in 7 children, is characterized by very profound psychomotor retardation, often associated with autistic features, and by S-Ado/SAICAriboside ratios between 1 and 2; the second one, diagnosed in one girl, is characterized by slight mental retardation and S-Ado/SAICAriboside ratios of ~ 4. Some patients also have muscular wasting and growth failure. The observation that the concentrations of SAICAriboside are similar in both types of patients suggests that it is the noxious compound and that its action can be counteracted by S-Ado. The deficiency of adenylosuccinase is marked in liver and kidney, partial in lymphocytes and fibroblasts, and absent in erythrocytes and granulocytes, indicating the existence of isozymes. Enzyme activity is also partially lost in muscle of the patients with growth retardation and muscular wasting. The latter might thus be explained by interruption of the purine nucleotide cycle. Recent studies have shown that activities with adenylosuccinate and SAICAR are lost in parallel in liver (29) and fibroblasts (Van den Bergh *et al.*, this symposium) of the profoundly retarded patients, but that the loss of activity with adenylosuccinate is more pronounced in the fibroblasts of the slightly retarded girl. This might provide an explanation for her higher S-Ado/SAICAriboside ratios.

Patients with growth retardation have been shown to benefit from the administration of adenine (10 mg/kg per day), together with allopurinol (5-10 mg/kg per day) to inhibit its conversion to insoluble 2,8-OH-adenine. This treatment might increase nucleotide levels in tissues.

ACKNOWLEDGMENTS

Work in the authors' laboratory was supported by the Fund for Medical Scientific Research (Belgium), and by the Belgian State - Prime Minister's Office for Science Policy Programming. G. Van den Berghe is Director of Research of the Belgian National Fund for Scientific Research.

REFERENCES

1. Van Waarde A. Operation of the purine nucleotide cycle in animal tissues. *Biol Rev* **63** : 259-98, 1988
2. Lowenstein J, Tornheim, K. Ammonia production in muscle : the purine nucleotide cycle. *Science* **171** : 397-400, 1971
3. Stayton MM, Rudolph FB, Fromm HJ : Regulation, genetics and properties of adenylosuccinate synthetase : a review. *Curr Top Cell Regul* **22** : 103-41, 1983
4. Ratner S. Argininosuccinases and adenylosuccinases. *The Enzymes, 3rd ed, vol 7* : 167-97, 1972
5. Casey PJ, Lowenstein JM. Purification of adenylosuccinate lyase from rat skeletal muscle by a novel affinity column. *Biochem J* **246** : 263-9, 1987
6. Zielke CL, Suelter CH. Purine, purine nucleoside and purine nucleotide aminohydrolases. *The Enzymes, 3rd ed, vol 4* : 47-78, 1971
7. Ogasawara N, Goto H, Yamada Y, Watanabe T, Asano T. AMP deaminase isozymes in human tissues. *Biochim Biophys Acta* **714** : 298-306, 1982
8. Sabina RL, Morisaki T, Clarke P, Eddy R, Shows TB, Morton CC, Holmes, EW. Characterization of the human and rat myoadenylate deaminase genes. *J Biol Chem* **265** : 9423-33, 1990
9. Morisaki T, Sabina RL, Holmes EW. Adenylate deaminase. A multigene family in humans and rats. *J Biol Chem* **265** : 11482-6, 1990
10. Van den Berghe G, Van Pottelsberghe C, Hers HG. A kinetic study of the soluble 5'-nucleotidase of rat liver. *Biochem J* **162** : 611-6, 1977
11. Bontemps F, Vincent MF, Van den Bergh F, van Waeg G, Van den Berghe G. Stimulation by glycerate 2,3-bisphosphate : a common property of cytosolic 5'-nucleotidase in rat and human tissues. *Biochim Biophys Acta* **997** : 131-4, 1989
12. Bontemps F, Van den Berghe G, Hers HG. Evidence for a substrate cycle between AMP and adenosine in isolated rat hepatocytes. *Proc Natl Acad Sci USA* **80** : 2829-33, 1983
13. Goodman MN, Lowenstein JM. The purine nucleotide cycle. Studies of ammonia production by skeletal muscle in situ and in perfused preparations. *J Biol Chem* **252** : 5054-60, 1977
14. Aragon JJ, Lowenstein JM. The purine nucleotide cycle. Comparison of the levels of citric acid cycle intermediates with the operation of the purine nucleotide cycle in rat

skeletal muscle during exercise and recovery from exercise. *Eur J Biochem* **110** : 371-7, 1980
15. Moss KM, McGivan JD. Characteristics of aspartate deamination by the purine nucleotide cycle in the cytosol fraction of rat liver. *Biochem J* **150** : 275-83, 1975.
16. Rognstad R. Sources of ammonia for urea sythesis in isolated rat liver cells. *Biochim Biophys Acta* **496** : 249-54, 1977
17. Krebs HA, Hems P, Lund P, Halliday D, Read WWC. Sources of ammonia for mammalian urea synthesis. *Biochem J* **176** : 733-7, 1978
18. Van den Berghe G, Bronfman M, Vanneste R, Hers HG. The mechanism of adenosine triphosphate depletion in the liver after a load of fructose. A kinetic study of liver adenylate deaminase. *Biochem J* **162** : 601-9, 1977
19. Tornheim K, Pang H, Costello CE. The purine nucleotide cycle and ammoniagenesis in rat kidney tubules. *J Biol Chem* **261** : 10157-62, 1986
20. Fishbein WN, Armbrustmacher VW, Griffin JL. Myoadenylate deaminase deficiency : a new disease of muscle. *Science* **200** : 545-8, 1978
21. Sabina RL, Swain JL, Holmes EW. Myoadenylate deaminase deficiency. In *The Metabolic Basis of Inherited Disease,* 6th ed (Scriver CR, Beaudet AL, Sly WS, Valle D, eds) McGraw-Hill, New York, pp 1077-84, 1989
22. Ogasawara N, Goto H, Yamada Y, Nishigaki I, Itoh T, Hasegawa I. Complete deficiency of AMP deaminase in human erythrocytes. *Biochem Biophys Res Commun* **122** : 1344-9, 1984
23. Ogasawara N, Goto H, Yamada Y, Nishigaki I, Itoh T, Hasegawa I. Park KS. Deficiency of AMP deaminase in erythrocytes. *Hum Genet* **75** : 15-8, 1987
24. Hers HG, Van den Berghe G. Enzyme defect in primary gout. *Lancet* **1** : 585-6, 1979
25. Van den Berghe G, Hers HG. Abnormal AMP deaminase in primary gout. *Lancet* **2** : 1090, 1980
26. Jaeken J, Van den Berghe G. An infantile autistic syndrome characterised by the presence of succinylpurines in body fluids. *Lancet* **2** : 1058-61, 1984
27. Van den Berghe G. Disorders of purine and pyrimidine metabolism. In *Inborn Metabolic Diseases. Diagnosis and treatment* (Fernandes J, Saudubray JM, Tada K, eds) Springer-Verlag, Berlin, pp 455-74, 1990
28. Jaeken J, Wadman SK, Duran M, van Sprang FJ, Beemer FA, Holl RA, Theunissen PM, De Cock P, Van den Bergh F, Vincent MF, Van den Berghe G. Adenylosuccinase deficiency : an inborn error of purine nucleotide synthesis. *Eur J Pediatr* **148** : 126-31, 1988
29. Van den Bergh F, Vincent MF, Jaeken J, Van den Berghe G. Radiochemical assay of adenylosuccinase : demonstration of parallel loss of activity toward both adenylosuccinate and succinylaminoimidazole carboxamide ribotide in liver of patients with the enzyme defect. *Analyt Biochem* **193** : 287-91, 1991

IMP DEHYDROGENASE AND GTP AS TARGETS IN HUMAN LEUKEMIA TREATMENT

George Weber

Laboratory for Experimental Oncology, Indiana University School of Medicine, Indianapolis, IN 46202-5200, USA

INTRODUCTION

GTP has multi-faceted cellular functions. Apart from its role in metabolism, in biosynthesis of RNA, proteins, biopterins, UTP and tubulin, GTP is an intricate part of signal transduction mechanisms, production of c-GMP and adenylates, G-protein action and expression of ras oncogene family. Guanylates are indispensable in DNA biosynthesis, since from GDP dGDP is formed and then dGTP, which is rate-limiting as it is the smallest pool among the dNTPs[1,2] (Fig. 1). Curtailing GTP and dGTP pools is an important chemotherapeutic objective. GTP de novo biosynthesis is governed by IMP dehydrogenase (EC 1.1.1.205), the rate-limiting enzyme[1,2]. GTP pools are influenced by the activity of GPRT (guanine-hypoxanthine phosphoribosyltransferase, EC 2.4.2.8), the salvage enzyme, which can recycle guanine to GMP in one step. The significance of GTP in cancer biochemistry and chemotherapy was highlighted by the discovery that IMP dehydrogenase activity increased in a transformation- and progression-linked fashion in rat hepatomas of different growth rates[1,2]. IMP dehydrogenase activity increased in all murine and 4 human cancer cell lines and was particularly high in rapidly proliferating neoplastic cells such as leukemic cells[1-3].

By application of the molecular correlation concept we discovered the relationship of the behavior of key enzymic activities with growth rate and malignancy of tumors and suggested that the increased activity of IMP dehydrogenase confers selective advantages to cancer cells[1,3]. Therefore, this enzyme should be a target of chemotherapy[3]. The approaches of enzyme-pattern-targeted chemotherapy[1] suggested that for successful therapy activities of both a key enzyme of de novo pathway (IMP dehydrogenase) and of salvage pathway (GPRT) should be inhibited.

The molecular correlation concept and enzyme-pattern-targeted chemotherapy are the biochemical basis of a new treatment for leukemia[1,4]. The experimental results are straightforward. They reach from the discovery in cancer cells (1975) of increased activity of IMP dehydrogenase[2] and the suggestion that it should be a sensitive target to chemotherapy[3] to a clinical trial of the inhibitor, tiazofurin, in end-stage leukemic patients (1987)[4-6]. R. K. Robins produced a number of compounds for inhibition of this enzyme, among them tiazofurin[7]. Tiazofurin is a pro-drug which must be metabolized in sensitive cells to TAD, an analog of NAD, (Fig. 2)[8] which inhibits IMP dehydrogenase at 0.1 μM in extracts of human leukemic cells[9].

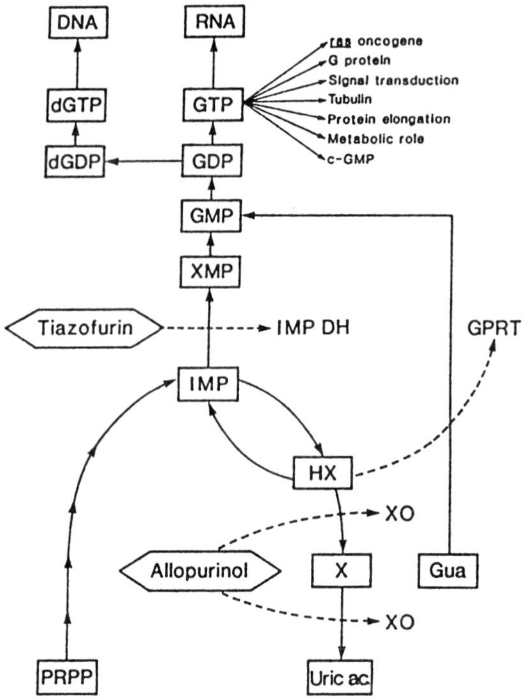

Fig. 1 Impact of tiazofurin and allopurinol in purine metabolism. Interrupted arrows point to enzymic targets.

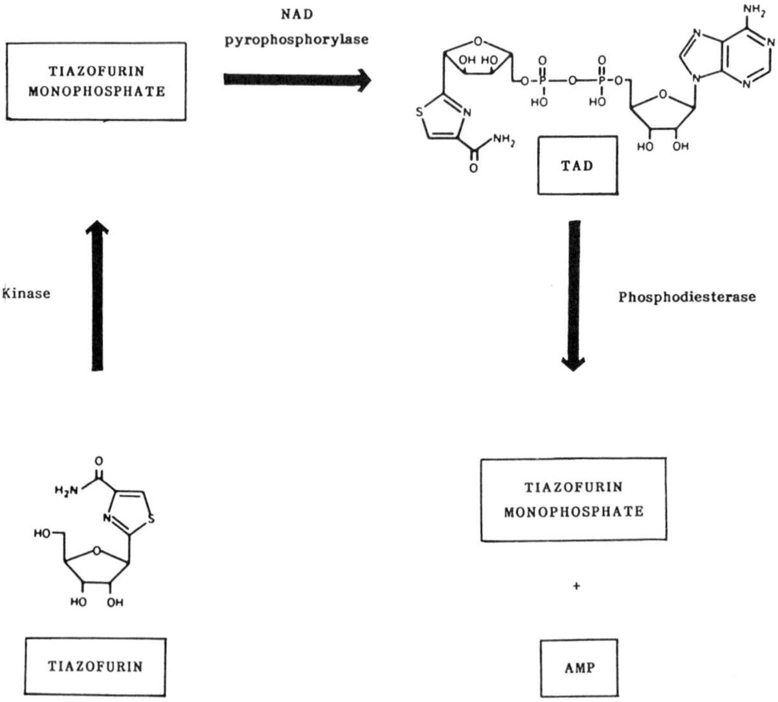

Fig. 2 Biochemistry of TAD formation and degradation.

TABLE 1. HUMAN MYELOCYTIC LEUKEMIA SHOULD BE A SENSITIVE TARGET FOR TIAZOFURIN

1. IMP DH activity : 5- to 20-fold increased
2. IMP DH activity : TAD has higher affinity (0.1 μM) to enzyme than NADH (150.0 μM)
3. IMP DH activity : strongly inhibited by TAD concentrations available in blast cells of patients
4. TAD accumulation : 20- to 30-fold higher than in normal leukocytes
5. GPRT activity : 3- to 6-fold increased
6. Incubation with tiazofurin : decreased GTP concentration
7. Tiazofurin inhibited murine leukemias (P-388, L-1210)

RESULTS AND DISCUSSION

We determined in human leukemia peripheral or bone marrow cells incubated with labeled tiazofurin that leukemic cells made 20- to 30-fold more TAD than normal leukocytes[10]. This should provide selective chemotherapy and spare normal bone marrow cells. The reasons for considering human myelocytic leukemia in blast crisis a sensitive target for tiazofurin are summarized in Table 1.

However, in human leukemic cells the activity of the guanine salvage enzyme GPRT is 100-fold higher than that of IMP dehydrogenase[11]. Therefore, enzyme-pattern-targeted chemotherapy required that GPRT activity also be inhibited. For this purpose we used allopurinol which inhibits xanthine oxidase activity, decreases uric acid formation and elevates plasma hypoxanthine levels (Fig. 1). Hypoxanthine in concentrations higher than 60 μM inhibits over 90% of GPRT and guanine salvage activities[11] (Table 2).

In tiazofurin treatment protocol the drug is given in a 1-h infusion with a starting dose of 2200 mg/m^2 per day. Allopurinol is given in 100

TABLE 2. ALLOPURINOL MODULATES TIAZOFURIN ACTION IN LEUKEMIC PATIENTS

1. <u>Tiazofurin</u>: anti-leukemic by decreasing IMP DH and GTP in leukemic cells
2. <u>Allopurinol</u>: modulates tiazofurin impact by increasing plasma hypoxanthine which inhibits GPRT and salvage
3. <u>Plasma hypoxanthine</u> (> 60 μM) is required for 90% inhibition of GPRT in leukemic cells
   Low hypoxanthine (< 20 μM): poor response
   In a <u>highly sensitive</u> patient: high hypoxanthine (> 60 μM) up-modulated tiazofurin impact, resulting in tumor-lysis syndrome
4. Increase in allopurinol dose permitted lowering tiazofurin dose
5. <u>Modulation of hypoxanthine level</u> in combination with tiazofurin: 83% response in CGL in BC

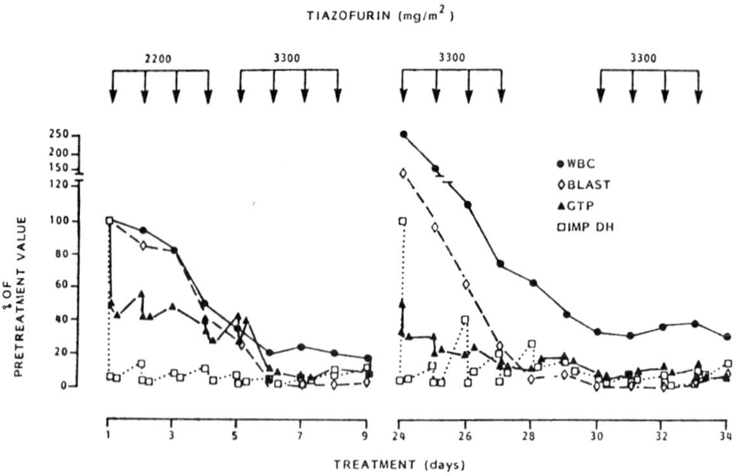

Figure 3 Changes in hematological and biochemical parameters during two consecutive courses of tiazofurin treatment in patient 4. The values of WBC, blast cells, GTP pools and IMPDH activity are expressed as a percentage of pretreatment values. The pretreatment value of WBC was 6.6 X 10^9/liter; blast cells, 4.0 X 10^9/liter; GTP pools, 344.3 nmol/10^9 leukemic cells; and IMPDH activity, 12.2 nmol/h/mg protein.

mg pills 6 to 8 times a day for a total of 600 to 800 mg. If IMP dehydrogenase activity is not decreased by 90% and GTP concentration by 80% in 3 days, tiazofurin dose is escalated to 3300 mg/m²/day[6,7]. We determined that tiazofurin dose need not be escalated when the patient relapses; treatment can be restarted at the previously effective dose level. Toxicity can be readily treated by standard medical procedures[7]. The protocol calls for 15 daily infusions, then the patient returns every 3 weeks for further treatment with 10 infusions each time. In responding patients IMP dehydrogenase activity rapidly decreases ($t_{\frac{1}{2}}$ = 30 min), GTP concentration declines ($t_{\frac{1}{2}}$ = 2-3 days), blast cells clear from the periphery and bone marrow is normo-cellular with induced differentiation of leukemic cells[6,7]. Typical therapeutic response and retreatment pattern are shown in Figure 3.

TABLE 3. PATIENTS' CHARACTERISTICS, TOTAL DOSE, NUMBER OF DAYS OF INITIAL TIAZOFURIN TREATMENT, OUTCOME AND DURATION OF RESPONSE IN PATIENTS WITH CHRONIC GRANULOCYTIC LEUKEMIA IN BLAST CRISIS

Patient	Sex/Age	Disease	Total dose of tiazofurin (mg/m²)	No. of days	Outcome	Response duration (mos.)	Total survival (mos.)
Responders							
1	M/27	CGL	41,800	11	Complete response	3	4
2	F/20	CGL	22,000	8	Complete response	6	6
3	M/44	CGL	24,200	11	Complete response	7	13
4	M/56	CGL	46,200	6	Complete response	6	13
5	M/46	CGL	4,400	2	Complete response	5	6
6	M/37	CGL	33,300	15	Complete response	3	4+
7	F/58	CGL	20,900	7	Hematologic improvement	10	16
8	F/65	CGL	46,200	15	Hematologic improvement	1	3
9	F/55	CGL	39,600	13	Hematologic improvement	1	3+
10	M/22	CGL	39,600	13	Antileukemia effect	<1	3
Non-responders							
11	M/28	CGL	42,900	15	No response		2+
12	M/30	CGL	33,000	15	No response		4+
13	M/49	CGL	19,800	9	No response		2+
Unevaluable for hematological effect							
14	M/27	CGL	2,200	1	Unevaluable		1
15	M/51	CGL	6,600	3	Unevaluable		<1

TABLE 4. TIAZOFURIN DOWN-REGULATES ONCOGENE EXPRESSION

Cell	Origin	Oncogene	Assay	Ref.
K-562	Human CML	c-*myc*	Dot-blot	13
		c-Ki-*ras*	Dot-blot	
HL-60	Human APL	c-*myc*	Northern-blot	14
		C-Ki-*ras*	Northern-blot	
Hepatoma 3924A	Rat hepatoma 3924A	c-*myc*	Dot-blot	15
		c-Ha-*ras*	Dot-blot	
Patient No. 27	CML in BC	c-Ki-*ras*	Northern-blot	12
		c-*myc*	Northern-blot	

APL = acute promyelocytic leukemia; CML = chronic myelogenous leukemia; BC = blast crisis.

In 30 consecutive patients there was a 55% response rate in 24 evaluable patients[4]. However, in a subgroup of 15 patients in the series with chronic granulocytic leukemia in blast crisis 83% of 13 evaluable patients responded (Table 3).

As expected, sensitivity to this drug varies, depending on pharmacodynamic, enzymic and metabolic circumstances in the patient and his leukemic cells. In a highly sensitive patient with two tiazofurin infusions tumor-lysis syndrome occurred and the patient returned to chronic phase within 5 days[12]. A single tiazofurin infusion (2200 mg/m^2) on days 1 and 2 decreased IMP dehydrogenase activity ($t\frac{1}{2}$ = 30 min), GTP concentration ($t\frac{1}{2}$ = 6 h) and expression of *ras* ($t\frac{1}{2}$ = 8 h) and c-*myc* ($t\frac{1}{2}$ = 38.5 h) oncogenes in leukemic cells. No further tiazofurin was given because on days 3 and 4 chemotherapeutic impact became evident in a tumor-lysis syndrome and blast cells were cleared from the periphery by day five. The decreases in IMP dehydrogenase activity, GTP concentration, and expression of c-Ki-*ras* oncogene were early markers of the successful chemotherapeutic impact of tiazofurin in a patient with chronic granulocytic leukemia in blast crisis[12]. The down-regulation of oncogene expression is in line with our earlier work in K-562, HL-60 and hepatoma cells[13,14] (Table 4).

Tiazofurin treatment entails 1. chemotherapy, 2. induced differentiation, 3. down-regulation of blast crisis oncogene.

TABLE 5. NOVEL TREATMENT OF LEUKEMIA: 6 PARADIGMS

1. BIOCHEMICALLY-TARGETED AND MONITORED DRUG PROTOCOL
2. ALLOPURINOL: BIOLOGICAL RESPONSE MODIFIER
3. DOSE ESCALATIONS: NOT NEEDED AT EACH RELAPSE
4. TARGETS: PRIMARY: IMPDH
   SECONDARY: GTP
   TERTIARY: *ras*, *myc*
5. 1-H INFUSION LOWERS TOXICITY VS 10-MIN BOLUS OR CONTINUOUS I.V.
6. SELECTIVITY: CONVERSION OF PRO-DRUG (TIAZOFURIN) TO ACTIVE METABOLITE (TAD) IN LEUKEMIC CELLS

This biochemically targeted and biochemically monitored therapy provides 6 paradigms in cancer chemotherapy (Table 5).

Current evidence suggests that on the basis of IMP dehydrogenase activity, inhibition of the enzyme by TAD and cytotoxicity studies in tissue culture and in animals, among the human solid tumors ovarian and colon carcinomas might be candidates for treatment with tiazofurin.

ACKNOWLEDGMENTS

Supported by USPHS Outstanding Investigator Grant CA-42510 to G.W.

REFERENCES

1. G. Weber, Biochemical strategy of cancer cells and the design of chemotherapy: G.H.A. Clowes Memorial Lecture, Cancer Res. 43:3466 (1983).
2. R. C. Jackson, H. P. Morris and G. Weber, IMP dehydrogenase: A proliferation and malignancy-linked enzyme, Nature 256:331 (1975).
3. G. Weber, N. Prajda and R. C. Jackson, Key enzymes of IMP metabolism: Transformation- and proliferation-linked alterations in gene expression, Advan. Enzyme Regul. 14:3 (1976).
4. G. Weber, M. Nagai, Y. Natsumeda, H. Nakamura, J. N. Eble, H. N. Jayaram, W. Zhen, E. Paulik, R. Hoffman and G. Tricot, Regulation of de novo and salvage pathways, Advan. Enzyme Regul. 31:45 (1991).
5. G. J. Tricot, H. N. Jayaram, E. Lapis, Y. Natsumeda, Y. Yamada, C. R. Nichols, P. Kneebone, N. Heerema, G. Weber and R. Hoffman, Biochemically directed therapy of leukemia with tiazofurin, a selective blocker of IMP dehydrogenase activity, Cancer Res. 49:3696 (1989).
6. G. Tricot, H. N. Jayaram, G. Weber and R. Hoffman, Tiazofurin: Biological effects and clinical uses, Intl. J. Cell Cloning 8:161 (1990).
7. R. K. Robins, Nucleoside and nucleotide inhibitors of inosine monophosphate (IMP) dehydrogenase as potential antitumor inhibitors, Nucleosides and Nucleotides, 1:35 (1982).
8. H. N. Jayaram, Biochemical mechanisms of resistance to tiazofurin, Advan. Enzyme Regul. 24:67 (1986).
9. Y. Yamada, Y. Natsumeda, Y. Yamaji and G. Weber, Kinetic properties and TAD inhibition of IMP dehydrogenase in leukemic leukocytes, Leukemia Res. 13:179 (1989).
10. H. N. Jayaram, K. Pillwein, C. R. Nichols, R. Hoffman and G. Weber, Selective sensitivity to tiazofurin of human leukemic cells, Biochem. Pharmacol. 35:2029 (1986).
11. G. Weber, H. N. Jayaram, E. Lapis, Y. Natsumeda, Y. Yamada, Y. Yamaji, G. J. Tricot and R. Hoffman, Enzyme-pattern-targeted chemotherapy in human leukemia, Advan. Enzyme Regul. 27:405 (1988).
12. G. Weber, M. Nagai, Y. Natsumeda, J. N. Eble, H. N. Jayaram, E. Paulik, W. Zhen, R. Hoffman and G. Tricot, Tiazofurin down-regulates expression of c-Ki-ras oncogene in a leukemic patient, Cancer Commun. 3:51 (1991).
13. E. Olah, Y. Natsumeda, T. Ikegami, Z. Kote, M. Horanyi, J. Szelenyi, E. Paulik, T. Kremmer, S. R. Hollan, J. Sugar and G. Weber, Induction of erythroid differentiation and modulation of gene expression by tiazofurin in K-562 leukemia cells, Proc. Natl. Acad. Sci. U.S.A. 85:6533 (1988).
14. M. Nagai, Y. Natsumeda and G. Weber, Tiazofurin action in HL-60 cells, Submitted for publication (1991).
15. E. Olah, Z. Kote, Y. Natsumeda, Y. Yamaji, G. Jarai, E. Lapis, I. Financsek and G. Weber, Down-regulation of c-myc and c-Ha-ras gene expression by tiazofurin in rat hepatoma cells, Cancer Biochem. Biophys. 11:107 (1990).

GUANINE RIBONUCLEOTIDE DEPLETION INHIBITS T CELL ACTIVATION

Jennifer S. Dayton,* Laurence A. Turka,§ Craig B. Thompson,† Beverly S. Mitchell*§

Departments of Pharmacology,* Medicine,§
Microbiology/Immunology,† University of Michigan, and the Howard Hughes Medical Institute, Ann Arbor, Michigan 48109

INTRODUCTION

Mizoribine is an immunosuppressive drug (4-carbamoyl-1-ß-D-ribofuranosylimidazolium-5-olate, also known as bredinin) which has been demonstrated to be beneficial in organ transplantation. In animal models: canine renal allografts, and rat cardiac and partial lung allografts, mizoribine is synergistic with cyclosporine in prolonging survival[1,2]. Mizoribine has been proven to be clinically useful for human renal transplantation as part of two or three drug therapy[3]. The results are comparable or superior to azathioprine; however mizoribine does not cause myelosuppresion or hepatotoxicity. Mizoribine is a nucleoside antibiotic isolated from a soil fungus, *Eupenicillum brefeldianum*. Mizoribine is phosphorylated to mizoribine monophosphate by adenosine kinase and the monophosphate is postulated to inhibit inosine monophosphate dehydrogenase (EC1.2.1.14)[4,5]. The predicted depletion of guanine ribonucleotides may block T cell proliferation by preventing signal transduction or act at some more distal point in the activation pathway.

MATERIALS AND METHODS

Reagents. Mizoribine (mol wt 259) was a gift of the Toyo Jozo Company, Tokyo, Japan. Ionomycin was obtained from CalbiochemBehring Corp. (San Diego, CA). The ionomycin was suspended in dimethylsulfoxide at 1mg/ml and stored at -20°C. PMA, guanosine, and 8-aminoguanosinewere purchased from Sigma Chemical Co. (St. Louis, MO). The PMA was reconstituted as a 1 mg/ml solution in dimethylsulfoxide. Fresh working dilutions of PMA (at 5 mg/ml in PBS) were prepared as needed for each experiment.

Isolation and characterization of peripheral blood T lymphocytes. Peripheral blood mononuclear cells were isolated from venous blood by density gradient centrifugation using Ficoll-Hypaque. T cells were purified from the mononuclear cells by negative selection using a cocktail of monoclonal antibodies, as previously described[6].

Cell Culture. Cells were cultured at a density of 1 x 10^6/ml in RPMI-1640 containing 10^5 U/liter penicillin, 100 mg/liter streptomycin, and 10% fetal calf serum. PMA was used at a final concentration of 3 ng/ml, ionomycin at 125 ng/ml, guanosine at 50 mM, 8-aminoguanosine at 100mM. The drugs were added simultaneously with the stimulating agents unless indicated otherwise. When used, anti-CD3 antibody was immobilized on plastic culture dishes using a 1mg/ml solution[7].

Purine and Pyrimidine Metabolism in Man VII, Part B
Edited by R.A. Harkness *et al.*, Plenum Press, New York, 1991

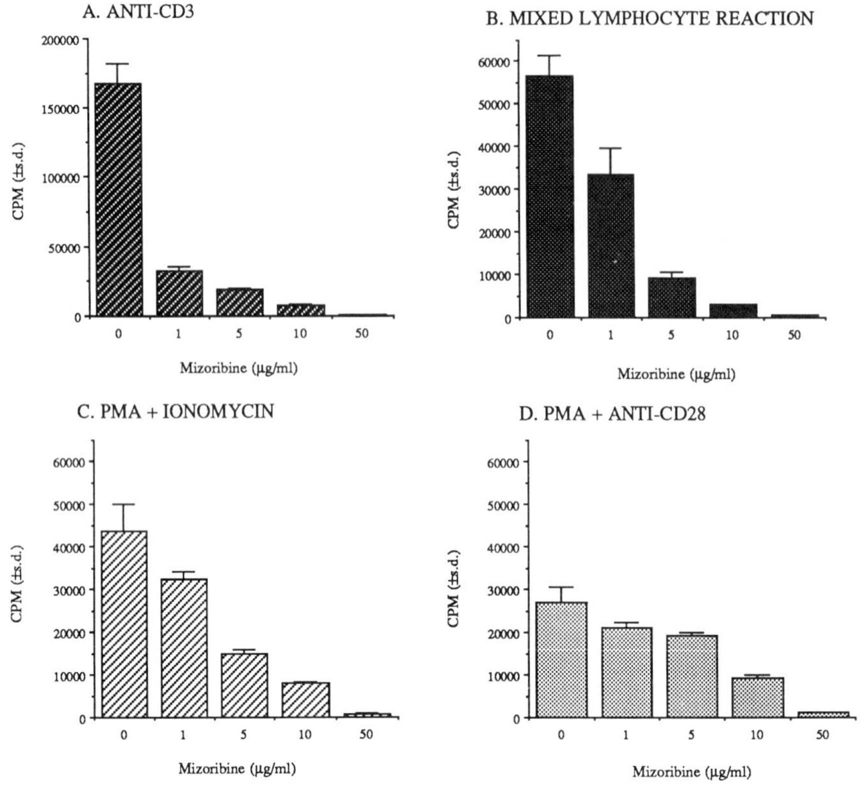

Figure 1. Effect of mizoribine on mitogen or alloantigen-dependent T cell proliferation

Measurement of ATP and GTP. Cellular nucleotide pools were quantitated from perchloric acid extracts by HPLC using a Partisil-10 SAX anion exchange column (Whatman, Clifton, NJ) as previously described[8]. Ribonucleoside triphosphates were separated using a linear gradient from 80% buffer A (0.002 M $NH_4H_2PO_4$, pH 2.8) to 90% buffer B (0.75 M $NH_4H_2PO_4$, pH 3.9) over 30 minutes at a flow rate of 2 ml/min. Total intracellular nucleotide levels were calculated by comparing areas of peaks absorbing at 254 nm to the areas generated by nmol amounts of pure standards using a 3390A Hewlett Packard integrator.

RESULTS AND DISCUSSION

Effect of mizoribine upon T lymphocytes stimulated by mitogen or alloantigen. Figure 1 shows the effect of mizoribine upon T cells induced to proliferate by A) immobilized antibodies to CD3, B) the mixed lymphocyte reaction, C) PMA and ionomycin, and D) PMA and monoclonal antibodies to CD28, an alternate pathway for T cell activation. In each case mizoribine causes a dose-dependent decrease in DNA synthesis at concentrations achieved in patients receiving the drug (peak-5mg/ml, trough-1mg/ml when receiving 1-3mg/kg/day)[3].

Effect of mizoribine upon GTP levels in T lymphocytes. The predicted consequence of inhibition of IMP dehydrogenase would be a depletion of intracellular guanine ribonucleotides. We measured GTP pools in order to determine whether mizoribine caused depletion of guanine ribonucletides in purified T cells. Mizoribine causes a dose-dependent depletion of GTP in purified T cells stimulated to proliferate by PMA and ionomycin Figure 2, open squares. In order to determine whether the inhibition of proliferation was causally related to the depletion of GTP we added guanosine to cultures treated with increasing concentrations of mizoribine. Guanosine can be converted to guanine by PNP and the guanine to GMP by HGPRT to provide an alternate source of guanine ribonucleotides independent of de novo purine biosynthesis. Figure 2 solid squares demonstrates that exogenous guanosine effectively blocks the GTP depletion caused by mizoribine.

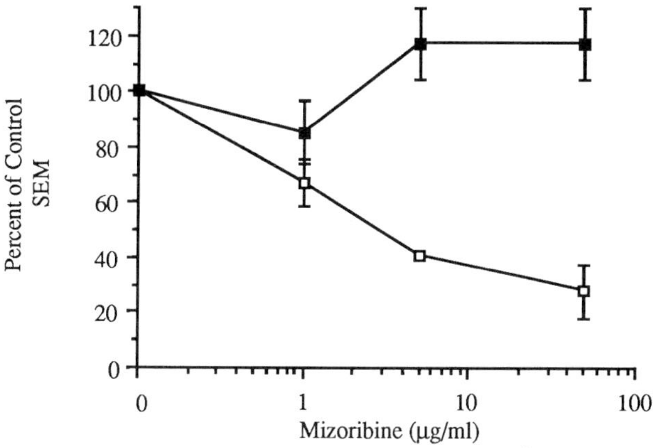

Figure 2. Effect of mizoribine on GTP levels (Control-open squares, guanosine-solid squares).

Effect of exogenous guanosine upon T cell proliferation in the presence of mizoribine. Since guanosine prevented the depletion of GTP we hypothesized that it would also reverse the inhibition of proliferation. The effect of guanosine upon T cell proliferation induced by and anti-CD3 in Figure 3. The addition of guanosine, solid squares, reverses the inhibition of DNA synthesis up to approximately 5mg/ml mizoribine.

Effect of mizoribine upon gene expression essential for T cell proliferation. Having demonstrated that mizoribine affects cells in S phase, we examined the effects of mizoribine upon T cell function prior to S phase. We performed a series of Northern blots to examine the effect of mizoribine on gene expression in T cells. In data not shown, c-myc expression was markedly increased by PMA + Ionomycin at 6 hours and declined at 24 hours, but, the presence of mizoribine did not affect c-myc expression. Similarly, IL2 steady-state RNA levels increased at 6 hours and were further elevated by 24 hour. The presence of 10mg/ml of mizoribine did not prevent IL2 expression. Thus, mizoribine does not appear to interfere with the initial processes of T cell activation.

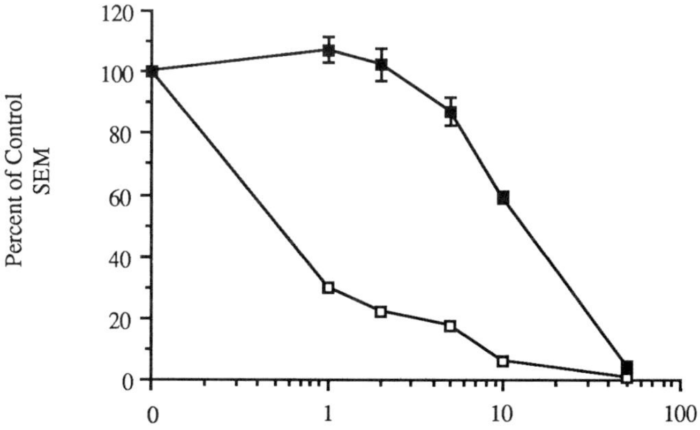

Figure 3. Effect of guanosine on mizoribine-induced inhibition of T cell proliferation mediated by CD3 (Control-open squares, guanosine-solid squares).

Mizoribine blocks the movement of cells from G1 to S phase of the cell cycle. Cell cycle analysis of DNA content was performed to determine whether T cell activation was inhibited by guanine nucleotide depletion prior to or following commitment to S phase and DNA synthesis. Table 1 shows data obtained from DNA histograms of mononuclear cells stimulated by immobilized monoclonal antibodies to CD3 for 48 hours. The presence of 2mg/ml mizoribine caused a decrease in the number of cells in S, and G2/M phases relative to control. The addition of guanosine reversed this effect. Therefore, mizoribine appears to have its effect following all events related to signal transduction and to act specifically at the G1/S interface.

Table 1. Effect of Mizoribine on Cell Cycle Progression

Condition	Go/G1	S+G2+M
A. anti-CD3	67	33
B. anti-CD3 + Guanosine	55	45
C. anti-CD3 + 2mg/ml MZ	79	21
D. anti-CD3 + 2mg/ml MZ+ Guanosine	64	36

In conclusion, mizoribine inhibits T lymphocyte proliferation by causing depletion of intracellular guanine ribonucleotides. Depletion of guanine nucleotides does not prevent early activation steps, such as IL-2 and IL-2 receptor expression, but interferes with a GTP-dependent event at the transition from G1 to S phase.

REFERENCES

1. H. Amemiya, S. Suzuki, S. Niiya, H. Watanabe, and T. Kotake, Synergistic effect of cyclosporine and mizoribine on survival of dog renal allografts, Transplantation. 46:768-771 (1988).
2. S. Suzuki, T. Hijioka, I. Sakakibara, and H. Amemiya, The synergistic effect of cyclosporine and mizoribine on heterotropic heart and partial-lung transplantation in rats, Transplantation. 43:743-744 (1988).
3. S. Takahara, T. Fukunishi, Y. Kokado, Y. Ichikawa, M Ishibashi, S. Nagano, and T. Sonoda, Combined immunosuppression with low-dose cyclosporine, mizoribine, and prednisolone, Transplant. Proc. 20:147-151 (1988).
4. K. Mizuno, M. Tsuijno, M. Takada, M. Hayashi, K. Atsumi, K. Asano, and T. Matsuda, Studies of Bredinin, Isolation, characterization and biological properties, J. Antibiot. (Tokyo) 27:775-782 (1974).
5. H. Koyama, and M. Tsuji, Genetic and biochemical studies on the activation and cytotoxic mechanism of bredinin, a potent inhibitor of purine biosynthesis in mammalian cells, Biochem. Pharm. 32:3527-3553 (1983).
6. L.A. Turka, J.A. Ledbetter, K. Lee, C.H. June, and C.B. Thompson, CD28 is an inducible T cell surface antigen that transduces a proliferative signal in CD3+ mature thymocytes. J. Immunol. 144:1646-1653 (19).
7. J.A. Ledbetter, C.H. June, L.S. Grosmaire, and P.S. Rabinovitch, Crosslinking of surface antigens causes mobilization of intracellular ionized calcium in T lymphocytes, Proc. Natl. Acad. Sci. USA. 84:1384 (1987).
8. Y. Sidi, J.L. Hudson, B.S. Mitchell, Effects of Guanine Ribonucleotide Accummulation on the Metabolism and Cell Cycle of Human Lymphoid Cells, Cancer Research, 45:4940-4945 (1985).

RAISED IMP-DEHYDROGENASE ACTIVITY IN THE ERYTHROCYTES OF A CASE OF PURINE NUCLEOSIDE PHOSPHORYLASE (PNP) DEFICIENCY

G Morgan (1), S Strobel (1), C Montero (2), JA Duley (2), PM Davies (2), HA Simmonds (2)

Host Defence Unit, Institute of Child Health, London (1).
Purine Research Laboratory, Guy's Hospital (2)

INTRODUCTION

The presenting features of PNP deficiency may be neurological, usually a mild non-progressive spastic diplegia, or related to immunodeficiency which predominantly affects T-lymphocytes. Patients usually die of infection or other manifestations of the immunodeficiency including lymphoma and transfusional graft versus host disease. The disorder may be correctable by bone marrow transplant(1). The lymphotoxicity is probably directly related to the inability to degrade deoxyguanosine with accumulation of dGTP in T-cells. It has been proposed that the neurological deficits may be an indirect consequence of the defect:in the absence of PNP the next step in the purine salvage cycle involving hypoxanthine-guanine phosphoribosyl-transferase (HPRT) is unable to operate due to lack of substrate(2). Low erythrocyte GTP levels have been found in PNP deficient subjects, a finding which may reflect a similar inability of the brain (also heavily dependent on a functional salvage cycle) to sustain GTP at levels compatible with normal physiological function(2).This paper reports studies in a PNP deficient individual with spastic diplegia where elevated GTP levels were identified.

PATIENTS + METHODS

Case History: A six year old male child of consanguinous parents was noted to have spastic diplegia at the age of 18 months which was non-progressive. He also suffered from recurrent respiratory infections and reversible airways obstruction before presenting with adenovirus Type 4 pneumonia requiring ventilation. He was treated with intravenous ribavirin 300mg tds for 2 days and then 150mg tds for a further 8 days, which was discontinued 42 days prior to purine enzyme and metabolite studies. In addition he received a blood transfusion 8 weeks prior to study and treatment at the time of blood sampling included corticosteroids, ceftazidime, theophylline and salbutamol. Biochemical investigations: Purine and pyrimidine concentrations in body fluids and enzyme activity in lysed erythrocytes were analysed by methods described previously(2,3). IMP dehydrogenase was measured using a specially developed assay adapted to HPLC (in preparation). Methods employed for the intact cell studies using radio-labelled substrates have also been reported(3). Immune function: T-lymphocyte subpopulations were identified by direct immunofluorescent staining and analysis by flow cytometry. Detailed analysis of the sub-populations of the total T-lymphocytes detected by monoclonal antibody to CD3 antigen was performed by double staining for helper/inducers (CD4), suppressor/cytoxics (CD8) and for evidence of activation (CD25, CD71, DR).

In addition CD4 and CD8 cells were studied for the expression of the CD45RO antigen which is known to divide CD4 cells into a mature/memory/activated subpopulation (CD4+CD45RO+) and a naive/unactivated subpopulation (CD4+CD45 RO-). Functional assays were performed by stimulation of lymphocytes by mitogens and third party lymphocytes in mixed lymphocyte culture (MLC).

RESULTS

Immune function: Selected results of T-lymphocyte investigations are shown in Table 1. There was absolute lymphopenia (and neutropenia) as well as reduced percentages of T-lymphocytes. Of particular interest was the complete absence of the subset of CD4+ve cells which are thought to constitute the pool of naive (ie unprimed, unactivated) CD4 cells. The corresponding subset of CD8+ cells was reduced but present. Levels of CD25, 71, and DR expression of T-lymphocytes were uniformly increased indicating a high level of activation of the residual lymphocyte population. Biochemical investigations: Very low PNP and uric acid levels (attributed to the blood transfusion), plus high concentrations of the metabolites characteristic of PNP deficiency (inosine, guanosine, deoxyinosine and deoxyguanosine) were detected in plasma and urine (Table 2). An unusual finding was the grossly-elevated erythrocyte GTP concentration (152umol/l). Even more remarkable was the conversion by intact erythrocytes of 42% of [14C] hypoxanthine into GTP (normally undetectable), confirming GTP formation via IMP in vitro. Significantly, IMP-dehydrogenase activity, normally extremely low in the human erythrocyte, was increased thirty-fold (Table 2).

DISCUSSION

A tenable hypothesis for the appearance of symptoms attributable to abnormalitites during the second to fourth year of life in PNP deficient individuals, is that of a gradual lymphocytoxic effect of guanosine metabolites (dGTP). The neurological abnormalities are, however, less easy to explain on this basis as they appear to be non-progressive. We hypothesised that a functional HPRT deficiency may be the mechanism by which neurological damage may occur(2) - though failure of progression may imply that CNS toxicity may only occur at a single, undefined stage in ontogeny. The findings of the absence of the pool of naive CD4+ lymphocytes may indicate an increased sensitivity of these cells to dGTP accumulation. At least part of the immunodeficiency may be similarly caused at a specific stage in T-lymphocyte ontogeny. The gradual decline of memory/mature T-lymphocytes due to GTP accumulation would be exacerbated by the failure to recruit replacement cells from the naive/unactivated cell pool. The other immunological finding of considerable note is the high level of activation of the residual T-lymphocytes. This may have been secondary to the recurrent infections which had been a notable feature of the clinical course in preceding months. We excluded the possibility of transfusional GVHD by demonstrating the absence of 3rd party cells in the circulation, using locus specific mini satellite probes. The absence of CD4+45RO- and low levels of CD8+CD45RO- cells might be explicable on the basis that all T-lymphocytes have lost this phenotype on activation. This would provide an alternative to the hypothesis that naive (particularly CD4) mature cells have a differential sensitivity to the toxicity of dGTP. However, we have never observed the complete absence of CD4+45RO- cells in any other condition, including disorders in which high levels of T-lymphocyte activation are present.

The biochemical workup in this PNP deficient individual revealed several unusual features. Erythrocyte GTP levels were extremely high, in complete contrast to the 4 other cases we have studied(2). The incorporation of radiolabel from hypoxanthine into GTP was equally remarkable, since previous in vitro studies in PNP deficient erythrocytes had demonstrated the formation of inosine with IMP being the only nucleotide formed from radiolabelled hypoxantine(3). This indicated a futile cycle, interrupted in PNP deficiency.

TABLE 1 - IMMUNOLOGICAL INVESTIGATIONS

Total Lymphocytes 0.96 x 10^9/l

LYMPHOCYTES + SUBPOPULATIONS

T-Lymphocytes (CD3)	25% (60-85)
Helper-Inducer (CD4)	15% (30-60)
Naive (CD4+CD45RO-)	0 (15-30)
Memory (CD4+CD45RO+)	16% (15-30)

PROLIFERATION RESPONSES

PHA	absent
MLC	absent

ACTIVATION MARKERS

CD25	48 (<10%)*	12% (<5)**
CD4+25+	47 (<10%)	7% (<5)
CD3+DR+	48 (<10%)	12% (<5)
CD71	52 (<10%)	13% (<5)

* = % activation of subpopulation
** = % of circulating cells

TABLE 2 - BIOCHEMICAL INVESTIGATIONS

PARAMETER	ERYTHROCYTES	
PNP	580	(3000-7000 nmol/mg Hb/h)
IMPDH	2670	(4 - 183 pmol/mg Hb/h)
GTP	205	(66 +/- 9 umol/l)
GDP	33	(17 +/- 3 ")
NAD	143	(69 +/- 15 ")
NADP	69	(54 +/- 12 ")

	PLASMA (umol/l)	URINE (mmol/l)
Uric Acid	50	1.4
Inosine	16	6.4
Guanosine	1	4.8
Deoxyinosine	6	5.2
Deoxyguanosine	1	3.0

The PNP contributed by the transfused erythrocytes coupled with the raised endogenous PP-rib-P levels characteristic of the defect, could have contributed to the elevated GTP levels by enhancing the salvage of guanine. The unexpected incorporation of hypoxanthine into GTP, however could not be explained on this basis. This is probably related to a 30 fold increase in activity of IMP-dehydrogenase, the enzyme responsible for the first step in the conversion of IMP to GMP. Activity of this enzyme is extremely low in lysed erythrocytes and essentially non-functional in intact red cells. Consequently no radiolabel from hypoxanthine has ever been detected in erythrocyte GTP. IMPDH is an enzyme associated with proliferation and malignancy(4) and shows a considerable increment in nucleated cells in human neoplasia, including leukaemias, but RBC activity is not raised. Erythrocyte IMPDH levels do show a modest increment (6 fold) in HPRT deficiency where stabilisation by elevated NAD levels has been implicated(5). However, we found no increment in IMPDH in other PNP deficient RBCs, despite elevated NAD levels of comparable magnitude. Ribavirin, a guanosine analogue and IMPDH inhibitor and it is thus possible that activity had been induced in early erythrocyte progenitor by ribavirin therapy 7 weeks previously. Alternatively the demonstration of high levels of T-lymphocyte activation combined with functional HPRT deficiency may provide an explanation for this finding, with RBC IMPDH induction possibly occuring as a result of secreted lymphokines. The increased GTP levels are not at variance with our hypothesis of functional HPRT deficiency if it is extended to encompass the possibility that the CNS insult occurs at an undefined stage in ontogeny. Further studies are necessary to establish the reason for the elevated IMPDH in the RBCs of this case.

References

1. Stone TW, Simmonds HA, Purines: Basic + Clinical Aspects. Kluwer, London 1991.

2. Simmonds HA, Fairbanks LD, Morris GS, Morgan G, Watson AR, Timms P, Pugh B. CNS dysfunction and erythrocyte GTP depletion in purine nucleoside phosphorylase deficiency. Archives of Disease in Childhood 1987, 62, 586-589.

3. Fairbanks LD, Webster DR, Simmonds HA, Potter CF, Watson AR. 1984 Inosine formation from hypoxanthine by intact erythrocytes and fibroblasts of an immunodeficient child with purine nucleoside phosphorylase deficiency. Adv Exp Med Biol 165B: 167-171.

4. Weber G. Metabolic strategies in cancer chemotherapy. Biochem Soc Transactions 1990; 18: 74-78.

5. Pehlke DM, McDonald JA, Holmes EW, Kelley WN. Inosinic acid dehydrogenase activity in the Lesch Nyhan syndrome. J Clin Invest 1972; 51: 1398-1404.

GUANINE RIBONUCLEOTIDE METABOLISM AND THE REGULATION OF MYELOPOIESIS

Daniel G. Wright, Vincent F. LaRussa, Robert D. Knight, Jana M. Bednarek, and Mary A. Cutting

Department of Hematology, Walter Reed Army Institute of Research Washington, DC, USA

Introduction

Previously reported studies of human leukemia cell lines by ourselves and others have established that there is a relationship between the maintenance of intracellular guanine ribonucleotide pools and the regulation of myeloid progenitor cell maturation. In particular, we have shown that induced maturation of human myeloid leukemia cells is associated with a down-regulation of guanylate sysnthesis from IMP at the initial rate limiting step mediated by IMP-dehydrogenase, and with a consequent depletion of guanine ribonucleotides (1-3). Furthermore, we and others have shown that inhibitors of IMP-dehydrogenase (IMPD), such as mycophenolic acid, tiazofurin, or ribavirin, are potent inducers of the maturation of these cells (1,2,4,5), with specific activities for maturation induction that are directly related to those for IMPD inhibition (6). On the other hand, addition of exogenous guanosine or guanine, which may be salvaged for guanylate synthesis by-passing the pathway from IMP, has been shown to prevent the induced maturation of leukemia cells in culture to the extent that high intracellular guanine ribonucleotide concentrations are maintained (2,5).

In studies described here, we have analyzed guanine ribonucleotide metabolism in normal human marrow cell isolates variously enriched for very primitive or mature myeloid cells, and we have examined the effects of pharmacologic depletion and repletion of guanylates on the proliferation and terminal maturation of normal myeloid, clonogenic progenitor cells in vitro. Results of these studies support the conclusion that guanine ribonucleotide metabolism and the activity of IMPD in hematopoietic progenitors is an important metabolic determinant for the rgulation of both normal and abnormal myelopoiesis in man.

Methods

Bone marrow aspirate specimens were obtained from healthy normal volunteers and collected into syringes containing preservative-free heparin (100 U/mL). Marrow specimens were diluted in sterile Dulbecco's PBS, and light-density (d<1.070) nucleated marrow cells were separated by Percoll (Pharmacia) density gradient centrifugation (7). CD34+ cells were then separated from these marrow cell isolates using immunomagnetic beads linked to the anti-CD34 monoclonal antibody K6.1 (8). These CD34+ marrow cell isolates had the morphology of primitive blasts and were highly enriched with clonogenic progenitor cells (>20% of the cells formed hematopoietic colonies in methylcellulose cultures). Relatively dense (d>1.070) marrow cells were also isolated, and following dextran sedimentation and hypotonic lysis to remove erythroid cells, these cell isolates could be shown to be highly enriched for mature myeloid cells and myeloid precursors (91% myelocytes, metamyelocytes, bands, and segmented neutrophils) and to be devoid of clonogenic progenitors.

Measurements of purine ribonucleotide concentrations and the activity of biosynthetic pathways for ribonucleotide synthesis via salvage of radiolabelled hypoxanthine in normal marrow cells isolates were done using methods reported previously (1,2). Cultures of clonogenic myeloid progenitors with recombinant human GM-CSF with and without IMPD inhibitors (tiazofurin, ribavirin) were also done using methods described previously (7,9).

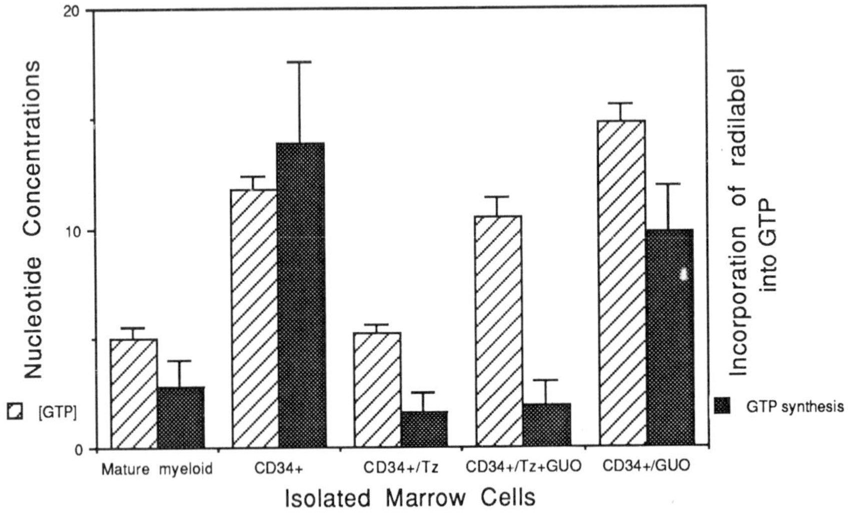

Figure 1. Intracellular GTP concentrations and guanylate synthetic activity via IMP from salvage of radiolabelled hypoxanthine in CD34+ cells vs. mature myeloid cells isolated from normal human marrow.

Results

As is illustrated in Figure 1, normal human CD34+ marrow cells, highly enriched with primitive hematopoietic progenitors, were found to have 3-5 fold higher intracellular concentrations of GTP and levels of guanylate synthetic activity than did marrow cells that were enriched for mature myeloid elements. Levels of IMPD acitivity in extracts of CD34+ cells were also 3-5 fold higher than in isolates of mature myeloid cells from the marrow (not shown). Incubation of CD34+ cells with tiazofurin (10^{-6}M x 2 hrs) caused a substantial inhibition of guanylate synthesis (as measured by the incorporation of radiolabel from ^{14}C hypoxanthine into GTP and GDP) and of IMPD activity in cell extracts, as well as a depletion of [GTP] in the cells. Addition of 5×10^{-4}M guanosine to tiazofurin-treated cell cultures, however, reversed the depletion of [GTP], although it did not reverse the inhibition of IMPD or guanylate systhesis via IMP. Of interest, the addition of guanosine alone to cultures of CD34+ cells was found to increase intracellular [GTP] levels above those observed under control conditions.

Figure 2 illustrates the effects of pharmacologic depletion and repletion of [GTP] on the growth of clonogenic neutrophilic progenitors from CD34+ marrow cell isolates with recombinant human GM-CSF across a broad range of concentrations of this myelopoietic growth factor. As shown in panel A, tiazofurin (10^{-6}M) inhibited the clonal growth of these progenitors, while the addition of guanosine (5×10^{-4}M) reversed this effect. Moreover, the addition of guanosine by itself to cultures enhanced the growth of neutrophilic progenitors, recruiting more progenitors to undergo clonal expansion at all conecntrations of rhGM-CSF than were detected under control conditions. Moreover, when the clonal growth of progenitors was normalized for all culture conditions (e.g. expressed as percent of the maximum growth detected; panel B), it was apparent that tiazofurin shifted the dose response curve for rhGM-CSF-stimulated progenitor cell growth to the right, indicating a drug induced loss of responsiveness to GM-CSF, while guanosine reversed this effect and by itself increased the sensitivity of the CD34+ cells to the growth factor, shifting the GM-CSF dose response curve to the left.

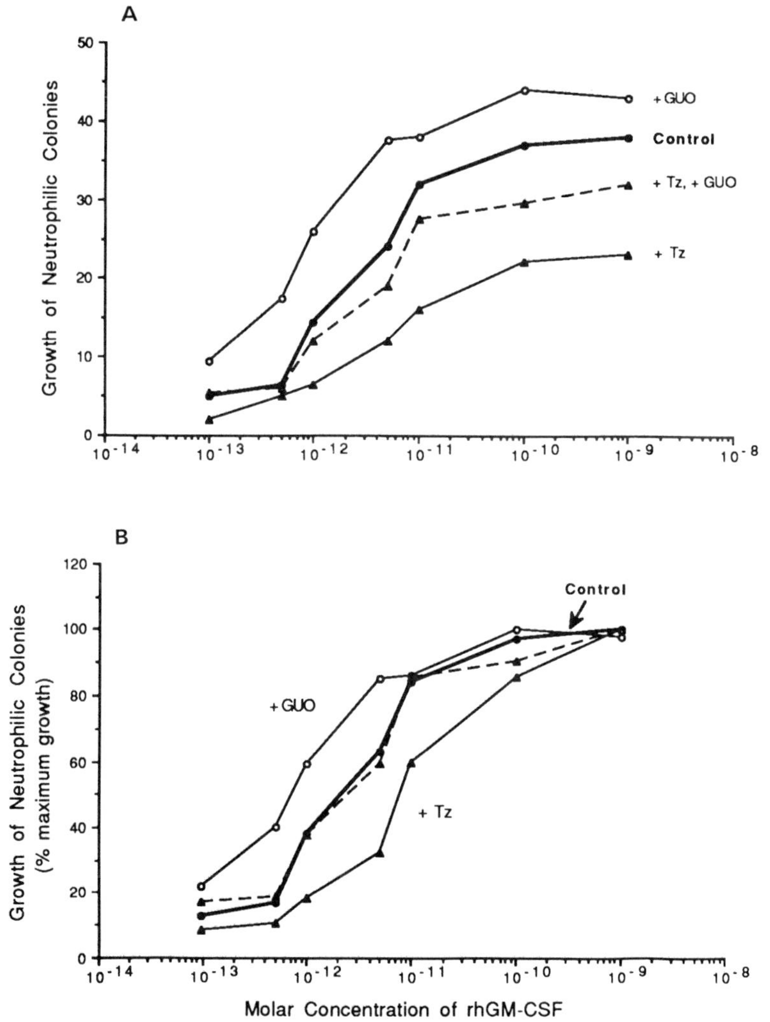

Figure 2. Clonal growth of neutrophilic colonies from CD34+ marrow cell isolates in methylcellulose cultures in response to rhGM-CSF. A. Numbers of colonies at day 14 of culture per 1000 CD34+ cells plated. B. Numbers of colonies expressed as percent of maximum growth.

In related studies, tiazofurin was also found to reduce the cellularity of neutrophilic colonies but to increase the proportion of mature neutrophils in the colonies (Figure 3), while the addition of guanosine again reversed these effects, and guanosine by itself promoted the growth of densely cellular colonies that were enriched with morphologically immature cells (Figure 3).

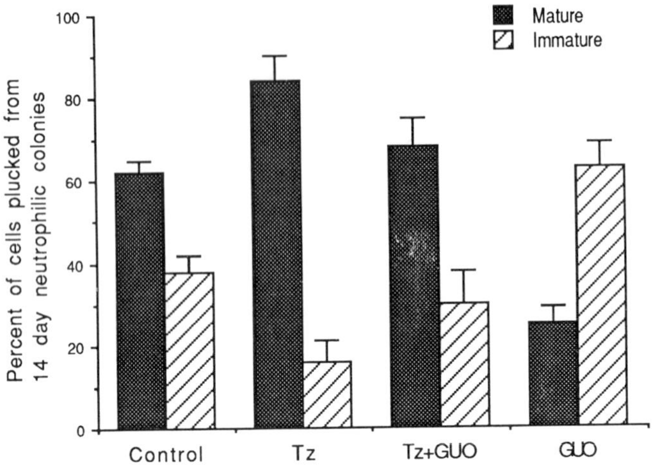

Figure 3. Morphology of cells plucked from 14 day neutrophilic colonies grown in methylcellulose cultures of CD34+ marrow cells with 100 pM rhGM-CSF, with or without tiazofurin (10^{-6}M), and with or without guanosine (5×10^{-4}M).

Conclusions

Results of our studies indicate that guanine ribonucleotide metabolism and the maintenance of intracellular [GTP] have a role in the regulation of normal myelopoiesis. Sepcifically, our studies show that the maturation of normal myeloid progenitors is associated with a down-regulation of IMP-dehydrogenase and with depletion of intracellular GTP pools. They also show that pharmacologic inhibition of IMPD promotes the terminal maturation of myeloid progenitors at the expense of progenitor cell proliferation and clonal expansion, while maintenance of high intracellular guanine ribonucleotide concentrations promotes progenitor cell proliferation and inhibits the terminal maturation of these cells.

References

1. D.L.Lucas, H.K.Webster, and D.G.Wright, J Clin Invest 72:1889 (1983).
2. D.G.Wright, Blood 69:334 (1987).
3. R.D.Knight, J.Mangum, D.L.Lucas, D.A.Cooney, E.C.Khan, and D.G.Wright, Blood 69:634 (1987).
4. J.A.Sokoloski, O.C.Blair, and A.C.Sartorelli, Cancer Res 46:2314 (1986).
5. J.Yu, V.Lemas, T.Page, J.D.Connor, and A.L.Yu, Cancer Res 49:5555 (1989).
6. D.L.Lucas, R.K.Robins, R.D.Knight, and D.G.Wright, Biochem Biophys Res Comm 115:971 (1983).
7. R.C.Meagher, A.J.Salvado, and D.G.Wright, Blood 72:273 (1988).
8. S.W.Kessler, D.Vembu, and A.T.Black, Blood 70(Suppl 1):321 (1987).
9. D.G.Wright, V.F.LaRussa, A.J.Salvado, and R.D.Knight, J Clin Invest 83:1414 (1989).

PYRIMIDINE PATHWAYS: NEWS CONCERNING THE MECHANISM OF OROTIDINE-5'-MONOPHOSPHATE DECARBOXYLASE

Mary Ellen Jones

Department of Biochemistry and Biophysics
University of North Carolina
Chapel Hill, NC 27599-7260

I will present the recently established mechanism (1,2) for the reaction catalyzed by orotidylate decarboxylases (ODCases) and detail the characteristics of one altered ODCase that we have prepared using biotechnology. The modified protein shows the dramatic loss of activity one would expect to see when a site-specific mutation is made to change an amino acid residue that is essential for catalysis.

ODCase is the sixth and last enzyme of the *de novo* pyrimidine biosynthetic pathway (3). This protein is monofunctional in bacteria and fungi but in the animal family it is part of a bifunctional protein, UMP synthase (4,5), that also contains orotate phosphoribosyltransferase (OPRTase). The ODCase domain is approximately 29,000 daltons while OPRTase domain is about 23,000 daltons and UMP synthase is therefore nearly 52,000 daltons.

For the present study we have utilized one of two proteins to study the ODCase reaction mechanism namely a pure and stable yeast ODCase or the ODCase domain of mouse UMP synthase (6,7). Crystals of the yeast protein are being studied by Dr. Charles W. Carter, Jr. (8). We are now working to prepare human UMP synthase in amounts large enough to permit crystallization.

Two mechanisms have been suggested from chemical studies of the decarboxylation of orotic acid analogs by Beak and Siegel (9) and by Silverman and Groziak (10). Our new studies show that ODCases use the zwitterion mechanism proposed by

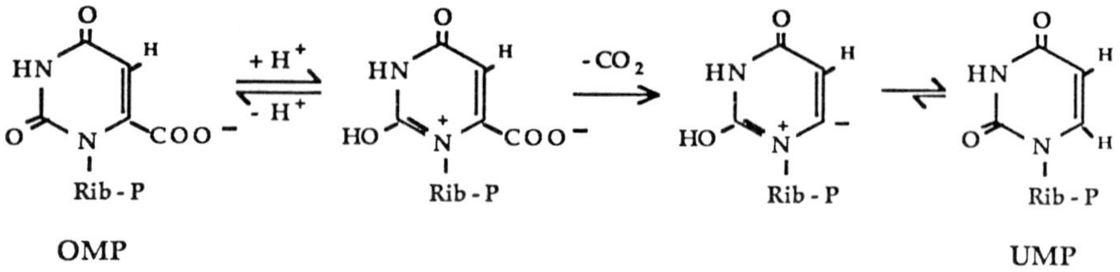

Scheme I. Proposed mechanism for decarboxylation of OMP by ODCase via non-covalent, zwitterionic intermediate.

Beak and Siegel (Scheme I). For most decarboxylations an electron sink is needed to stabilize the negative charge left on the C atom adjacent to the carboxylate-C of the substrate when the neutral product, CO_2, is released. Often the positively charged electron sink is a metal ion or the positively charged pyridinium-N of pyridoxal phosphate. In the case of ODCase it is N1 of the pyrimidine ring of OMP. A positive charge is generated on N1 of the orotate ring of orotidylic acid (OMP) when the enzyme protonates either the 2- or 4-keto group of OMP. On decarboxylation the negative charge remaining at C6 is momentarily stabilized by the positively-charged N1; the resulting structure is a ylid. The atomic stoichiometry of the ylid and UMP are identical. All that is required to go from ylid to UMP is loss of a proton at C2 and the addition of this or another proton to C6; one presumes the enzyme aids in this proton interchange. However, the crucial proton addition is the initial step where the proton is added to the 2-keto group to form the zwitterion.

A histidine residue would seem ideal for this role but there is no histidine residue that is invariant in the 17 known ODCase structures. There is however a highly conserved sequence of about 20 amino acids in ODCases that has 6 identical amino acids (F, D, K, D, I and T). The lysine (K93 in yeast ODCase or K97 in the mouse ODCase domain) of this sequence seemed like a possible candidate to donate the essential proton to the 2-keto group of the orotate ring to form the transition-state zwitterion (Scheme I). Enzymes increase the rate of chemical reactions by a factor between

10^{10} to 10^{20}; therefore, if an amino acid essential in the chemistry of the catalysis is altered, a factor much greater than a 10^2 reduction would be expected in the rate of the catalysis.

To test whether yeast K93 or mouse K97 is the proton donor for ODCase we have made analog (or mutant) proteins with the aid of biotechnology such that codons for other amino acids replaced the normal lysine codon at K93 (yeast) or K97 (mouse) in the respective cDNA of vectors pGU2 (6) or $pODC_{tac}$ (7). Since there is 90% identity between the amino acid sequences of the human and mouse ODCase domain (5) and about 53% identity between the amino acid sequence of the yeast protein (7) and either the human or mouse proteins, the yeast ODCase and the mouse ODCase domain are good representatives for the human protein.

Using the mouse ODCase domain the wild type lysine (K97) codon was changed to a codon for either arginine, histidine, glutamine, asparagine, cysteine, serine, aspartate, tyrosine, threonine, glycine, alanine, valine, isoleucine, leucine, methionine, or a stop codon in 16 altered proteins (11). All of these proteins were produced in mutant *E. coli* cells that lack ODCase using the appropriately modified $pODC_{tac}$. All altered cDNA's were sequenced in order to select cells containing a plasmid with an altered cDNA (12). The proteins with sequences that produced the amino acid substitutions listed above were all inactive. One mutant also had the second lysine codon. This plasmid produced ODCase in the *E. coli* homogenate that had the same specific activity as is produced by the normal (wild-type) cDNA. This "silent" mutation served as a nice control to indicate that our techniques were good.

Although the mouse ODCase protein is greatly enriched using the E. coli expression system it is rather hard to obtain pure mouse ODCase domain using this system. Pure protein is necessary in order to quantitatively compare the activity of the kinetic characteristics of the altered protein with the normal protein. We decided therefore to select one of the inactive amino acid codons and produce an altered yeast ODCase using the yeast pGU2 plasmid. A cysteine residue was selected since Hartman and his coworkers (13) have used a cysteine to replace lysine. The cysteine can be chemically modified with Br-ethylamine to yield a lysine analog with a thioether replacing one - CH_2-group.

The inactive yeast K93C protein was obtained as a 90% pure protein. However, by lengthening the enzyme assay, by putting in a maximal amount of protein and by using ^{14}C-OMP of the highest possible specific activity, we could estimate that, if the altered protein has activity, it is < 1.5 x 10^{-6} nmol/min/mg protein; the wild-type protein produces 75 nmol CO_2/min/mg protein. Thus the activity of the altered protein is reduced by at least 5 x 10^7. This large loss of activity is what one would expect if the side chain of an amino acid residue that is essential for catalysis has been modified.

In changing a cysteine residue to the thioether analog there are several limiting conditions. The first of these is that there must not be an essential cysteine residue in the wild-type protein. This is the case for yeast ODCase since the greatest loss of activity we have observed using Br-ethylamine after urea denaturation followed by renaturation with wild-type protein was 36%. Treatment of the wild-type protein either with Br-ethylamine or with urea followed by renaturation did not cause a significant loss of ODCase activity (3 and 17% respectively). These two treatments also did not reactivate the K93C mutant. However when the 93C-protein was treated with Br-ethylamine after urea denaturation and subsequently renatured, the activity of this protein increased to 1% of the activity of wild-type protein given the same treatment. This is a 10^5 recovery of activity at a minimum and demonstrates that this lysine residue is essential for enzyme activity. We believe the K93 residue protonates the orotate ring of orotidylic acid to yield the zwitterion.

An experiment was done to establish that the three dimensional structure of the K93C protein had not been seriously changed. Since the protein is inactive, we cannot test activity to measure OMP binding. However we can test the ability of nucleotides to bind since dilute yeast ODCase is a monomer and does not dimerize unless a ligand that binds to the ODCase active site is present (6). The rapid equilibrium between the monomer and dimer (relative to the separation method used) leads to a single protein peak sedimenting as a 45 kilodalton protein, ie a mixture of monomer and dimer, when 900 μM UMP or when 900 mM OMP was added to wild-type, K93, protein. With the K93 protein the OMP was all converted to UMP in fractions of the solution through which ODCase had passed. The K93C protein does not hydrolyze OMP so one would expect the amount of OMP needed to yield 100% dimer to be 10^2 X K_m for OMP (14) or about 70 μM. Fifty μM OMP changed better than two-thirds the K93C protein to dimer.

We belive K93C is an altered yeast protein that binds substrate and product normally. K93C can be made active again by the converting the 93C residue to a lysine analog with Br-ethylamine. The sum of the data suggest that K93 may be the amino acid sidechain that donates a proton to a keto group of the orotate ring of OMP to form the zwitterion essential for the reaction.

REFERENCES

1. Acheson, S. A., Bell, J. B., Jones, M. E. and Wolfenden, R., 1990, Orotidine 5'-monophosphate decarboxylase catalysis: kinetic isotope effects and the state of hybridization of a bound transition state analog, Biochemistry, 29:3198.
2. Smiley, J. A., Paneth, P., O'Leary, M. H., Bell, J. B., and Jones, M. E., 1991, Investigation of the enzymatic mechanism of yeast orotidine 5'-monophate decarboxylase using ^{13}C kinetic isotope effects, Biochemistry, 30:6216.
3. Jones, M. E., 1980, Pyrimidine nucleotide biosynthesis in animals: genes, enzymes and regulation of UMP biosynthesis, Annu Rev. Biochem., 49:253.
4. McClard, R. W., Black, M. J., Livingstone, L. R. and Jones, M. E., 1980, Isolation and initial characterization of the single polypeptide that synthesizes uridine 5'-monophosphate from orotate in Ehrlich ascites carcinoma, Biochemistry, 19:4699.
5. Suttle, D. P., Bugg, B. Y., Winkler, J. K., and Kanalas, J. J., 1988, Cloning and nucleotide sequence for the complete coding region of human UMP synthase, Proc. Natl. Acad. Sci. U.S.A., 85:1754.
6. Bell, J. B. and Jones, M. E., 1991, Purification and characterization of yeast orotidine 5'-monophosphate decarboxylase overexpressed from plasmid pGU2, J. Biol. Chem., 266 (in press).

7. Ohmstede, C.-A., Langdon, S. D., Chae, C. B., and Jones, M. E., 1986, Expression and sequence analysis of a cDNA encoding the orotidine 5'-monophosphate decarboxylase domain from Ehrlich ascites uridylate synthase, J. Biol. Chem., 261:4276.

8. Bell, J. B., Jones, M. E. and Carter, C. W. Jr., 1991, Crystallization of yeast orotidine 5'-monophosphate decarboxylase complexed with 1-(5'-phospho-β-D-ribofuranosyl) barbituric acid, Proteins: structure, function and genetics, 9:143.

9. Beak, P. and Siegel, B., 1976, Mechanism of decarboxylation of 1,3-dimethylorotic acid, J. Am. Chem. Soc., 98:3601.

10. Silverman, R. B. and Groziak, M. P., 1982, Model chemistry for a covalent mechanism of action of orotidine 5'-phosphate decarboxylase, J. Am. Chem. Soc., 104:6434.

11. Smiley, J. A., 1991, Aspects of the catalytic mechanism of orotidylate decarboxylase, Ph.D. Thesis for Univ. North Carolina at Chapel Hill, N. C.

12. Hutchison, III, C. A., Nordeen, S. K., Vogt, K., and Edgell, M. H., 1986, A complete library of point-substitution mutations in the glucocorticoid response element of mouse mammary tumor virus, Proc. Natl. Acad. Sci. U.S.A., 83:710.

13. Smith, H. B., Latimer, F. W., and Hartman, F. C., 1988, Subtle alterations of the active site of ribulose bisphosphate carboxylase/oxygenase by concerted site-directed mutagenesis and chemical modification, Biochem. Biophys. Res. Commu., 152:579.

DIHYDROPYRIMIDINE DEHYDROGENASE DEFICIENCY IN A HUTTERITE NEWBORN

K.J. Adolph*, E. Fung*, D.R. McLeod@, K. Morgan• and F.F. Snyder*@#

@ Department of Paediatrics, # and Medical Biochemistry University of Calgary, Calgary, Alberta, Canada. * Biochemical Genetics Laboratory, Alberta Children's Hospital, Calgary Alberta, Canada • Departments of Epidemiology and Biostatistics and Medicine, McGill University, Montreal, Quebec, Canada

INTRODUCTION

We have previously described three inherited biochemical disorders in the Hutterite Brethren of Western Canada. These include hypophosphatasia, mucopolysaccharidosis IVA (Morquio Syndrome)[1] and methylmalonic acidemia[2]. A urine specimen from the patient described in this paper came to the Biochemical Genetics Laboratory for a routine metabolic screen. A voluntary screening program for methylmalonic acidemia is in place for the Hutterite Brethren of Alberta and other infants at high risk.

Dihydropyrimidine dehydrogenase catalyzes the reduction of the C5,6 double bond in the pyrimidine bases uracil and thymine in a NADPH dependent reaction. This enzyme catalyzes the first reaction in the degradation of uracil to beta-alanine and thymine to beta-aminoisobutyric acid, the other products being carbon dioxide and ammonia. No clear picture has yet emerged regarding clinical findings in this disorder[3,4,5]. Although the clinical presentation may be unrelated to the enzyme deficiency, four cases have had late onset, atypical seizures. Patients have also presented with 5-fluorouracil toxicity[6,7].

MATERIALS AND METHODS

Dihydropyrimidine dehydrogenase activity was assayed in 50 mM phosphate, pH 7.4, 400 μM NADPH, 100 μM [^{14}C]-uracil (55 mCi/mmol) or thymine (55 mCi/mmol), and 0.4 - 1.2 mg protein/ml lymphocyte lysate for 30 min at 37°C. Substrate and product were separated on silica gel thin layers and radioactivity in substrate and products counted.

Urine samples were derivatized and examined by gas chromatography-mass spectrometry as previously described[8]. Pyrimidine bases were quantitated by reverse phase HPLC[9] and ribonucleotides by anion exchange HPLC[10] essentially as described. Interleukin-2 dependent T lymphocytes were cultured from the proband and his parent's peripheral blood[11] for enzyme assay and nucleotide analysis.

Cousin relationships and kinship and inbreeding coefficients were computed using Pedpack[12], a package of programs for pedigree analysis which was kindly provided by Alun Thomas and Elizabeth Thompson. The inbreeding coefficient of an individual is equal to the kinship coefficient of his parents.

RESULTS AND DISCUSSION

A male infant was born at 27 weeks gestation after a pregnancy complicated by the premature rupture of membranes. The child was discharged with mild retinopathy of prematurity and this has subsequently resolved at 8 months. As part of a metabolic screen at 1 month, organic acids were examined primarily to assess methylmalonic acid levels. Gas chromatography-mass spectrometry revealed the presence of significant amounts of two compounds not normally present in urine having mass spectra identical to thymine and uracil. Reverse phase HPLC quantitation of pyrimidine bases in the patient's urine on two occasions showed marked excretion of both uracil, 630 and 1300 μM, and thymine, 350 and 780 μM. Control urines had less than 5 μM uracil or thymine. Plasma analysis also revealed the presence of uracil, 60 μM, and thymine, 45 μM, compared to less than 1 μM in control plasma. These observations suggested the likelihood of a defect in pyrimidine base catabolism.

Dihydropyrimidine dehydrogenase activity was examined by radiochemical assay in T-lymphocyte cultures from the patient. There was no detectable activity even upon prolonged assay with either base substrate, <0.01 nmole/hr/mg protein (U), as compared to controls, 7.5 ± 2.5 and 5.9 ± 2.1 U for uracil and thymine respectively. The activity was significantly reduced in the mother's T-lymphocytes, 1.2 and 1.1 U, for uracil and thymine respectively, consistent with carrier status for the enzyme deficiency. The activity in the father's T-lymphocytes was at the lower limit of the control range, 5.2 and 3.7 U, for uracil and thymine, respectively. The results are consistent with an autosomal recessive or X-linked recessive disorder.

A deficiency of dihydropyrimidine dehydrogenase might result in the expansion of pyrimidine nucleotide pools and a deficit in products of the degradative pathway. The impact of dihydropyrimidine dehydrogenase deficiency on cellular nucleotide pools was examined. A decrease in the degradative pathway for uracil could conceivably cause an increase in the uridine and possibly cytidine ribonucleotides. Nucleotides were examined in cultured T-lymphocytes by anion exchange HPLC and no differences were evident for NAD, uridine, cytidine, adenosine or guanosine nucleotides between control and patient. The UXP/AXP ratio was 0.23 and 0.32 for two separate harvests of the patient's cells as compared to 0.24 ± 0.06 for controls. Thus we have not observed any change in pyrimidine ribonucleotide pools. Examination of the concentration of the terminal products of this pathway would also be of interest. The possibility of administering beta-alanine or a precursor distal to the enzyme block may be considered as rational metabolite therapy in this disorder.

The Hutterite Brethren divided into three Leuts following their migration to North America in the 1870's and are a genetically isolated population. In Alberta there are only Dariusleut and Lehrerleut and since 1918 there have been few instances of marriage between these Leuts. We maintain a computerized genealogical database of the Hutterite Brethren with information dating back to the 1700's. The inbreeding coefficient of the patient was 0.0615 which is close to the inbreeding coefficient of an offspring of first-cousins. The nearest relationship of the parents was first cousins once-removed (a kinship coefficient of 0.03125); the additional kinship was due to more distant relationships. For example, the parents were full and half-cousins in 310 and 29 different ways, respectively, for relationships up to seventh-cousins once-removed. We estimate that the average kinship coefficient of Lehrerleut couples was about 0.047[13,14]. Therefore, the proband has a relatively large inbreeding coefficient. The kinship coefficient of the patient's parents was above the 75th percentile for 500 Lehrerleut couples who had a child during 1971-81[14]. The Hutterite population would seem to be an ideal population in which to test the hypothesis that dihydropyrimidine dehydrogenase deficiency is a benign enzyme deficiency by implementation of a broader screening program. Notice has also been given to oncologists for awareness of enhanced pharmacogenetic susceptibility to 5-fluorouracil toxicity among the Hutterite Brethren.

ACKNOWLEDGEMENTS

This work has supported by the Alberta Hereditary Disease program and the Medical Research Council of Canada (MT-11009) and the Canadian Genetic Diseases Network (Network for Centres of Excellence).

REFERENCES

1. Lowry RB, Snyder FF, Wesenberg, RL, Machin GA, Applegarth DA, Morgan K, Carter RJ, Toone JR, Holmes TM and Dewar RD. Morquio Syndrome (MPS IVA) and hypophosphatasia in a Hutterite kindred. Am J Hum Genet 22:463-475 (1985).
2. Fowlow SB, Holmes TM, Morgan K and Snyder FF. Screening for methylmalonic aciduria in Alberta: A voluntary program with particular significance for the Hutterite Brethren. Am J Med Genet 22:513-519 (1985).
3. Bakkeren JAJM, De Abreu RA, Sengers RCA, Gabreels FJM, Maas JM, Renier WO. Elevated urine, blood and cerebrospinal fluid levels of uracil and thymine in a child with dihydrothymine dehydrogenase deficiency. Clinica Chim Acta 140:247-256 (1984).
4. Berger R, Stoker-de Vries SA, Wadman SK, Duran M, Beemer FA, de Bree PK, Weits-Binnerts JJ, Penders TJ and van der Woude JK. Dihydropyrimidine dehydrogenase deficiency leading to thymine-uraciluria. An inborn error of pyrimidine metabolism. Clinica Chim Acta 141:227-234 (1984).
5. Suttle DP, Becroft DMO, Webster DR. Hereditary orotic aciduria and other disorders of pyrimidine metabolism. In "The Metabolic Basis of Inherited Disease", Scriver CR et al. McGraw Hill 6th Edition, 1119-1121 (1989).
6. Tuchman M, Stoeckler JS, Kiang DT, O'Dea RF, Ramnaraine KL and Mirkin BL. Familial pyrimidinemia associated with severe fluorouracil toxicity. N Eng J Med 313:245-249 (1985).

7. Diasio RB, Beavers TL and Carpenter JT. Familial deficiency of dihydropyrimidine dehydrogenase. Biochemical basis for familial pyrimidinemia and severe 5-fluorouracil-induced toxicity. J Clin Invest 81:47-51 (1988).

8. Dias VC, Fung E, Snyder FF, Carter RJ and Parsons HG. Effects of medium chain triglyceride feeding on energy balance in adult humans. Metabolism 39:887-891 (1990).

9. Mably ER, Fung E and Snyder FF. Genetic deficiency of purine nucleoside phosphorylase in the mouse. Characterization of partially and severely enzyme deficient mutants. Genome 32:1026-1032 (1989).

10. Hodges SD, Fung E, McKay DJ, Renaux BS and Snyder FF. Increased activity, amount and altered kinetic properties of IMP dehydrogenase from mycophenolic acid resistant neuroblastoma cells. J Biol Chem 264:18137-18141 (1989).

11. Adolph KJ, Kucey MT, Hodges SD, Carter RJ and Snyder FF. Interleukin-2 dependent T lymphocytes for the diagnosis and investigation of inherited metabolic disorders. Clin Chim Acta 173:147-156 (1988).

12. Thomas A. Pedpack: User's Manual. Technical report no. 99, Department of Statistics, University of Washington, Seattle, 30pp (1987).

13. Thompson EA and Morgan K. Recursive descent probabilities for rare recessive lethals. Annals of Human Genetics 53:357-374 (1989).

14. Pearce WG, MacKay JA, Holmes TM, Morgan K, Fowlow SB, Shokeir MHK and Lowry RB. Autosomal recessive juvenile cataract in Hutterites. Ophthalmic Paediatrics and Genetics 8:119-124 (1987).

SUPERACTIVE UMP HYDROLASE: CAUSE OR CONSEQUENCE OF HAEMOLYTIC ANAEMIA?

J.A. Duley and H.A. Simmonds

Purine Research Laboratory, Clinical Science Laboratories
UMDS Guy's Hospital
London, U.K.

INTRODUCTION

Pyrimidine-5'-nucleotidase (UMP hydrolase 1: UMPH) is a degradative enzyme which is widely distributed in human tissues, including erythrocytes. A deficiency of the enzyme presents clinically as non-sperocytic haemolytic anaemia characterised by marked basophilic stippling in the erythrocyte (1). Grossly elevated levels of pyrimidine nucleotides (normally low to undetectable) are found in the erythrocytes. Conversely, UMPH has been reported to be raised two- to three-fold in a kindred with an hereditary haemolytic anaemia associated with grossly elevated adenosine deaminase (ADA) and low ATP (2).

Similarly, raised activity of enzymes of the de novo pyrimidine synthetic pathway have been reported (3) in patients suffering from congenital hypoplastic anaemia - Diamond-Blackfan Syndrome (DBS) - where ADA was also found to be raised (4). However, the proportion of DBS patients exhibiting raised erythrocyte ADA has been shown to be as low as a third, and thus not a useful diagnostic parameter for the syndrome (5). In contrast, adenine phosphoribosyltransferase (APRT), has been reported to be higher in young red cells (6) and frequently raised in renal failure (7), but was found to be normal in the above anaemias.

We investigated erythrocyte UMPH activity in 5 patients referred with haemolytic anaemia, and compared these results with 2 patients referred for possible Diamond-Blackfan syndrome, as well as anaemias resulting from vitamin/iron deficiency and secondary to renal failure. We also measured UMPH in erythrocyte fractions separated by density gradient centrifugation. The results indicate that elevation of this enzyme may be associated with specific forms of anaemia and it is proposed that this association may compromise red cell stability.

METHODS

Patients - Heparinised blood was collected from 7 patients, 2 of whom were DBS patients and 5 were suffering from haemolytic anaemia. Blood from 8 patients with transient anaemias resulting from vitamin or iron deficiencies, and from 5 with anaemia secondary to renal failure was also obtained as part of their normal clinical work-up.

Enzyme assays - Packed red blood cells were washed twice in saline, with careful removal of the top fifth layer containing reticulocytes, white blood cells and platelets. UMPH was assayed using undialysed erythrocyte lysates by the method of Hopkinson et al (8), with UMP as the substrate. The reaction was stopped by the addition of trichloroacetic acid, and the supernatant back-extracted into water-saturated ether following centrifugation. A high performance liquid chromatography (HPLC) method was developed to quantify the assay products. 10ul of the extract was injected into a 125x4.5mm ODS-2 column eluted isocratically at 1ml/min with 40mM ammonium acetate buffer containing 5mM tetrabutyl ammmonium acetate, pH2.8. Erythrocyte nucleotides, APRT and ADA were assayed as described (9). Measurement of activity in erythrocytes separated by density centrifugation using discontinuous Percoll gradients was also performed (6). Erythrocyte enzyme activity in the different patient groups was compared using non-parametric assessment of significance.

RESULTS

Under the HPLC conditions used, the uridine product eluted at a retention time of 2.5min, clearly separated from UMP at 4.0min (Figure 1). Analyses showed that during incubation of undialysed lysate with substrate, in addition to uridine, an increase in UTP, AMP and ADP occurred, concomitant with a decrease in the ATP present. UMPH activity (Table 1) in all 5 patients was not low, but rather was elevated, thereby excluding UMPH deficiency. The pattern of ADA activity was not consistent, being raised in 3 of the cases, and similarly APRT was also high in 3 of the 5 (not shown). UMPH activity was high in one two Diamond-Blackfan patients, but normal in the second. ADA was normal in both patients, while APRT was raised in the second.

Table 1. UMPH activity in patients with haemolytic anaemia or Diamond-Blackfan Syndrome.

Patients	UMPH activity nmol/h/mg Hb	
Haemolytic anaemia:		
LH.	31	
TW.	40	
PD.	72	
BN.	25	
ZK.	26	(median= 31)*
Diamond-Blackfan:		
M.M.	10	
O.F.	44	
Controls:		
(n=13)	6-20	(median= 12)
Anaemia:		
(n=8)	9-21	(median= 14)
Renal failure:		
(n= 5)	11-28	(median= 16)

*significantly different from control group

The UMPH activity in patients with transient anaemias was not significantly different from that of the controls, and although UMPH in patients with anaemia secondary to renal failure was slightly raised, this was not significant. The APRT and ADA activities in the patient groups were grossly raised in some cases, but no consistent pattern was noted (not shown).

Comparison of UMPH and APRT activities in erythrocytes of controls following fractionation by density gradient centrifugation showed that while APRT activity was higher in less dense (younger) cells, there was no apparent change in UMPH between the different fractions.

The erythrocyte nucleotide patterns of the 5 haemolytic anaemia patients were found to be unremarkable in 2 of these patients, whereas the erythrocytes of one subject exhibited a greatly decreased ATP concentration, compared with control bloods stored for the same period of time, and in the remaining 2 patients there were unusually high concentrations of erthrocyte CDP-choline. None of the 5 haemolytic anaemia patients showed any evidence of the grossly raised UTP and CTP characteristic of UMPH deficiency, thus supporting the lysate findings.

DISCUSSION

Study of seven patients with anaemia, five of whom were referred with haemolytic anaemia for suspected UMPH deficiency and two queried for Diamond-Blackfan Syndrome showed that rather than being deficient, UMPH was elevated up to 4-fold above the normal range. Activities of other enzymes such as APRT and ADA, while frequently raised, showed no consistent pattern. A kindred with a dominantly inherited haemolytic anaemia associated with grossly elevated ADA was reported by Valentine et al (2), where UMPH was raised 3-fold. Although elevated levels of pyrimidine-synthesising enzymes were found by Zielke et al (3) in DBS, the activity of UMPH was not measured. These authors suggested that that consequent pyrimidine deregulation may interfere with erythropoiesis. The same comment could apply to our own findings. In support of the above observations, red cell nucleotide patterns showed no characteristic changes, apart from two patients with elevated CDP-choline levels. These resembled a case reported by Paglia et al (10) and appear to have a quite different metabolic defect that will be the subject of a separate report.

In this study we were unable to find any correlation between induction of UMPH and anaemia <u>per se</u> or the age of the red cell population or the presence of coexistent renal disease - factors frequently associated with raised APRT. Alternatively, our findings in the patients with haemolytic anaemia support that concept that the induction, or stabilisation, of UMPH may be associated with an underlying mechanism involving an <u>increased flux</u> through pyrimidine pathways. Haemolytic anaemia thus may be the consequence of abnormal pyrimidine metabolism affecting cellular stability, e.g., through an effect of aberrant CDP- and UDP- lipid and sugar metabolism on membrane lipid and glycoprotein synthesis.

REFERENCES

1. Valentine, W.N., Fink, K., Paglia, D.E., Harris, S.R., and Adams, W.S. Hereditary haemolytic anaemia with human erythrocyte pyrimidine 5'-nucleotidase deficiency. <u>J.Clin.Invest.</u> 54:866 (1974)

2. Valentine, W.N., Paglia, D.E., Tartaglia, A.P., and Gilsanz, F., Hereditary haemolytic anaemia with increased red cell adenosine deaminase and decreased adenosine triphosphate. <u>Science</u> 195:783 (1977)

3. Zielke, H.R., Ozand, P.T., Luddy, R.E., Zinkham, W.H., Schwartz, A.D., and Sevdalian, D.A., Elevation of pyrimidine enzyme activities in the RBC of patients with congenital hypoplastic anaemia and their parents. Br.J.Haematol. 42:381 (1979)

4. Glader, B.E., and Backer, K., Elevated red cell adenosine deaminase activity: a marker of disordered erythropoiesis in Diamond-Blackfan anaemia and other haematological diseases. Br.J.Haem. 68:165 (1983)

5. Whitehouse, D.B., Hopkinson, D.A., Pilz, A.J., and Arredondo, F.X., Adenosine deaminase activity in a series of 19 patients with the Diamond-Blackfan Syndrome. Adv.Exper.Med.Biol. 195A:85 (1986)

6. Micheli, V., Ricci, C., Taddeo, A., Gili, R., Centrifugal fractionation of human erythrocytes according to age: comparison between ficoll and percoll density gradients. Quad.Sclavo.Diagn. 21:236 (1985)

7. Banholzer P., Grobner, W., Loffler, W., Reiter, S., and Zollner N., Adenine phosphoribosyltransferase (APRT) activity in patients with nephrolithiasis or renal failure. Adv.Exper.Med.Biol. 165A:27 (1984)

8. Hopkinson, D.A., Swallow, D.M., Turner, V.S., and Aziz, I., Evidence for a distinct deoxypyrimidine 5'-nucleotidase in human tissues. Adv.Exper.Med.Biol. 165A:535 (1984)

9. Simmonds H.A., Duley J.A., and Davies P.M., Analysis of purines and pyrimidines in blood, urine and other physiological fluids. Chapter 25, In Techniques in Diagnostic Human Biochemical Genetics: A Laboratory Manual, Hommes F.A. (Ed); pp397-424. NY:Wiley-Liss, (1991).

10. Paglia, D.E., Valentine, W.N., Nakatani, M., and Rauth, B.J., Selective accumulation of cytosol CDP-choline as an isolated erythrocyte defect in chronic hemolysis. Proc.Natl.Acad.Sci 80:3081 (1983)

NAD SYNTHESIS IN HUMAN ERYTRHOCYTES: STUDY OF ADENYLYL TRANSFERASE ACTIVITIES IN PATIENTS BEARING PURINE ENZYME DISORDERS

Sylvia Sestini, Monica Pescaglini, Claudia Magagnoli, Gabriella Jacomelli, Marina Rocchigiani and H. Anne Simmonds

Department of Molecular Biology, University of Siena, Italy, *Purine Research Laboratory, Guy's Hospital, London.U.K.

INTRODUCTION

The role of NAD in human erythrocytes is completely devoted to its function of cofactor in glicolysis and pentose shunt, mantaining the oxidoreductive state of the cells, since they lack mono- and poly- ADP ribosylations, and then its availability is absolutely necessary for cell survival.

It is well known that human erythrocytes need the preformed pyridine rings of Nicotinamide (NAm) or Nicotinic Acid (NA) to accomplish NAD synthesis, through two different pathways, the amidated (from NAm) or deamidated one (from NA). Both involve, first, two phosphoribosyl-transferases, NAmPRT and NAPRT, synthesizing the respective monunucleotides, then the adenylyltransferases, acting on Nicotinamide Mononucleotide (NMN-AT) and on Nicotinate Mononucleotide (NAMN-AT), and leading to the synthesis of the dinucleotides, NAD and NAAD. The latter is amidated to NAD by NAD synthetase, utilizing glutamine as amine donor, while the synthesis of NADP is catalyzed by NAD kinase.

The two routes seem to proceed without any connections, no deamidating activity for NAm or NMN, having been demonstrated, despite several investigations conducted by us and by other researchers.

The existence of all the enzymes leading to NAD in human red blood cells had been supposed on the basis of experiments conducted incubating intact cells in the presence of ^{14}C precursors,[1,2] but only few of them have been identified and studied.

In particular, no information was available about NMN-AT, whose presence in an anucleated cell, like human erythrocyte, is very peculiar. In fact, adenylyltransferase has been reported to be located in the nucleus in eukariotic cells[3] and in yeast,[4] and to be associated with poly-ADPR polymerase activity;[5] such enzyme has been demonstrated in rat liver nuclei, where it correlates with DNA synthesis,[6] and in other mammalian tissues.[7] Enucleated cells of the human line D 98/AH 2 have been shown not to possess a complete amidated pathway, lacking NMN-AT.[8] Whether NMN- and NAMN- AT activities are related to an unique enzyme or not, is still an open question.

Table 1. Kinetic characteristics of NMN- and NAMN- AT in crude lysates.

	NMN-AT	NAMN-AT
$Km_{(NMN) or (NAMN)}$	0.300 mM	0.139 mM
$Km_{(ATP)}$	0.100 mM	0.495 mM
Vmax	350 nmoles/(h·g Hb)	550 nmoles/(h·gHb)

Some characteristics of both NMN- and NAMN- AT have been investigated in red cell lysates using a new method developed in our laboratory, for the assay of NAMN-AT in human red blood cells[9] and then adapted to NAMN-AT.

The relationships between purine and pyridine metabolism have also been investigated in our laboratory, in the aim of clarifying the connections between the two metabolisms.[10] In fact, it is well known, that the synthesis of NAD is strictly connected with purine metabolism through ATP and PPRibP availability. ATP is necessary both as an energy source in most reactions leading to NAD(P), and as the adenylyl moyety donor for dinucleotides, on the other hand, PPRibP is the common substrate of phosphoribosyltransferases of both purines and pyridines.

Recently, elevated NAD levels have been observed in the erythrocytes of patients bearing some inherited purine defects, as Hypoxanthine-Guanine Phosphoribosyltransferase (HGPRT) and Purine Nucleoside Phosphorylase (PNP) deficiences.[11] Such findings have been referred to alterations in the synthesis, more than in the breakdown of NAD,[12] but the connections between the purine defects and the high NAD levels are under investigation.

To evaluate the involvement of adenylyltransferases in purine disorders, associated or not with altered NAD levels, the activities of both NMN- and NAMN- AT have been assayed in the erythrocytes of patients bearing HGPRT and PNP deficiences, and also in patients bearing Adenosine Deaminase (ADA) and Adenine Phosphoribosyltransferase (APRT) deficiences, in which no altered NAD levels have been detected.

METHODS

Erythrocyte Preparation

Erytrocytes from healthy voluntary donors and from patients were separated by centrifugation from freshly drown heparinized blood, and whashed twice with TRIS 310 mOsM. The cells were lysed by adding 9 volumes of hypotonic TRIS buffer 20 mOsM in ice, gently shaking, and centrifugated at 20,000xg for 30 min to discard stroma.

Adenylyltransferase activities

NMN-AT activity was assayed incubating, in a final volume of 125 µl, a lysate volume corresponding to 2-3 mg of Hemoglobin, 2.5 mM ATP, 6 mM NMN, 10 mM $MgCl_2$ and 50 mM TRIS-HCl buffer pH 7.6, for 30 min at 37°C. The reaction was stopped by adding 20 µl of 40% trichloracetic acid (TCA) centrifuging to remove precipitated proteins. TCA was removed from the clear supernatant by extraction with water-saturated diethyl-ether till pH above 5.0; 50 µl of extract were processed by HPLC.

NAMN-AT activity was assayed using the same conditions, NMN being replaced by NAMN, at the same concentration.

Table 2. NMN- and NAMN- AT activities in erythrocytes of patients bearing some inherited purine defects. Data are expressed as nmoles of formed product in one hour per gram of Hemoglobin.

	n°	NMN-AT	NAMN-AT		n°	NMN-AT	NAMN-AT
controls	1C	282	449	HGPRT$^-$	1	261	339
	2C	304	444		2	221	354
	3C	264	378		3	186	367
	4C	373	389		4	331	333
	5C	305	373		5	285	267
	6C	299	459				
	7C	225	326		Mean±SD	257±50	331±34
	8C	219	313				
	9C	210	408	PNP$^-$	6	350	350
	10C	353	454		7	228	327
	11C	299	331				
	12C	235	285	ADA$^-$	8	256	336
	13C	252	338		9	206	381
	14C	212	390		10	0	99
	15C	309	457		11	373	389
	16C	271	345		12	457	634
	17C	300	405				
					Mean±SD	258±156	373±169
	Mean±SD	277±46	384±54				
				APRT$^-$	14	391	567

HPLC separation

HPLC separation of products from the respective substrates was carried out in a Millipore Waters Assoc. (Harrow U.K.) automated system, with the method described elsewhere.[13]

RESULTS

The apparent kinetic characteristics of both NMN- and NAMN- AT are reported in Table 1.

The activity of both enzymes in the erythrocytes of patients and controls are shown in the Table 2. In the group of healthy controls, the activity of NMN- is lower than NAMN-AT, the former ranging from 210 (lower value) to 373 (higher value) nmoles/h/g Hb, and the latter from 313 to 475 nmoles/h/g Hb. No significant alterations have been shown in HGPRT$^-$, PNP$^-$ and in four of the five ADA$^-$ patients. The APRT$^-$ subject showed a marked increase in the activity of both enzymes.

DISCUSSION

In normal subjects, the activity of NAMN- is higher than NMN- AT, though the affinity constants are of the same magnitude. among the observed pathologies, the implication of both AT activities seems to be restricted to one ADA$^-$ patient, and to the APRT$^-$ subject. The undetectable NMN- and the very low NAMN-AT activity found in patient n°10 (ADA$^-$) is without explanation at the moment. The APRT deficient patient had normal erythrocyte content of NAD, while NADP content was in the upper values of the range, probably due to the high activity of both adenylyltransferases.

These results let us hypothesize two possible explanations: first, that the endocellular regulation of adenylyltransferases, and not their maximal activity, may be responsible for the altered NAD levels, observed in some purine disorders; second, that the formation of dinucleotides is not the key step which may lay elsewhere, probably in the phosphoribosyltransferase reactions. Nevertheless it is noteworthy to observe that both enzymes show parallel behaviours, showing parallel alterations, thus supporting the hypotesis that they are an unique enzyme.

REFERENCES

1. J. Preiss and P. Handler, Intermediates in the synthesis of diphosphopyridine nucleotide from Nicotinic acid, J. Am.Chem. Soc. 79:4246 (1957)
2. I. G. Leder and P. Handler, Synthesis of Nicotinamide Mononucleotide by human erythrocytes in vitro, J. Biol. Chem., 189:889 (1951)
3. G. H. Hogeboom and W. C. Schneider, The synthesis of diphospho pyridine nucleotide by liver cell nuclei, J. Biol. Chem., 197:611 (1952)
4. P. Natalini, S. Ruggieri and G. Magni, Nicotinamide mononucleotide adenylyltransferase. Molecular and enzymatic properties of the homogeneous enzyme from Baker's yeast. Biochemistry, 25:3725 (1986)
5. S. Ruggieri, L. Gregori, P. Natalini, A. Vita, M. Emmanuelli, N. Raffaelli, and G. Magni, Evidence for an inhibitory effect exerted by yeast NMN adenylyltransferase on poly(ADP-ribose) polymerase activity, Biochemistry, 29:2501 (1990)
6. M. E. Haines, I. R. Johnston, A. P. Mathias and D. Ridge, Synthesis of Nicotinamide Adenine Dinucleotide and poly (adenine diphosphate ribose) in various classes of rat nuclei, Biochem. J., 115:881 (1969)
7. T. Kato and O. H. Lowry , Distribution of enzymes between nucleous and cytoplasm of single nerve cell bodies, J. Biol. Chem., 248:2044 (1973)
8. M. Rechsteiner and V. Catanzarite, The biosynthesis and turnover of Nicotinamide Adenine Dinucleotide in enucleated culture cells, J. Cell. Physiol., 84:409 (1974)
9. S. Sestini, P. Lusini, V. Micheli, M.L. Ceccuzzi and C. Ricci, Evidence for NMN Adenylyltransferase in human erythrocytes. Preliminary observations, Ital J. Biochem., 37:63A, (1988)
10. V. Micheli, S. Sestini and C. Ricci, Purine and pyridine production in human erythrocytes, Arch. Biochem. Biophys. 244:454 (1986)
11. H. A. Simmonds, L. D. Fairbanks, G. S. Morris, D. R. Webster and E.H. Harley, Altered erythrocyte nucleotide patterns are characteristics of inherited disorders of purine or pyrimidine metabolism, Clin. Chim. Acta, 171:197 (1988)
12. V. Micheli, H. A. Simmonds and C. Ricci, Regulation of Nicotinamide-Adenine Dinucleotide synthesis in erythrocytes of patients with hypoxanthine-guanine phosphoribosyl transferase deficiency and a patient with phosphoribosyl pyrophosphate synthetase superactivity, Clin. Science, 78:239 (1990)
13. M. Rocchigiani, S.Sestini, V. Micheli, M. Bari and H. A. Simmonds, NAD synthesis in human erythrocytes: determination of the activities of some enzymes, 7th International/3rd European Joint Symposium on PURINE AND PYRIMIDINE METABOLISM IN MAN, Bournemouth, 30th June-5th July

PYRIDINE NUCLEOTIDE METABOLISM: PURINE AND PYRIMIDINE INTERCONNECTIONS

Vanna Micheli, Carlo Ricci, Sylvia Sestini, Marina Rocchigiani, Monica Pescaglini, Giuseppe Pompucci

Dipartimento di Biologia Molecolare, Universita' di Siena
Siena, Italia

STATE OF THE ART

The role of the pyridine coenzymes NAD and NADP in oxidation-reduction reactions, leading to energy production and to reductive synthesis, is well documented, and the characteristics of the proteins involved have been extensively studied. Since the balance between the oxidised and reduced form of pyridine coenzymes modulates both catabolic and synthetic pathways, their differing levels in different tissues, organs and cellular compartments indicate metabolic differentiations[1]. In addition to this role, pyridine coenzymes take part in a number of other cellular processes occurring in prokaryotes or eukaryotes or both[2]. These processes include the utilization of NAD for: protein ADP-ribosylations, mainly involved in the mechanism of action of bacterial toxins, affecting protein synthesis or some processes mediated by cAMP and G-proteins; poly-ADPribose synthesis, involved in the regulation of DNA repair and replication and in cellular differentiation; DNA ligase reactions, active in prokaryotes. NAD(P) involvement in the production of cytotoxic compounds and in phagocytosis process has also been described[2]. Such findings revealed the versatility of the NAD(P) molecule in cell function, suggesting that its biological role is not yet fully appreciated.

Pyridine nucleotide precursors, nicotinic acid (NA) and nicotinamide (NAm), have been described to produce several effects, not correlated with normal vitamin function (enhancement of cAMP availability, vasodilation, lowered mobilization of fatty acids, reduction of circulating cholesterol and lypoproteins[3,4,5]). The effectiveness of NAm in the therapy of psychiatric diseases has also been reported in connection with altered tryptophan metabolism and disturbances of serotonine-mediated neurotransmittance[6]. Moreover, the therapeutic employment of NAm in insuline dependent diabete type 1 has been reported to induce islet B-cells regeneration through poly(ADPR)synthetase inhibition and NAD level restoration[7].

There are several indications on the metabolic interconnections taking place between pyridine, purine and pyrimidine nucleotides. PRPP synthetase inhibition by NAD, NADH and NADPH has been reported[8], which might have a physiological role on the whole nucleotide metabolism, together with the inhibition by ADP, AMP, GDP and 2,3DPG. A direct influence of pyrimidine nucleotides on PRPP production has also been reported, mainly referred to magnesium ion sequestration[9]

Pyridine coenzymes are known to regulate some branch points in purine metabolism, mainly through the balance between the oxidised and reduced forms. NAD/NADH value modulating IMP dehydrogenase reaction and consequently guanyl nucleotide availability is one of the best known examples which received increasing interest in the last years, owing to its involvement in tumours and in their possible therapy by tiazofurin and its metabolites[10]. NADP(H) levels are in turn responsible for GMP conversion to IMP. Regulation on purine metabolism can occur also through direct modulation of non-oxidative enzymes, as described for HGPRT, which is regulated by NADH/NAD ratio in rat heart[11]. Modification in the activity of xanthine dehydrogenase[12] and xanthine oxidase activities[13] by NAm in rat serum and liver has also been reported. Hyperuricemia has been described in man following the oral administration of NA, referred to the alteration in the renal handling of uric acid. NA treatment was also associated with PRPP depletion in erythrocytes and fibroblasts and with alterations in purine de novo synthesis[14].

Purine and Pyrimidine Metabolism in Man VII, Part B
Edited by R.A. Harkness *et al.*, Plenum Press, New York, 1991

The metabolism of pyrimidine nucleotides is dependent on NAD(H) level for the dihydroorotic dehydrogenase reaction, while NADPH availability regulates the pyrimidine catabolism through the formation of the dihydropyrimidines. The association between the increased urinary excretion of 5-ribosyluracil and N-methylpyridone-5-carboxamide, a pyridine catabolite, has been reported in mentally retarded children[16]. Up to date there is a limited amount of studies concerning this field.

Both purine and pyrimidine deoxynucleotide production depends on NADPH, thus connecting DNA metabolism with the oxidative state of pyridine coenzymes.

Nucleotide pathways of synthesis and interconversion show important connections and common effectors: inorganic phosphate; PRPP as the donor of the ribosyl moiety; glutamine as amide group donor; ATP , which plays a major role in synthetase reactions, in nucleotide phosphorylation and as the adenylyl donor in the dinucleotide synthesis.

Pyridine nucleotide synthesis has been studied in experimental animals and in man by several authors, included our group. After the early finding on the stimulation of NAD synthesis by nicotinamide administration to rats[17], possible connections with purine nucleotide synthesis were identified[18]. More recently we investigated NAD metabolism in human erythrocytes, focusing on the synthetic pathways and on their regulation by purine nucleotides[19].

Pyridine nucleotide synthesis in man may occur in liver and kidney from tryptophan, via quinolinic acid, or from preformed pyridine rings, namely NA and NAm. The former pathway is not very effective, and the latter is used in most tissues. This route proceeds through the cytosolic formation of the mononucleotides, followed by the synthesis of adenine-dinucleotides, occurring in cell nucleus, and by phosphorylation of NAD to NADP in the cytosol (Fig. 1). NAD metabolism is known to occur through the splitting of NAm by NAD glycohydrolase (yielding ADPribose, or possibly transglycosidation products), or by poly(ADPR)synthetase, with different significance and subcellular localization. Hydrolysis of the pyrophosphate binding by NAD pyrophosphatase may occur, originating AMP and NMN. Both NAm and NMN may be recycled for NAD synthesis through "pyridine cycles", though not all of them are active in the humans. The catabolic fate of NMN is to be dephosphorylated by 5'-nucleotidase; NAm nucleoside is substrate for purine phosphorylase, yielding free NAm, which in turn can be deamidated to NA. Pyridine ring is not broken down in eukaryotes; in mammals NAm is methylated by S-adenosylmethionine to N-1-methylnicotinamide and excreted in the urine. Also 2-pyridone- and 4-pyridone-carboxamide, NA, NAm, NAm-N-oxide and the 6-hydroxyderivatives of both NA and NAm can be found in human urines[20].

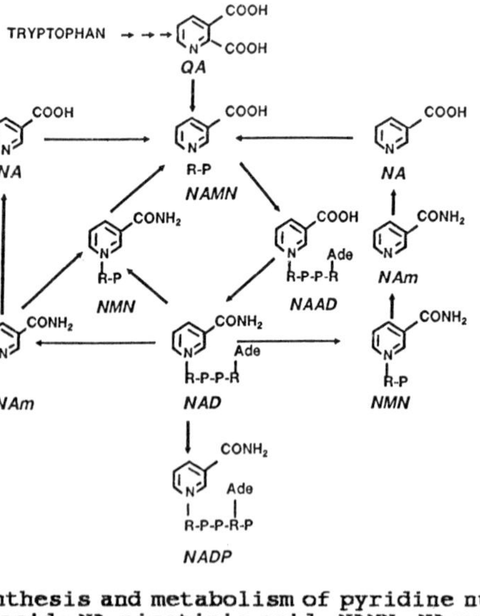

Fig.1. Pathways of synthesis and metabolism of pyridine nucleotides
QA: quinolinic acid; NA:nicotinic acid; NAMN: NA mononucleotide; NAAD: NA adenine dinucleotide; NAm: nicotinamide; NMN: NAm mononucleotide; NAD: NAm adenine dinucleotide; NADP:NAm adenine dinucleotidephosphate

No systematic study has been conducted on the significance of NAD(P) level alteration in cells and tissues and on the regulation of their endocellular turnover, except that related to vitamin deficiency (pellagra or pellagra-like syndromes). There are several sparse reports on this subject, many of which postulate a close relationship between purine and pyridine metabolism.

Altered NAD levels have been found in human erythrocytes in association with purine enzyme deficiencies, namely PNP and HGPRT, and with PRPP synthetase superactivity[21]. In the latter two disorders we found altered NAD synthesis rather than breakdown[22]. Lowered NAD levels have also been found in ATP-depleted erythrocytes due to pyruvate kinase deficiency[23] associated with disaggregation of PRPP synthetase subunits. Moreover, increased NAD levels have been reported in sickle cell disease[24] and lowered levels in tumours[25], while erythrocyte NADP levels and synthesis rate have been shown to be increased in G6PD deficiency[26].

Some effects of purine nucleotides on different steps of pyridine nucleotide synthesis have been described, such as inhibition of adenylyl-transferase by ITP and 6-mercaptopurine riboside triphosphate, competing with ATP[27].

Human erythrocytes, though a limited biological system, may provide useful informations on the metabolic connections occurring among nucleotides, without any interference of nucleic acid and protein turnover. Nucleotide metabolism, though all the *de novo* synthesis are lacking, is very active in these cells, which also exert a function of transport of nucleosides and bases to different tissues[28]. The contribution of pyrimidines to nucleotide metabolism in these cells seems to be very poor, since their levels are virtually undetectable, except for UDP sugars, though the Orotidine-phosphoribosyltransferase/OMP decarboxylase pathway is active.

Alterations in the eryhthrocyte nucleotide content may be not only of diagnostic value[29], but also an useful tool in the understanding of the mutual effects of the different sorts of nucleotides. The interactions among separate nucleotide pools seems to be of particular interest also in the light of their different turnover rate and stability; the latter proved to be very low for adenine nucleotides but very high for NAD, NADP and UDPG within the human erythrocyte[30]. Moreover, the investigation of pyridine nucleotide metabolism in human erythrocytes may provide a possible approach to an hitherto unknown "NAD pathology".

OUR CONTRIBUTION

Our previous investigations on NAD synthesis in normal human erythrocytes identified some effectors of this process[31] and the different roles of the two precursors, NA and NAm[32]. The present paper illustrates some reciprocal interactions occurring on the early steps of the synthesis of pyridine and purine nucleotides. The study has been conducted on the incorporation of precursors by intact cells and on the lysate activities of the phosphoribosyltransferases (PRT) leading to the conversion of adenine (Ade), hypoxanthine (Hyp), NA and NAm into nucleotides: APRT, HPRT, NAPRT and NAmPRT, respectively (E.C.:2.4.2.7.; 2.4.2.8.; 2.4.2.11.; 2.4.2.12). The competition for the common substrate PRPP and the key role played by ATP are the main results.

METHODS

Nucleotide production was determined on intact washed erythrocytes incubated at 37°C for one hour in an isotonic PRPP generating medium supplied with inorganic phosphate (Pi) and with the appropriate ^{14}C precursor, as described[31].

HPRT, APRT, NAPRT and NAmPRT activities were tested on crude lysates by radiochemical methods associated with TLC or HPLC separation of products, as described elsewhere[31,33]. Partially purified preparations (PPE) [34] were also used (about 20-fold purification for APRT and HPRT and 63-fold purification for NAPRT).

RESULTS AND DISCUSSION

Competition among precursors present in the incubation mixtures of intact erythrocytes was checked: NA did not affect significantly the amount of nucleotides produced from either Ade and Hyp, while either purine base, particularly Ade, lowered the production of total nucleotides from NA (Fig.2). Such effect is likely to be referred to the competition for the common substrate PRPP, or to direct inhibition by metabolites on NAPRT activity, or to both mechanisms.

TAB. 1. Apparent kinetic parameters of purine and pyridine phosphoribosyltransferases in crude lysates. Km values are expressed as M^{-6} and Vmax as nmoles/h·mg_{Hb}.

	Km_{PRPP}	Km_{base}	Vmax
APRT	74.9	Ade 120.0	33.9
HPRT	3.9	Hyp 13.1	212.0
NAPRT	23.1	NA 6.7	0.8
NAmPRT	52.8	NAm 15.1	0.015

Kinetic constants of purine and pyridine PRTs in crude lysated are summarised in table 1, showing striking differences among the maximal activities, but not among the Kms for PRPP . Partially purified APRT, HPRT and NAPRT confirmed the results in crude lysates.

The effect of the following purine and pyridine compounds was tested on NAPRT activity both in crude lysates and in PPE: Ade, adenosine, AMP, ADP, ATP, Hyp, inosine, IMP, GMP, GTP, NAm, NAMN, NAD, NADH. Appreciable inhibitory effect was exerted only by NAMN, which is the reaction product, and by NAm, suggesting a further interaction between the two precursors. Adenine mono- di- and triphosphate nucleosides displayed different extents of inhibition (Fig.3A). Neither Ade nor Hyp exerted any direct effect, supporting the hypothesis that the observed decrease of NAMN synthesis in intact cells in the presence of the purine bases may depend on the competition for PRPP. The particular role of ATP, in association with Pi, on NAPRT activity has been reported [19]: Pi lowers the Km for PRPP and increased the Vmax; the addition of ATP furtherly lowers the Kms both for NA and PRPP, also decreasing the Vmax and thus causing inhibition. The incubation of erythrocytes in the presence of Ade in our conditions yields ADP and ATP beside AMP[19], suggesting that inhibition of NAPRT activity might also contribute to explain our results.

NAmPRT activity in crude lysates was slightly increased by low ATP concentrations and decreased by concentrations close to the endocellular ones (Fig.3A), thus suggesting that the endocellular ATP levels modulate the activity of this enzyme.

The effect of the following pyridine compounds was tested on APRT and HGPRT activities both in lysate and PPE: NA, NAMN, NAAD, NMN, NAD, NADH, NADP, NADPH. APRT activity was slightly inhibited by NAMN, NAAD, NADP and NADPH while neither NAD nor NADH showed any appreciable effect (Fig.3B). HPRT activity was not influenced by any of the above listed compounds; only NAAD and NMN showed a mild inhibition. The lack of appreciable effect by both NAD and NADH is in contrast with what reported for rat heart [11], suggesting that the regulation of this enzyme activity in the erythrocytes may be different from that in respiring cells.

Present data suggest that ATP plays a major role in regulating the conversion of pyridine precursors into their nucleotides in human erythrocytes, while pyridine compounds exert a poor direct effect on purine

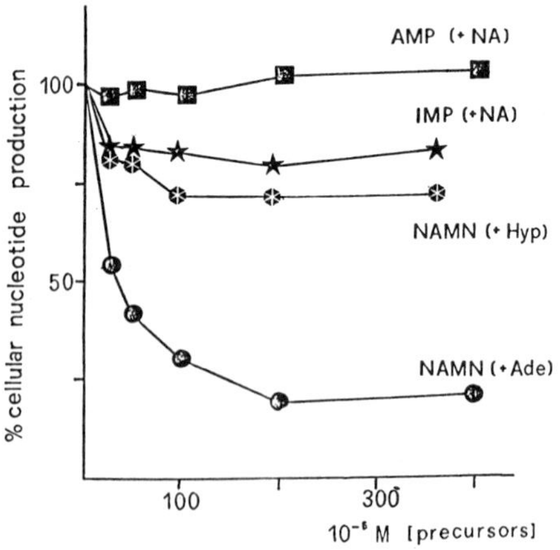

Fig.2 Nucleotide production from ^{14}C- Ade,- Hyp and -NA by red blood cells incubated with each precursor alone and in the presence of increasing concentrations of the others (unlabeled). Values expressed as % of production with no added compound.

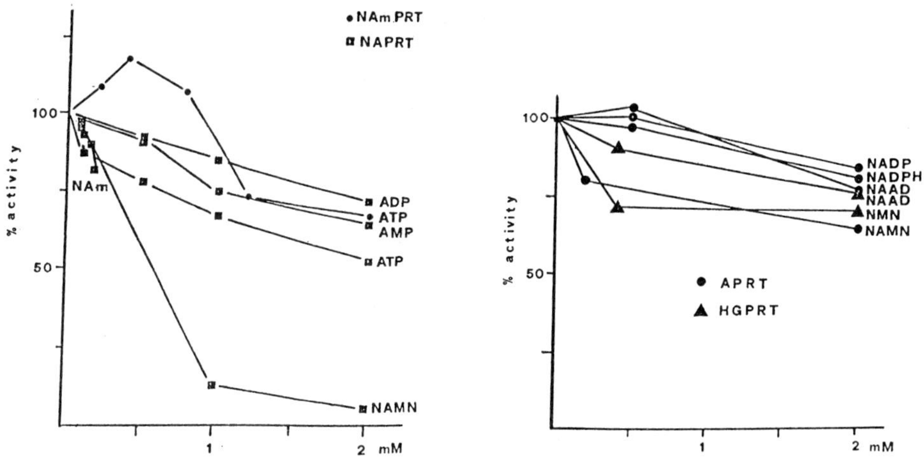

Fig.3. Effect of some purine and pyridine compounds on NAPRT and NAmPRT activities (A) and on APRT and HGPRT activities (B) in human erythrocyte lysates.

base conversion into nucleotides. PRPP can be effectively utilized for pyridine nucleotide synthesis, as demonstrated in human erythrocytes in vivo and in vitro[14], and the competition with the other pathways utilizing PRPP, in physiological conditions, is probably dependent on the modulation of enzyme characteristics by Pi and ATP.

AKNOWLEDGEMENTS

This research was supported by Italian MURST (60% and 40% funds)

REFERENCES

1. H. Sies, Nicotinamide nucleotide compartmentation, in "Metabolic compartmentation" H. Sies ed., Academic Press, London (1982)
2. "The pyridine nucleotide coenzymes", J.Everse, B.Anderson, K. You eds., Academic Press, New York (1982)
3. P.Lusini, C.Ricci, Metabolism of cAMP in the liver of rats treated with nicotinamide, Ital J Biochem 32:152 (1983)
4. M.F. Sorrell, H.Baker, D.J. Tuma, D.Frank, A.J.Barak, Potentiation of ethanol fatty liver in rats by chronic administration of nicotinic acid Biochem Biophys Acta 450:231 (1976)
5. J.P.Kane, M.J.Malloy, P.Tun, N.R.Phillips, D.D.Freedman, M.L.Wliams, J.S.Rowe, R.J.Havel, Normalization of LDL in heterozygous familial hypercholesterolemia with a combined drug regimen, New Engl J Med 304:251 (1981)
6. W.Blom, G.B.Van den Berg, J.G.M. Huymans, J.A.R. Sanders- Woudstra, Successful nicotinamide treatment in an autosomal dominant behavioral and psychiatric disorder, J Intern Metab Dis 8:107 (1985)
7. P.Pozzilli, N.Visalli, G.Ghirlanda, R.Manna, B.Andreani, Nicotinamide increases C peptide secretion in patients with recent onset typel diabetes, Diabet Med 6:568 (1985)
8. I.H.Fox, W.N.Kelley, Phosphoribosylpyrophosphate in man: biochemical and clinical significance, Ann Int Med 74:424 (1971)
9. N.A Lachant, C.R. Zerez, K.R. Tanaka, Pyrimidine nucleotides impair PRPP synthetase subunit aggregation by sequestering magnesium. A mechanism for the decreased PRPP synthetase activity in hereditary erythrocyte 5'nucleotidase deficiency, Biochem Biophys Acta 994: 81 (1989)
10. G.Weber, Critical issue in chemotherapy with tiazofurin, Adv Enz Regul 29:75 (1989)

11. K.Ravid, P.Diamant, Y.Avi-Dor, Regulation of the salvage pathway of purine nucleotide synthesis by the oxidation state in rat heart cells, Arch Biochem Biophys 229:632 (1984)
12. O.R.Affonso, M.F.Lemos, E.Mitidieri, Effect of nicotinamide on the serum xanthine dehydrogenase activity during fasting, Acta Biol Med Germ34:1607 (1975)
13. A.Di Stefano, M.Pizzichini, E.Marinello, Nicotinamide and liver xanthine oxidase, Adv Exp Med Biol 122B:189 (1980)
14. S.L.Gershon, I.H.Fox, Pharmacologic effects of nicotinic acid on human fibroblast purine metabolism, J Lab Clin Med 84:179 (1974)
15. J.A.RBoyle, K.O.Raivio, M.A.Becker, J.E.Seegmiller, Effects of nicotinic acid on human fibroblast purine biosynthesis Biochim Biophys Acta 269:179 (1972)
16. E.W.Lis, R.Bijon, A.W.Lis, The urinary excretion of 5-ribosyluracil and Nmethyl-2pyridone-5-carboxamide in mentally retarded children, J Cell Biol 31:150A (1966)
17. A.Bonsignore, C.Ricci, Nicotinamide and synthesis of pyridine nucleotides in the liver, Scientia Medica Ital VI: 1 (1958)
18. G.Pompucci, V.Micheli, M.Bari, Effetti della nicotinamide sulla sintesi "in vivo" dei nucleotidi purinici, Boll Soc It Biol Sper XLVII, 22:804 (1972)
19. V.Micheli, S.Sestini, M.Rocchigiani, M.Pescaglini, C.Ricci, Nucleotide synthesis in the human erythrocyte: correlation between purines and piridines, Biomed Biochim Acta 46: S268 (1987)
20. R.Fumagalli, Pharmacokinetics of NA and some of its derivatives, in "Metabolic effects of NA and its derivatives", K.F.Gey, L.A.Carlson eds.,Hans Huber Publishers Bern (1971)
21. H.A.Simmonds, D.R.Webster, J.Wilson, S.Lingham, An X- linked syndrome characterized by hyperuricaemia, deafness, and neurodevelopmental abnormalities, The Lancet 10 : 68 (1982)
22. V.Micheli, H.A.Simmonds, C.Ricci Regulation of NAD synthesis in erythrocytes of patients with HGPRT deficiency and a patient with PRPP synthetase superactivity, Clin Sci, 78: 239 (1990)
23. C.R.Zerez, K.R. Tanaka, Imaired NAD synthesis in pyruvate kinase deficiency human erythrocytes: a mechanism for decreased total NAD content and a possible secondary cause of Hemolysis Blood 69 : 99 (1987)
24. C.R.Zerez, N.A.Lachant, S.J.Lee, K.R.Tanaka, Decreased erythrocyte NAD redox potential and abnormal pyridine nucleotide content in sickle cell disease, Blood, 71, 512; (1988)
25. G.E.Glock, T.McLean, Levels of oxidized and reduced diphosphopyridine nucleotide and triphosphopyridine nucleotide in tumours, Biochem J 65: 413 (1957)
26. A.Bonsignore, G.Fornaini, G.Segni, A.Fantoni, Biosynthesis of pyridine coenzymes in erythrocytes of subjects with a case history of favism, Ital J Biochem 10: 212 (1961)
27. M.R.Atkinson, J.R.Jackson, P.K.Morton, Substrate specificity and inhibition nicotinamide mononucleotide adenylyltransferase of liver nuclei: possible mechanism of effect of 6- mercaptopurine on turnover growth, Nature 192: 946 (1961)
28. E.H.Harley, S.Sacks, P.Berman, Source and fate of circulating pyrimidine, Adv Exp Med Biol 195A: 109 (1986)
29. H.A.Simmonds, L.D.Fairbanks, G.S.Morris, D.R.Webster, E.H.Harley, Altered erythrocyte nucleotide patterns are characteristic of inherited disorders of purine or pyrimidine metabolism , Clin Chim Acta 171: 197 (1988)
30. H.A.Simmonds, V.Micheli, P.M.Davies, M.B.McBride, Erythrocytes nucleotide stability and plasma hypoxanthine levels: is 4°C really the best short-term storage temperature? Clin Chim Acta 192: 121 (1990)
31. V.Micheli, S.Sestini, C.Ricci, Purine and pyridine nucleotide production in human erythrocytes, Arch Biochem Biophys 244: 454 (1986)
32. V.Michheli, H.A.Simmonds, S.Sestini, C.Ricci, Importance of nicotinamide as an NAD precursor in the human erythrocyte, Arch Biochem Biophys 283: 40 (1990)
33. M.Rocchigiani, H.A.Simmonds, M.Bari, C.Ricci, NAmPRT assay in human red blood cells, Ital Biochem Soc Trans 1: 297 (1990)
34. J.Niedel, L.S.Dietrich, Nicotinate phosphoribosyltransferase in human erythrocytes. Purification and properties, J Biol Chem 240: 3500 (1973)

NAD SYNTHESIS IN ADA DEFICIENT ERYTHROCYTES OF THE OPOSSUM

DIDELPHIS VIRGINIANA

Nicholas C. Bethlenfalvay[1], Joseph E. Lima[2] and
Joseph C. White[2]

Departments of Primary Care[1] and Clinical Investigation[2],
Fitzsimons Army Medical Center, Aurora, CO 80045 USA

INTRODUCTION

In the human red cell, ATP is an absolute requirement for the production of NAD. Thus, in pyruvate kinase deficient human erythrocytes NAD concentrations are significantly reduced and NAD synthesis is impaired to a degree that is dependent on ATP concentration of those cells (1). SCID is characterized by variable pancellular deficiency of adenosine deaminase (ADA). Red cells of SCID patients may have low ATP and high 2'dATP concentrations (2) but there are no reports on NAD or 2'dNAD content and synthesis in the se cells. D. virginiana is North America's only native marsupial. Red cells of this species have low (1.5 nmoles/mg hg/h) ADA activity and high concentrations of 2'dATP resembling human SCID erythrocytes (3). In this study we have determined the concentration of NAD and investigated its synthesis in opossum red cells having variable ATP content.

METHODS

Nine adult opossums were studied, heparinized blood was obtained by cardiocenthesis.

Synthesis of Nucleotides in Whole Cells: Saline-washed red cells were used "as is" or after depleting ATP while raising 2'dATP content by incubating them in 1 mM 2'dADO for 1 h. The cells were then suspended (15% PCV) in a TRIS/HCl buffered physiologic electrolyte solution (Pi 1 mM). Suspensions of 1 ml were supplemented to contain 100 nmoles of either ^{14}C NA (54.8 mCi/mmole) or ^{14}C NAm (41.2 mCi/mmole) and incubated under an atmosphere of N_2 for 5 hours at 37°C.

Activities of Enzymes of NAD Synthesis in Hemolysates: The method of Zerez et al. (4) was used, modified as follows: ^{14}C ATP (10 mCi/mmole) was 1 mM and ^{3}H 2'dATP (30 mCi/mmole) was 1 mM. Using ^{3}H 2'dATP as a potential 2'deoxyadenyl donor, provided concentrations of NA mononucleotide and of NAm mononucleotide were 0.03 - 5.0 mM.

Biochemical Analyses: Red cell NAD content and metabolic intermediates in whole cells and in lysates were quantitated by reverse phase HPLC-radiochromatography (5,6). Purine nucleotide concentrations were measured by ion-pair exchange HPLC (7).

Purine and Pyrimidine Metabolism in Man VII, Part B
Edited by R.A. Harkness *et al.*, Plenum Press, New York, 1991

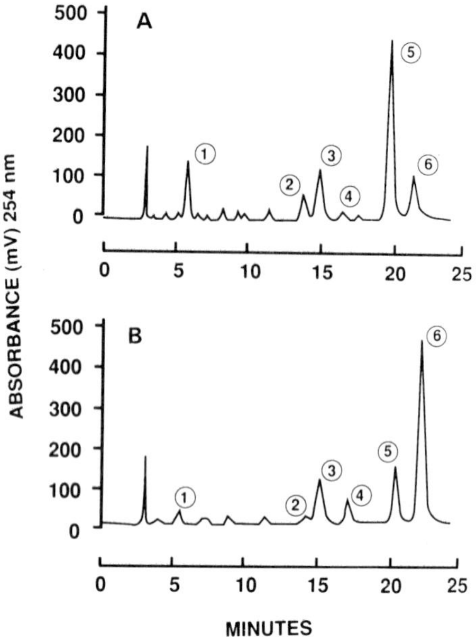

Fig. 1

Ion pair-exchange HPLC pattern of purine and pyrimidine nucleotides in opossum red cells. (A) high ATP, (B) low ATP phenotype. 1. NAD, 2. ADP, 3. GTP, 4. dADP, 5. ATP, 6. dATP

RESULTS AND DISCUSSION

Despite the narrow range of their ADA activity (3), red cell ATP concentrations vary in individual animals ranging 0.4-1.3 mM. 2'dATP levels reveal an inverse relationship of 1.2-0.3 mM (Fig. 1 A-B, Fig. 2A). Characterization of (d)ADO kinase(s) may clarify the cause for these observations. NAD(T) content in red cells of individual animals reveals a linear relationship to ATP concentrations (Fig. 2B). NAD(T) concentrations in opossum red cells are higher than in human erythrocytes and reach levels observed in P. falciparum infected (4) and PNP deficient human red cells (8).

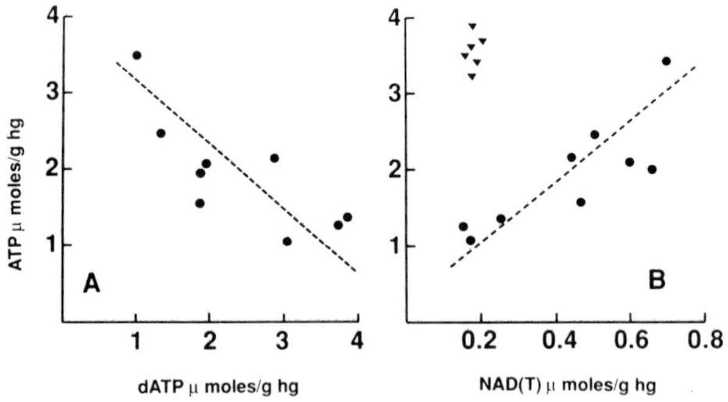

Fig. 2 Correlation of (A) ATP vs dATP and (B) NAD(T) vs ATP concentrations in red cells of nine opossums. Closed triangles denote NAD(T) concentrations in human red cells (n=6)

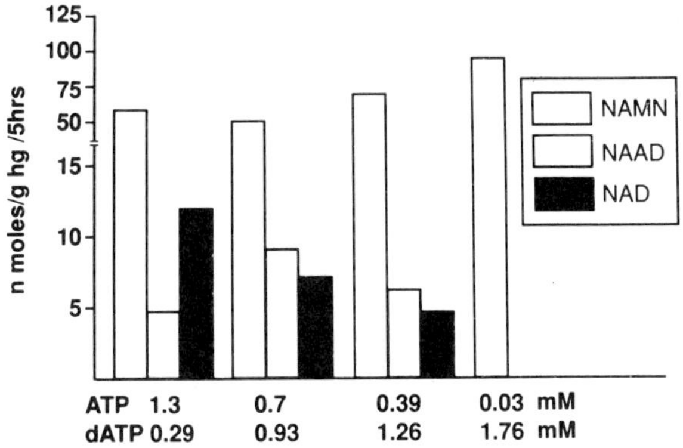

Fig. 3 Effect of "zero-time" ATP and 2'dATP concentrations in opossum red cells on pyrimidine nucleotide synthesis in-vitro

The capacity to synthetize NAD was positively correlated with ATP content in intact cells (Fig. 3). Because of the instability of ATP ($T_{1/2}$ 6h) in metabolizing cells in-vitro (6) only 5 h incubations were done. 2'dNAD was not detected, nor was its synthesis observed in intact cells. In the absence of 2'dATP, activities of NAPRT, NAAD pyrophosphorylase and of NAD synthetase were of the same order of magnitude as was determined in human red cell lysates (4). No activity of NAmPRT, NAD pyrophosphorylase or of NAm deamidase were detected. 2'dATP did not serve as 2'd adenyl donor to NAMN and to NAmMN as provided substrates under the conditions of these experiments.

It will be of interest to study ADA deficient human SCID red cells in regard to NAD content prior to and after ADA enzyme replacement (9) or ADA gene therapy.

ACKNOWLEDGEMENTS

This research was wholly supported by grant 80/650 from the Department of Clinical Investigation, Fitzsimons Army Medical Center. We thank Ms. C. Montoya for preparation of the manuscript.

REFERENCES

1. C. R. Zerez and K.R. Tanka: Impaired Nicotinamide Adenine Dinucleotide Synthesis in Pyruvate Kinase-Deficient Human Erythrocytes: A Mechanism for Decreased Total NAD Content and a Possible Secondary Cause of Hemolysis, Blood 69:999 (1987).
2. M. S. Coleman, J. Donofrio, J.J. Hutton, L. Hahn, A. Daoud, B. Lampkin and J. Dyminski: Identification and Quantitation of Adenine Deoxynucleotides in Erythrocytes of a Patient with Adenosine Deaminase Deficiency and Severe Combined Immunodeficiency, J. Biol. Chem. 253:1619 (1978).
3. N. C. Bethlenfalvay, C. Chadwick and J.E. Lima: Studies on the Energy Metabolism of Opossum Didelphis Virginiana Erythrocytes-IV. Red Cells Have Low Adenosine Deaminase Activity and High Levels of Deoxyadenosine Nucleotides, Life Sci. 44:963 (1989).

4. C. R. Zerez, E.F. Roth, S. Schulman and K. Tanaka: Increased Nicotinamide Adenine Dinucleotide Content and Synthesis in Plasmodium falciparum-Infected Human Erythrocytes, Blood 75:1705 (1990).
5. G. Leoncini, E. Buzzi, M. Maresca, M. Mazzei and A. Balbi: Alkaline Extraction and Reverse-Phase High-Performance Liquid Chromatography of Adenine and Pyridine Nucleotides in Human Platelets, Anal. Biochem. 165:379 (1987).
6. N. C. Bethlenfalvay, J.C. White, E. Chadwick and J.E. Lima: Studies on the Energy Metabolism of Opossum (D. Virginiana) Erythrocytes: V. Utilization of Hypoxanthine for the Synthesis of Adenine and Guanine Nucleotides In Vitro, J. Cell. Physiol. 143:563 (1990).
7. N. C. Bethlenfalvay, J.C. White, E. Chadwick and J.E. Lima: Studies on the Energy Metabolism of Opossum (D. Virginiana) Erythrocytes - VI. DE NOVO Purine Nucleotide Biosynthesis is Limited to the Final Steps of the Pathway In Vitro, Comp. Biochem. Physiol. 97B:193 (1990).
8. H. A. Simmonds, L.D. Fairbanks, G.S. Morris, G. Morgan, A.R. Watson, P. Timms and B. Singh: Central Nervous System Dysfunction and Erythrocyte Guanosine Triphosphate Depletion in Purine Nucleoside Phosphorylase Deficiency, Arch. Dis. Child. 62:385 (1987).
9. C. Bory, R. Boulieu, G. Souillet, C. Chantin, M.O. Rolland, M. Mathieu and M. Hershfield: Comparison of Red Cell Transfusion and Polyethylene Glycol-Modified Adenosine Deaminase Therapy in an Adenosine Deaminase-Deficient Child: Measurement of Erythrocyte Deoxyadenosine Triphosphate as a Useful Tool, Ped. Res. 28:127 (1990).

PYRIDINE NUCLEOTIDE METABOLISM: PURIFICATION AND PROPERTIES OF NMN ADENYLYLTRANSFERASE FROM HUMAN PLACENTA

Monica Emanuelli, Nadia Raffaelli, *Silverio Ruggieri, *Paolo Natalini and Giulio Magni

Istituto di Biochimica, Università di Ancona, Italy and *Dipartimento di Biologia MCA, Università di Camerino, Italy

INTRODUCTION

In mammalian cells the main biosynthetic pathway leading to the formation of NAD involves the enzyme NMN adenylyltransferase (NMNAT), which catalyzes the following reaction:

$$ATP + NMN \longleftrightarrow NAD^{+} + PPi$$

It has long been known the function of NAD as electron carrier in various biological oxidation-reduction systems, but quantitative measurements of NAD cellular utilization have indicated that it is mostly consumed to form ADP-ribose for the covalent modification of proteins by attachment of ADP-ribose in either monomeric or polymeric form[1]. Yet studies in vitro have shown that an increase in DNA strand breaks formation activates poly(ADP-ribose)polymerase (ADPRP) so that the rate of formation of ADP-ribose can exhaust intracellular NAD pool and thereby interfere with NAD-dependent enzymes in intermediary metabolism. It is not known how cells control the formation of NAD so as to satisfy the needs of oxidation-reduction reactions and of ADP-ribosylation reactions. It would seem likely that NMNAT must play an important role in this control and its nuclear location appears very significant. In fact the presence of NMNAT in the nucleus could allow: (i) a direct transfer of NAD from NMNAT to ADPRP, without the equilibration with NAD in solution; (ii) a control of NAD synthesis by ADP-ribosylation of NMNAT; (iii) a regulation of poly(ADP-ribose) synthesis mediated by a possible inter-action between the two enzymes. In order to better understand how NMNAT is involved in pyridine nucleotide cycle regulation in man we purified and characterized the enzyme, for the first time, from a human source. This report describes our purification procedure and major molecular and catalytic properties of NMNAT isolated from human placenta.

METHODS

Assay. Enzyme activity was routinely tested by a continuous spectrophotometric coupled enzyme assay[2]. A different

Purine and Pyrimidine Metabolism in Man VII, Part B
Edited by R.A. Harkness *et al.*, Plenum Press, New York, 1991

assay procedure, based on HPLC techniques[3], was adopted when the kinetic of the reaction was studied in the backward direction. One enzyme unit is defined as the amount of enzyme which catalyzes the synthesis of 1 µmol of NAD per minute at 37 °C.

Purity The purity of the final enzyme preparation was assessed by analytical polyacrylamide gel electrophoresis carried out in homogeneous buffer system (0.015 M sodium borate, pH 9.0). Protein bands were detected by silver staining. To correlate the stained band with the enzyme activity duplicates of non denaturating gel, loaded with 0.100 units of adenylyltransferase activity, were run simultaneously at 4°C. One gel was fixed and stained for proteins, while the second gel was sliced into 2-mm sections and placed in 250 µl of 100 mM potassium phosphate buffer, containing 40 mM $MgCl_2$, pH 7.4 for one hour at 4 °C.The enzyme activity was assayed on 150 µl of supernatant.

Molecular weight determinations The molecular weight of the purified enzyme was determined by gel filtration on a Superose 12 FPLC column by elution with 50 mM potassium phosphate buffer, pH 7.5, containing 1.0 mM DTT, 1 mM $MgCl^2$, 0.5 mM EDTA, 0.5 M KCl at a flow rate of 0.5 ml/min. For the estimation of subunit molecular weight gel electrophoresis under denaturating conditions was conducted according to Laemmli in 15 % polyacrylamide gels[4].

Isoelectrofocusing. Isoelectrofocusing was conducted on a 110-ml LKB 8100 ampholine column filled with a linear, 0-65 % (w/v), gradient of glycerol containing 1 % (w/v), pH 3.5-10, ampholine carrier ampholytes.

Amino Acid Analysis. Samples of protein were hydrolyzed for 45 min at 155 °C in 6 N HCl, in sealed evacuated tubes, as described by Hare[5]. Cysteine was determined as cysteic acid after performic acid oxidation. The analysis was performed on a Chromakon 500 (Kontron Instruments) amino acid analyzer and the amino acids were detected after post-column reaction with o-phthalaldehyde.

Kinetic analysis. Kinetic parameters were determined on homogeneous preparation of NMNAT. The effect of pH on maximal velocity was studied at 37 °C by using a 50 mM Bis-Tris and Tris-HCl buffering mixture. The activity was determined according to[3].

RESULTS AND DISCUSSION

The details of the purification procedure will be published elsewhere. The results of a typical purification procedure starting from 400 g of human placental tissue are summarized in Table I. All steps except FPLC analysis were performed

Table I. Purification of NMNAT from human placenta

Step	total act. (Units)	purification (x-fold)	specific act. (U/mg)
crude extract	4867	-	0.0004
pH 5.0 fraction	4554	6	0.0027
Phenyl Sepharose	4560	76	0.0317
Green A	3664	1364	0.573
Hydroxyapatite	2198	3338	1.402
TSK Phenyl-5PW	1099	22548	9.47

at 4 °C. The pure enzyme, stored at 4 °C in high ionic strenght buffers, was stable for several months retaining 80 % of its original activity. Repeated freezing and thawing led to inactivation of the enzyme. The maintenance of activity is, in any storage condition, strictly dependent on the presence of high KCl concentrations, whereas the stability exhibited by the enzyme purified to homogeneity from yeast seems to be independent upon KCl presence[2]. The final enzyme preparation, subjected to 5 % polyacrylamide gel electrophoresis under non-denaturating conditions, showed one diffuse band associated with the enzyme activity (R_m=0.25). The molecular weight of the native enzyme, estimated by gel filtration as described under "Methods", was 132,000 ± 10,000, in agreement with the value found for the enzyme from bull testis (unpublished results). Upon denaturating polyacrylamide gel electrophoresis the purified enzyme ran as a single protein band of 33,000 mol. wt. A similar SDS-polyacrylamide gel electrophoresis pattern, obtained after incubation of the enzyme both in the presence and in the absence of 2-mercaptoethanol, revealed the absence of disulfide linkages between the four subunits. The activity and pH profile of an eluate from an isoelectric focusing column showed that three catalytically active components with isoelectric pH values of 4.7, 5.7 and 6.6 were present. This behaviour, in agreement with that observed both for yeast[2] and chicken enzyme[6], can be ascribed to aggregation of the enzyme molecules, as observed for other nuclear proteins[7] and is consistent with the excess of aspartic and glutamic acid over the other aminoacid residues. The behaviour of the enzyme activity at different [H^+] was consistent with a plateau ranging from pH 6.0 to pH 9.0. NMNAT exhibits linear kinetics with respect to NMN, ATP, NAD and PPi; our K_m values are one order of magnitude lower than those determined for NMNAT obtained from from other sources[2,8] (Table II). K_m value for NaMN is higher than that for NMN, suggesting that the amido pathway is predominant in man. Kinetic experiments conducted by analyzing the enzyme catalyzed reaction in both directions are consistent with a sequential mechanism. To establish whether substrate addition and product release were random or ordered product inhibition studies were carried out. Product inhibition results suggest an Ordered Bi-Bi mechanism, in agreement with the stereochemical analysis reported by Lowe[9]. NMNAT activity has been assayed in the presence of several nucleotides, nucleosides and bases, including both physiological intermediates of the biochemical pathways and some of their analogs, with special emphasis to the pyridine derivatives. Only ADP-ribose and β-NMNH were found to significantly inhibit the enzyme activity; the anomer of NMN (α-NMN) was uneffective both as inhibitor and substrate, showing that the nucleotidyl transfer is strongly

Table II

Substrate	Michaelis Constant (mM)
ATP	0.023
NMN	0.038
NaNM	0.180
NAD^+	0.067
PPi	0.125

stereospecific. None of the different monosaccharides tested seem to modulate the catalytic properties of the enzyme. Even though the level of ADP-ribose necessary for substantial inhibition of NMNAT appears higher than the levels reported for mammalian cells in vivo the inhibition exerted by ADP-ribose towards NAD (reverse reaction) deserves some attention. Further studies should be undertaken in order to investigate on the possibility that NMNAT might play a role in metabolic events associated to ADP-ribosylation reactions.

REFERENCES

1. M.Rechsteiner, D.Hillyard and B.M.Olivera, Magnitude and significance of NAD turnover in human cell line D98/AH2, Nature, 259: 695 (1976).
2. P.Natalini, S.Ruggieri, N.Raffaelli and G.Magni, Molecular and enzymatic properties of the homogeneous enzyme from baker's yeast, Biochemistry, 25: 3725, (1986).
3. V.Stocchi, L.Cucchiarini, M.Magnani, L.Chiarantini, P.Palma and G.Crescentini, Simultaneous extraction and reverse-phase high-performance liquid chromatographic determination of adenine and pyridine nucleotides in human red blood cells, Anal. Biochem., 146: 118, (1985).
4. U.K.Laemmli, Cleavage of structural proteins during the assembly of the head of bacteriophage T4, Nature, 277: 680 (1970).
5. P.E.Hare, Subnanomole-range amino acid analysis, Methods Enzymol., 47: 3, (1977).
6. W.Cantarow and B.D.Stollar, Nicotinamide mononucleotide adenylyltransferase, a nonhistone chromatin protein. Purification and properties of the chicken erythrocyte enzyme, Arch. Biochem.Biophys., 180: 26, (1977).
7. P.Adamietz, K.Klapproth and H.Hilz, Isolation and partial characterization of the ADP-ribosylated nuclear proteins from Ehrlich ascites tumor cells, Biochem. Biophys. Res. Commun., 91: 1232, (1979).
8. A.M.Ferro and L.Kuehl, Adenosine triphosphate: nicotinamide mononucleotide adenylyltransferase of pig liver. purification and properties, Bioch.Bioph.Acta, 410: 285 (1975).
9. G.Lowe and G.Tansley, The stereochemical course of nucleotidyl transfer catalysed by NAD pyrophosphorylase, Eur. J. Biochem., 132: 117, (1983).

NAD SYNTHESIS IN HUMAN ERYTHROCYTES: DETERMINATION OF THE ACTIVITIES OF SOME ENZYMES

Marina Rocchigiani, Sylvia Sestini, Vanna Micheli, Mario Bari and H. Anne Simmonds*

Department of Molecular Biology, University of Siena Italy, *Purine Research Laboratory, Guy's Hospital, London U.K.

INTRODUCTION

The oxido-reductive properties of pyridine coenzymes (NAD and NADP) are well known. In the last fifteen years different roles have been pointed out for NAD as substrate in ADP-ribosylation and other reactions involved in DNA repair and replication, and in protein synthesis. Such findings suggested that hitherto unrevealed functions and biological roles may exist for NAD.

Mammalian erythrocytes lack the "de novo" pathways for the synthesis of all nucleotides (purine, pyrimidine and pyridine) and need preformed precursors: namely purine and pyrimidine bases and the preformed pyridine ring, either nicotinic acid (NA) or nicotinamide (NAm). In human erythrocyte both NA or NAm can be utilized for pyridine coenzyme synthesis. In the so called "deamidated pathway" (Fig.1) nicotinic acid phosphoribosyltransferase (NA-PRT, E.C. 2.4.2.11) catalyzes the conversion of NA into its mononucleotide (NAMN), which is converted into the dinucleotide (NAAD) by NAMN adenylyltransferase (NAMN-AT, E.C. 2.7.1.23); NAD synthetase catalyzes NAAD amidation to NAD. In

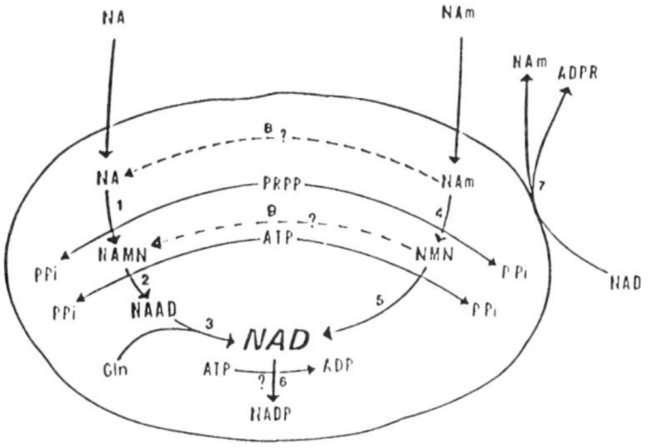

Fig. 1. Synthesis of pyridine nucleotide in human erythrocytes. Enzymes: 1) NA-PRT; 2) NAMN-AT; 3) NAD sintetase; 4) NAm-PRT; 5) NMN-AT; 6) NAD kinase; 7) NAD glycohydrolase; 8) NAm deamidase; 9) NMN deamidase.

the so called "amidated pathway", starting from NAm, the synthesis of the mononucleotide (NMN) is catalyzed by nicotinamide phosphoribosyl-transferase (NAm-PRT, E.C. 2.4.2.12) and the conversion of NMN into NAD is catalyzed by NMN adenylyltransferase (NMN-AT, E.C. 2.7.7.1). The synthetic pathway involve ATP availability at several steps both as a substrate (adenylyltransferase reactions) and as an energy donor (NAD synthetase reaction) or phosphate donor (NAD kinase reaction).

In the human red blood cell the existence or functional capacity of some of the above mentioned enzymes was in doubt until recently. An unphysiologically high Km value for NAm (0.1 M) has been reported for NAm-PRT;[2] while the AT activities had been postulated to be located exclusively in cell nuclei, thus supposing[3] that such activities were absent in human erythrocytes.

In the present study we describe an HPLC method developed to determine the activity of the four enzymes responsible for the first two steps of NAD synthesis on NA and NAm pathways in human red blood cells.

METHODS

Blood, obtained from normal adult volunteers of both sexes, was collected into heparin and centrifuged immediately (1,500xg, 10 min). The plasma, buffy coat and top fifth reticulocyte-rich layer of red blood cells were discarded and the remaining cells were washed twice in isotonic saline (1,500xg, 10 min).

Enzymatic Assayes

NAPRT. One volume of packed erythrocytes was lysed by adding 4 volumes of water, freezing and thawing twice in dry ice. Stroma were discarded by centrifugation (15 min, 9,000xg). The assay mixture contained: 25 mM Phosphate Buffer pH 7.4; 10 mM $MgCl_2$; 0.6 mM PRPP; 0.1 mM ^{14}C NA (19,990 dpm/nmole); lysate corresponding to 2-3 mg Hb. The reaction mixture was incubated for 30 min at 37°C in a shaking water bath. The incubation was stopped by the addition of 20 µl of 40% trichloracetic acid (TCA) while vortex mixing and then the mixture was centrifuged to remove precipitated proteins in a microfuge for 2 min. TCA was removed from the clear supernatant by extraction with water-saturated diethyl-ether till pH above 5.0; 50 µl of extract were processed by HPLC.

NAmPRT. One volume of packed erythrocytes was lysed by the addition of 10 volumes of hypotonic TRIS-HCl buffer (11 mM, pH 8.8). After 10 min at room temperature the lysate was centrifuged (9,000xg, 15 min) in a microfuge, and the membranes removed immediately. The assay mixture contained: 25 mM TRIS-HCl buffer pH 8.8; 5 mM $MgCl_2$; 0.5 mM PRPP; 0.1 mM ^{14}C NAm (33,600 dpm/nmole); lysate corresponding to 2-3 mg Hb. The reaction mixture was incubated for 120 min at 37°C in a shaking water bath. The incubation was stopped as described above (see NAPRT); 100 µl of extract were processed by HPLC.

NAMN- NMN- AT. One volume of packed erythrocytes was lysed by the addition of 9 volumes of hypotonic TRIS-HCl buffer (11 mM, pH 7.6). Stroma-free lysates were obtained by centrifugation in a microfuge at 9,000xg (15 min); membranes were removed immediately. The assay mixture contained: 50 mM TRIS-HCl buffer pH 7.6; 10 mM $MgCl_2$; 2.5 mM ATP; 6 mM NAMN or NMN; lysate corresponding to 2-3 mg Hb. Reaction mixture was incubated for 30 min at 37°C in a shaking water bath. The incubation was stopped by adding 10 µl of 40% TCA and the precipitated proteins were removed as described above; 50 µl of extract were processed by HPLC.

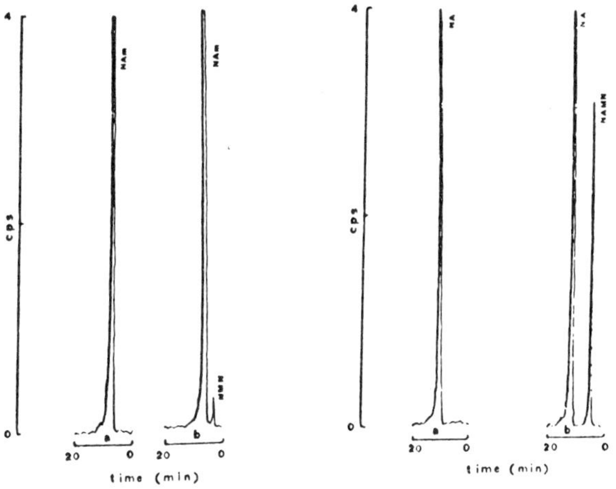

Fig. 2 Radioactivity traces of HPLC separation of: NMN (produced by NAmPRT) from NAm in the blank (a) and in the assay (b); NAMN (produced by NAPRT) from NA in the blank (a) and in the assay (b).

HPLC Method

The HPLC system used was a Millipore Waters Assoc. (Harrow U.K.) trimodule fully automated system, with a dual channel plot UV detector (254 and 280 nm), coupled with a radiodetector (Precision Radioactivity Monitor, Reeve Analytical, Glasgow, U.K.) with a heterogeneous flow cell of 200 µl capacity. When the radiodetector was used, the 280 nm channel recorder was disconnected. Each compound was identified on the basis of its retention time and 280/254 ratio. A reverse phase column (5 µm Sperisorb, 12.5 cm x 4.6 mm) was used. Substrates were separated from products, using a gradient of 0.1 M KH_2PO_4, containing 6 mM TBA, pH 5.5 (Solution A), and methanol (Solution B). The chromatographic conditions were as follows: initial conditions 96%A and 4%B; up to 10%B in 5 min; up to 12%B in 5 min; 3 min in isocratic mode at 12%B; up to 30%B in 2 min; 4 min in isocratic mode at 30%B; return to 96%A and 4%B in 1 min. The initial conditions were restored in 10 min; the flow rate was 1 ml/min.

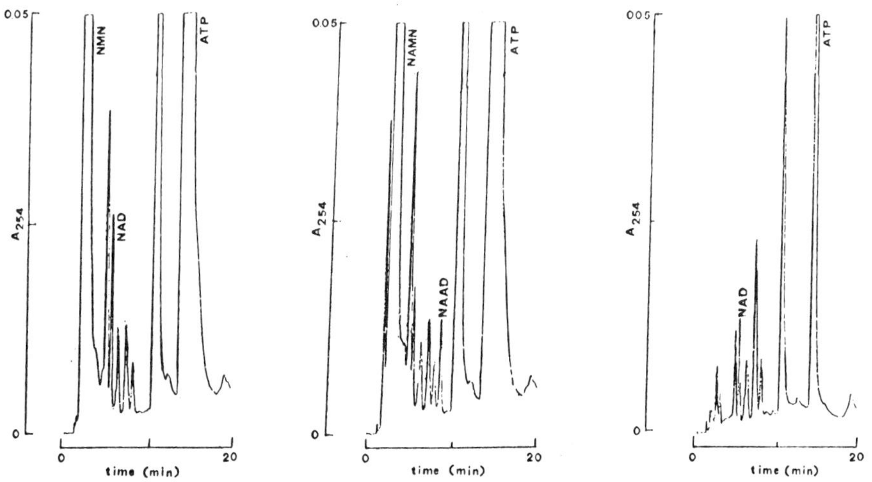

Fig. 3. UV traces at 254 nm of HPLC separation of NAD (produced by NMN-AT) from NMN; of NAMN (produced by NAMN-AT) from NAMN and the blank.

Table 1. Enzymatic activities on crude lysate

Enzyme	nmoles/h·g Hb ± SD	n
NA-PRT	1539±225	9
NAm-PRT	21±3.5	15
NAMN-AT	382±58	15
NMN-AT	279±50	15

RESULTS

Fig. 2 shows the chromatograms obtained injecting the protein free assay mixture and the respective blanks of the two phosphoribosyltransferases (NAPRT and NAmPRT). There is a striking difference in the activity of the two enzymes, NAPRT activity being about eigthy-fold higher than NAmPRT.

Fig. 3 shows the chromatograms obtained injecting the protein free assay mixtures and the respective blanks of the two adenylyltransferases. The activity values are quite similar.

The activities determined in normal adults of both sexes are reported in Table 1.

DISCUSSION

The data reported in this paper suggest that human erythrocyte posses all the enzymatic activities for NAD synthesis. The determination of NAD synthetase activity in hemolysate according to Zerez[4] was also achieved using a similar HPLC method performed in a System Gold Beckman,[5] obtaining results in accord with the enzymatic cycling assay described by this Author. The identification and quantification of enzymatic activities, which had not been detected previously in human red blood cells, is of particular interest. In fact adenylyltransferases and NAmPRT were considered not functioning in these cells.

The availability of sensitive and reliable methods to determine the activity of the enzymes synthesizing NAD is a useful tool for the study of the metabolism of pyridine nucleotides, also in connection with the altered erythrocyte NAD levels, found in metabolic disorders such as HGPRT and PNP deficiency and in PRPP synthetase superactivity.[6]

REFERENCES

1. H. B. White in "The Pyridine Nucleotide Coenzyme" J. Everse, B. Anderson, K. S. You, ed., Academic Press, N.Y. (1982)
2. J. Preiss and P. Handler, Enzymatic synthesis of nicotinamide nucleotide, J. Biol. Chem. 225:759 (1957)
3. A. Malkin and O.F. Denstedt, Synthesis of diphosphopyridine nucleotide in the erythrocyte, Can. J. Biochem. Physiol. 34:130 (1956)
4. C. R. Zerez, M.D. Wong and K.R. Tanaka, Blood 75:1576 (1990)
5. M. Pescaglini unpublished data
6. H. A. Simmonds, L. D. Fairbanks, G. S. Morris, D.R. Webster and E.H. Harley, Altered erythrocyte nucleotide patterns are characteristic of inherited disorders of purine and pyrimidine metabolism, Clin. Chim. Acta 171:197 (1988)

APPLICATIONS OF PRPP METABOLISM IN HUMAN ERYTHROCYTES

Eric Harley, Duncan Black, Peter Cole, Tony Marinaki and Rosemary Hickman*

Department of Chemical Pathology and *Department of Surgery, University of Cape Town Medical School Observatory, 7925 South Africa

INTRODUCTION

Hershko et al. demonstrated in 1967[1] that rabbit erythrocytes incubated in a high phosphate medium accumulate PRPP. Berman et al. in 1988[2] defined details of the precise conditions of phosphate concentration, pH and pO_2 which control PPRP accumulation in human erythrocytes. Under anoxic conditions xanthine dehydrogenase can be converted to xanthine oxidase which in the presence of hypoxanthine, and molecular oxygen as an electron acceptor, can form superoxide radicals[3]. If this mechanism contributes to reperfusion injury then removal of hypoxanthine from ischaemic tissue before exposure to oxygen might alleviate injury. In this paper, the ability of erythrocytes to accumulate PPRP under appropriate conditions in vitro is utilised to investigate whether such PPRP 'primed' erythrocytes can facilitate the removal of the hypoxanthine which accumulates in ischaemic tissues.

METHODS

Two preparations of human erythrocytes were used: control (unprimed) erythrocytes drawn from volunteers were immediately washed twice in cold 0.9% saline and resuspended in normal saline at a ratio of 1:2 v/v. PPRP-enriched (primed) erythrocytes were prepared by resuspending cells washed as above in 10 mM phosphate pH 7.0 at a ratio of 1:2 v/v and incubating for 2 hours at 37°C. Under these conditions erythrocyte PPRP levels rise from less than 5 µmoles/l intracellular water to typically about 90 µmoles/l.

Four groups of twenty male Long-Evans rats were treated as follows: in all groups the right kidney was removed; the left renal blood vessels were cannulated and occluded for 50 minutes. In group 1 the vessels were then repaired without perfusion; in group 2 the kidney was perfused with warm saline; in group 3 the kidney was perfused with warm control human erythrocytes, and in group 4 perfusion was with primed erythrocytes. After perfusion for 10 mins, the renal vessels in groups 2-4 were repaired. Perfusate in representative animals from groups 2-4 was collected, and trichloracetic acid (TCA) extracts were analysed by anion exchange HPLC[2]. Morbidity was assessed by measuring plasma creatinine levels on days 0, 1, 2, 4, 6, and 8 in surviving animals.

In order to compare hypoxanthine levels in perfused and non-perfused ischaemic kidneys, six rats were treated as above except that both kidneys were rendered ischaemic for 50 minutes. The right kidney was then removed

Purine and Pyrimidine Metabolism in Man VII, Part B
Edited by R.A. Harkness *et al.*, Plenum Press, New York, 1991

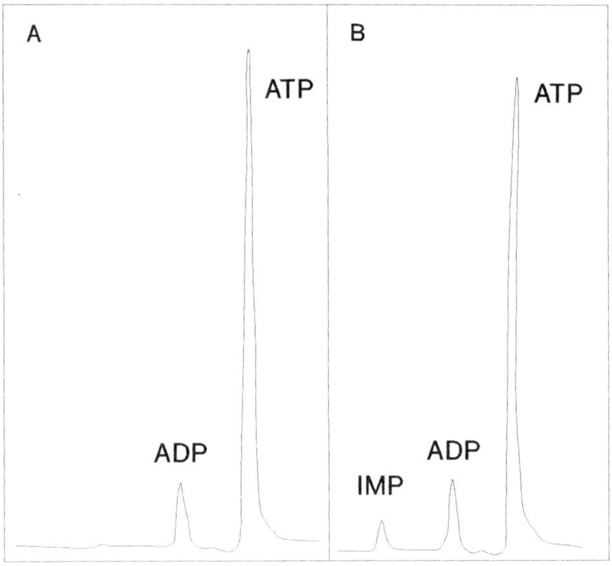

Figure 1

Anion exchange HPLC of a) unprimed, and b) primed erythrocyte extracts after passage through ischaemic rat kidneys.

Table I. Hypoxanthine content of ischaemic rat kidneys after perfusion with either primed or unprimed human erythrocytes

Perfusate		µg hypoxanthine/g kidney (dry wt)		
		Unperfused (right)	Perfused (left)	Difference
Unprimed erythrocytes	1	1269	1050	219
	2	580	340	240
	3	429	103	326
Primed erythrocytes	1	810	271	539
	2	1152	325	877
	3	1303	475	828

for measurement of hypoxanthine by freezing in liquid N_2. followed by TCA extraction and reverse-phase HPLC. The left kidney was perfused with either unprimed or primed erythrocytes (3 rats each) for 10 minutes before removal for hypoxanthine determination as above.

RESULTS AND DISCUSSION

In the HPLC traces from rats perfused with primed erythrocytes a peak of IMP was clearly visible but was absent or small in the unprimed samples (Fig. 1). Levels of IMP, expressed as a % of the ATP levels, were 0.25 in group 2 (saline), 0.60 in group 3 (unprimed erythrocytes) and 5.90 in group 4 (primed), demonstrating a much higher accumulation of IMP in the erythrocytes with elevated PPRP levels.

Comparison of hypoxanthine levels in the perfused and non-perfused kidneys (Table I) showed that although the absolute levels of hypoxanthine varied considerably between the six individual animals tested, there was a consistently greater difference in the hypoxanthine levels between perfused and unperfused kidneys in the primed group (mean 748 $\pm$ 149 µg/g dry weight) than in the unprimed group (mean 202 $\pm$ 46), consistent with the increased levels of IMP found in the PRPP-primed erythrocyte perfusate.

Despite the biochemical evidence of effective removal of hypoxanthine from ischaemic kidneys by primed erythrocytes there was no significant difference in survival between the four groups of rats in the first experiment, the mean survival being 35, 30, 30, and 30% respectively in the four groups. Neither was there significant variation in the serum creatinine profiles in these groups. This implies either that hypoxanthine, in acting as a substrate for free radical production by xanthine oxidase, does not play a significant role in reperfusion injury in rat kidney, or that the experimental design was not appropriate to detect it. The phenomenon whereby erythrocytes can sequester hypoxanthine so effectively under appropriate conditions is therefore still at the stage of a 'mechanism in search of a role', and it will be an intriguing search to discover what this role is.

ACKNOWLEDGEMENTS

This work was funded by the MRC and the University of Cape Town.

REFERENCES

1. A. Hershko, A. Razin, T. Shoshani, and J. Mager, Turnover of purine nucleotides in rabbit erythrocytes II. Studies in vitro. Biochim. Biophys. Acta, 149:59 (1967).
2. P.A. Berman, D.A. Black, L. Human, and E.H. Harley, Oxypurine cyles in human erythrocytes regulated by pH, inorganic phosphate and oxygen. J. Clin. Invest. 82:980 (1988).
3. E. De La Corte, and F. Stirpe, The regulation of rat liver xanthine oxidase, Biochem. J. 126:739 (1972).

A SYNDROME OF MEGALOBLASTIC ANEMIA, IMMUNODEFICIENCY, AND EXCESSIVE NUCLEOTIDE DEGRADATION

Theodore Page, William L. Nyhan, Alice L. Yu, and John Yu*

Department of Pediatrics, 0609
University of California, San Diego
La Jollà, CA 92093 USA

*Department of Molecular and Experimental Medicine
Research Institute of Scripps Clinic
La Jolla, CA 92037 USA

INTRODUCTION

Several defects of purine and pyrimidine metabolism have been associated with behavioral abnormalities. The most common and best studied is Lesch-Nyhan syndrome[1], with its characteristic aggressive, self-mutilating behavior. Deficiency of adenylosuccinate lyase[2] has been reported to be associated with infantile autistic behavior. Autistic behavior as well as seizures have been associated with thymine-uraciluria[3]. We report here a syndrome which involves behavioral abnormalities, seizures, and macrocytic anemia and which is associated with increased degradation of purine and pyrimidine nucleotides.

CASE HISTORY

The patient, a three-year-old white female, was first seen because of recurrent infections, developmental delay, and seizures. Upon examination, she was found to have mild immunodeficiency, macroycytic anemia, ataxia, and alopecia. Physical development and speech were notably delayed. IgG was borderline to low, and MCV was variable from 90 to 100 (normal for age and sex <88). A severe, recurrent sinus infection required surgical drainage. No abnormalities of amino acid or organic acid metabolism were identified by a routine metabolic screen and amino acid analysis. No unusual compounds were detected in plasma or urine by HPLC. All parameters of folic acid and B12 metabolism were found to be within normal limits. The most striking feature of her phenotype was her bizarre, and often aggressive behavior. She was hyperactive, with a short attention span, inappropriate verbalizations, and poor interaction with other children. Aggressive behavior took the form of pinching or scratching others, or biting toys. She would sometimes bang her head or poke at her eye with her finger. She had 2-3 seizures of 1-2 min duration per day. Initially, she was treated with IgG, folinic acid, depakote, and tegritol. Her sinus infection resolved and seizure activity decreased but there was no change in her MCV, behavior, or speech development. At this time, an investigation of her nucleotide metabolism in cultured fibroblasts was begun. Based on the findings of these studies, a trial with oral nucleotides was begun. Upon initiation of this treatment, an almost immediate improvement in speech, behavior, and cognitive function was seen. Speech became more understandable and appropriate and she seemed to pay more attention to her surroundings, and focus better on tasks. Her interactions and play with other children became appropriate, and her mother described her behavior as that of a normal child. MCV remained elevated. Seizure activity decreased markedly, such that she was taken off depakote (625 mg/day), and the dose of tegritol was gradually reduced from 500 to 50 mg/day, with the intention of eliminating it as well. However, an interuption in the supply of nucleotides caused a one week interuption in oral nucleotide therapy. During this time, seizure activity increased to >10 seizures per day. Her attention span became limited, and her frustration tolerance low. Verbalization and interaction with others deteriorated, and behavior became more aggressive. At that time she was returned to pretreatment doses of depakote and tegritol. Upon resumption of nucleotide therapy, these symptoms resided, and her condition prior to the interuption of therapy gradually returned.

MATERIALS AND METHODS

Incorporation studies were done as described earlier[4] for adenine, guanine, hypoxanthine, formate, uridine, and thymidine. For glycine and orotic acid incorporation studies, isotope (10 uCi/ml) was added to Minimal Essential Medium and cells were grown in 75 mm plates for 72 hr, harvested by trypsinization, and analyzed by HPLC as above. Incorporation experiments were done in duplicate. For the assay of individual enzymes, cultured fibroblasts were harvested in the log phase of growth and lysed at a concentration of approximately 1 mg/ml in a 0.10 sodium phosphate buffer, pH 7.2 containing 0.05 M magnesium chloride. Serial dilutions of this lysate were incubated with a 10 uM concentration of radiolabeled substrate for 1 hr at $37^{o}C$. For the assay of UMP synthetase 1 mM PRPP was added. For the assay of uridine kinase, 1 mM ATP was added. The assays were deproteinized and analyzed by HPLC. Assays were done in triplicate. Reported values do not necessarily represent maximum enzyme activities due to the low substrate concentration. Nucleoside inhibition of erythroid colony formation by bone marrow cells was done as described earlier[5].

RESULTS AND DISCUSSION

The incorporation of purine and pyrimidine precursors into nucleotides is shown in Table 1. Normal incorporation of glycine into purines, as well

Table 1. Incorporation of Precursors into Nucleotides

Precursor	Patient	Controls (n)
adenine	8160	9727 (4)
hypoxanthine	3308	2849 (4)
guanine	3392	3129 (4)
formate + AICAR	3688	3658 (4)
glycine	10071	8350 (2)
uridine	3469	8511 (4)
thymidine	1223	1027 (2)
orotic acid	5694	18315 (2)

Incorporation is reported in units of pmol/100 nmol purines/2 hr as normal excretion of uric acid by the patient indicate normal denovo purine synthesis. From the incorporation of adenine, hypoxanthine, and guanine into the various purine nucleotides it is clear that the activities of enzymes of purine nucleotide interconversion (i.e. adenylosuccinate synthetase, adenylosuccinate lyase, AMP deaminase, IMP dehydrogenase GMP, synthetase, GMP reductase, and the purine nucleotide mono- and diphosphate kinases) are comparable to those of normal controls (data not shown). Similarly, the production of normal amounts of UTP, CTP, and TTP from uridine indicates that the enzymes of pyrimidine nucleotide interconversion (i.e. CTP synthetase, thymidylate synthetase, and the pyrimidine mono- and diphosphate kinases) are comparable to normal controls. The only notable differences in these precursor incorporation studies was the low incorporation of orotic acid and uridine into pyrimidine nucleotides. This could reflect low activities of the synthetic enzymes or increased catabolism of the nucleotide products.

To study this question further, individual enzyme activities in dialyzed fibroblast lysates were measured (Table 2). To determine if a deficiency existed in the synthesis of pyrimidine nucleotides, the activities of UMP synthetase and uridine kinase were measured, and found to be normal. The only consistent difference between the patient and normal controls was a ten- to thirty-fold increase in the catabolism of UMP. When the rate of purine nucleotide catabolism was measured for comparison a similar elevation was found. Interestingly, the increased catabolism of purine nucleotides had no noticeable effect on the incorporation of purine precursors into purine nucleotides, whereas the increased catabolism of pyrimidine nucleotides appeared to result in a net decrease in pyrimidine nucleotide synthesis, as measured in cultured fibroblasts. To further study the metabolism of pyrimidine nucleotides in intact cells, the effect of pyrimidine nucleosides on erythroid colony formation was measured in the presence and absence of nucleosides (Table 3). Clearly, the patient's bone marrow cells show much less inhibition of colony formation in presence of thymidine and uridine than control bone marrow cells. This indicates that in the patient's cells, pyrimidine nucleosides are either transported or phosphorylated more slowly, or that the pyrimidine nucleotides, once formed, are degraded more rapidly. Again, it is interesting to note that this effect is present with thymidine, although intact fibroblasts show no decrease in the incorporation of thymidine into thymidine nucleotides.

On the basis of these results, it was decided to intiate pyrimidine nucleotide replacement therapy. In orotic aciduria, in which there is a known defect in pyrimidine synthesis associated with macrocytic anemia, pyrimidine nucleotide replacement therapy has been quite successful. The patient was started with 150 mg/kg/day each of UMP and CMP Plasma uri

Table 2. Activities of Catabolic Enzymes in Dialyzed Cell Lysates

Enzyme (substrate)	Patient	Controls (n)
5'Nucleotidase (UMP)	7.44	0.65 (4)
5'Nucleotidase (AMP)	9.64	0.80 (4)
Nucleoside Phosphorylase (uridine)	0.41	0.44 (4)
Nucleoside Phosphorylase (inosine)	5.72	4.24 (4)
Adenosine Deaminase	3.54	5.09 (4)
UMP Synthetase	1.56	2.17 (4)
Uridine Kinase	2.86	3.51 (2)

Enzyme activities are in nmol/min/mg protein

Table 3. Inhibition of Erythroid Colony Formation by Nucleosides

Nucleoside (concentration)	Patient	Control
None	347 (100%)	169 (100%)
Uridine (10 uM)	322 (94%)	166 (98%)
Uridine (50 uM)	320 (94%)	164 (97%)
Uridine (75 uM)	335 (98%)	56 (33%)
None	508 (100%)	328 (100%)
Thymidine (10 uM)	486 (95%)	251 (76%)
Thymidine (50 uM)	298 (59%)	0 (0%)
Thymidine (75 uM)	96 (18%)	0 (0%)

Colony formation in units of average colonies per plate

dine and erythrocyte UTP were monitored during therapy. Plasma uridine was undetectable before therapy and stayed in the range of 20-50 uM during therapy. Erythrocyte UTP was similarly undetectable before therapy, and was maintained in the range of 20-60 nmol/ml packed red cell during therapy. A general improvement in the patient's condition was noted, but MCV remained abnormally high. Because increased catabolism of purine nucleotides was also indicated, 75 mg/kg/day AMP was included, and the daily dose of pyrimidines was increased to 500 mg/kg/day. These measures produced no additional improvement, and MCV remained high.

At present, the precise metabolic basis of these symptoms remains unknown. The increase in nucleotidase activity could be the primary defect, or it could be a response to abnormal amounts of some as yet unidentified nucleotide. The fact that MCV remained high, even during nucleotide replacement therapy with adenine, cytidine, and uridine nucleotides might indicate a shortage of other nucleotides, perhaps deoxynucleotides. Alternatively, the amount of nucleotides used here may have been inadequate to maintain normal nucleotide levels in the presence of increased nucleotidase activity.

REFERENCES

1. M Lesch and WL Nyhan, A familial disorder of uric acid metabolism and central nervous system function, Am J Med 36:561 (1964)

2. J Jaeken and G Van den Berghe, An infantile autistic syndrome characterized by the presence of succinylpurines in body fluids, The Lancet, Nov 10:1058 (1984)

3. JAJM Bakkeren, RA De Abreu, RCA Sengers, FJM Gabreels JM Maas, and WO Renier, Elevated urine, blood, and cerebrospinal fluid levels of uracil and thymine in a child with dihydrothymine dehydrogenase deficinecy, Clin Chim Acta 140:247 (1984)

4. T Page, Purine nucleotide production in normal and HPRT$^-$ cells, Int J Biochem 21:1377 (1989)

5. J Yu, L Shao, J Vaughan, W Vale, and A Yu, Characteristics of the potentiation effect of activin on human erythroid colony formation in vitro, Blood 73:952 (1989)

ADENINE NUCLEOTIDE CATABOLISM IN THE ERYTHROCYTES OF URAEMIC PATIENTS

Maciej Marlewski, Ryszard T. Smoleński, Julian Swierczyński, Boleslaw Rutkowski* and Mariusz M. Żydowo

Departments of Biochemistry and Kidney Diseases*, Academic Medical School Gdansk, 80-211 Gdańsk, Dębinki 1, Poland

INTRODUCTION

Chronic renal failure is frequently associated with anaemia. Red blood cells (RBC) of these patients are characterised by several biochemical abnormalities such as increased glucose utilisation[1], accumulation of unusual purine and pyrimidine nucleotides[2] and elevated ATP and GTP concentrations[3]. Some of these biochemical abnormalities such as elevated ATP and GTP levels are reversed within a few days after successful renal transplantation[4]. This effect cannot be explained by the replacement of the erythrocytes by a new population of cells but could be a consequence of changes in the metabolic environment of the erythrocyte. This observation excludes also a possibility that a different age of erythrocytes is a primary cause of elevated nucleotide levels. However, a positive balance of nucleotide pool turnover could result also from excessive rate of nucleotide generation or inhibition of catabolic pathways.

Adenine nucleotides interconversions are directly linked to cellular energy metabolism. Any disturbances of this processes could therefore reduce erythrocyte viability promoting the development of anaemia. To obtain further informations concerning the nature of the abnormalities in the uraemic erythrocytes we studied fluxes through nucleotide catabolic pathways in the intact cells according to the concentration of accumulated products of these pathways. We found much faster nucleotide degradation in the uraemic erythrocytes both under basal and stimulated conditions which primarily resulted from accelerated flux through the AMP-deaminase.

METHODS

Blood was collected in heparinised tubes from healthy subjects (n=3) and from patients with advanced renal failure (n=4). The blood samples were centrifuged, the plasma and the top fifth layer removed. The erythrocytes were washed subsequently three times in the incubation buffer of the following composition: 10 mM HEPES (pH 7.4), 125 mM NaCl, 2.6 mM KCl, 1.2 mM KH_2PO_4, 1.2 mM $MgSO_4$, 1 mM $CaCl_2$ and 5.5 mM glucose. Washed erythrocytes were adjusted to hematocrit = 20 % and the incubations were carried out at 37°C. Some incubations contained erythro-9-(2-hydroxy-3-nonyl)adenine (EHNA) an adenosine deaminase inhibitor at 15 μM concentration. The glycolytic inhibitor - iodoacetic acid (IA) was present where indicated at 0.5 mM concentration. At the time indicated, 0.5 ml of the erythrocyte suspension was collected from the incubation flask and mixed with 0.25 ml of 1.3 M $HClO_4$. After removal of the protein precipitate by centrifugation, clear supernatant was neutralised with 2 M KOH. Nucleotide, nucleoside and base concentration was analysed using reversed-phase high performance liquid chromatographic procedure, as described previously[5].

RESULTS

Initial ATP remained relatively stable in the erythrocytes of both healthy subjects and renal failure patients. However, some increase in hypoxanthine concentration was observed (Fig. 1). The rate of hypoxanthine formation under basal conditions was six times greater in the erythrocytes of uremic patiens than in healthy subjects. Presence of EHNA did not attenuated hypoxanthine production or even some stimulating effect was observed in renal failure erythrocytes.

Incubation in the presence of IA caused a gradual decrease in ATP concentration at the faster rate in the renal failure erythrocytes (Fig. 2). A transient increase in ADP concentration was also observed. Interestingly, progressive increase in AMP concentration was similar in renal failure and healthy subjects erythrocytes (Fig. 2). Hypoxanthine was the predominant catabolite with IA alone (Fig. 3). Presence of EHNA caused a change in catabolite pattern without great influence on the total rate of degradation.

DISCUSSION

The most important finding of this study was the demonstration of the accelerated degradation of adenine nucleotides in the erythrocytes of patients with chronic renal failure. This was demonstrated either under basal conditions and after stimulation of the nucleotide catabolism by application of the glycolytic inhibitor.

We could confirm that basal nucleotide catabolism in the erythrocytes proceed via AMP-deaminase[6] as we have not observed any inhibition of hypoxanthine formation by EHNA (Fig. 1), in fact stimulating effect was observed in the renal failure erythrocytes. The flux through the AMP-deaminase pathway under basal conditions was six fold higher in the erythrocytes of the renal failure patients than in the healthy subjects.

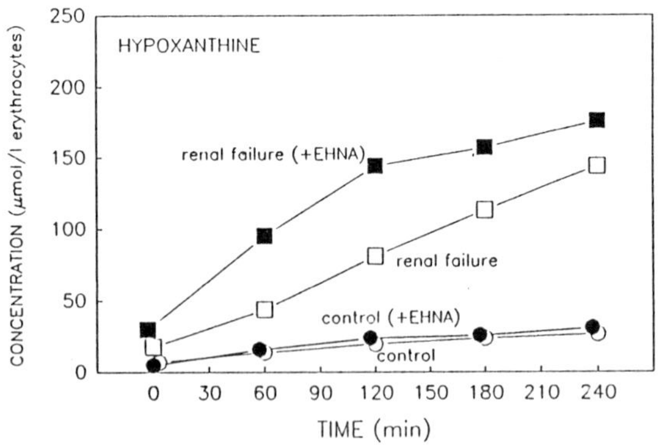

Fig.1
Production of hypoxanthine in the erythrocytes of uraemic patients and controls (healthy subjects) under basal conditions in the presence or absence of EHNA.

A striking observation demonstrated during the incubation in the presence of IA was the dissociation of nucleotide catabolic rate from the AMP concentration (Figs. 2,3). Purine catabolite production was 3 times greater in the erythrocytes of uraemic patients while the AMP concentration changes exhibited a similar pattern in both types of cells. An increased flux through 5'-nucleotidase was responsible for some of this observation, but this mainly resulted from the acceleration of the AMP-deaminase pathway (Fig. 3). However, no difference in the maximal AMP-deaminase activity was found in the haemolysates between normal and renal failure erythrocytes[7]. Our results must imply therefore an activation of AMP-deaminase. An allosteric effect of ATP[8] could be taken into consideration as its concentration was higher in renal failure erythrocytes.

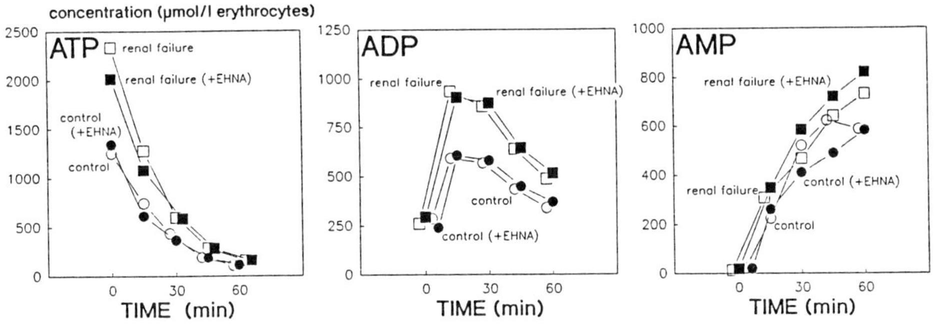

Fig.2

Changes in ATP, ADP and AMP concentrations in the erythrocytes of uraemic patients (squares) and healthy subjects (circles) in the presence of iodoacetate with (closed symbols) or without EHNA (open symbols).

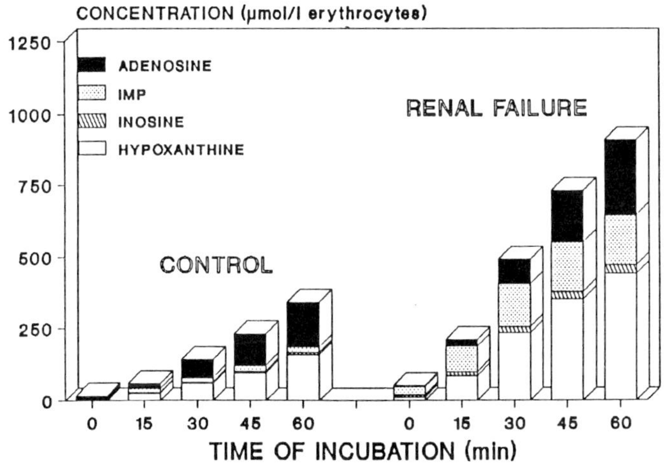

Fig.3

Purine catabolite production in the erythrocytes of uraemic patients and controls (healthy subjects) incubated in the presence of iodoacetate and EHNA allowing estimation of the flux through 5'-nucleotidase (adenosine) and AMP-deaminase (IMP, inosine and hypoxanthine).

The estimate of 5'-nucleotidase flux do not take into account possible recycling of adenosine via adenosine kinase. However, maximal capacity of this process[9] is very small in comparison with the rate of adenosine formation observed in our study. In addition it seems to be further inhibited by the depletion of ATP concentration.

An important conclusions relating to the mechanism of elevated nucleotide concentration in uraemic erythrocytes can be drawn from these experiments. An elevation of nucleotide pool could result from the inhibition of nucleotide degradation as was observed in AMP-deaminase deficient erythrocytes[10]. However, it could not apply to renal failure erythrocytes as just the opposite effect was demonstrated in our study. Therefore an enhanced nucleotide synthesis could be a primary cause of raised nucleotide concentration with faster nucleotide degradation as a secondary effect. Acceleration of the nucleotide pool turnover would have to be associated with increase of high energy phosphate consumption and acceleration of glucose utilisation which was demonstrated in uraemic erythrocytes[1]. However, definite confirmation of this hypothesis, and possible contribution of this mechanism to the reduction of erythrocyte number in uraemic blood require further detailed studies.

ACKNOWLEDGEMENT

We are indebted to Dr. H. Anne Simmonds for many stimulating discussions relating to these results. This work was supported by a grant from the Polish Committee for Scientific Research.

REFERENCES

1. D.E. Lichtman, and R.D. Miller, Erythrocyte glycolysis, 2,3-diphosphoglycerate, and adenosine triphosphate concentration in uraemic subjects: relationship to extracellular phosphate concentration, J Lab Clin Med. 76:267 (1970).
2. C.R. Angle, M.S. Swanson, S.J. Stohs, R.S. Markin, Abnormal erythrocyte pyrimidine nucleotides in uremic subjects, Nephron. 39:169 (1985).
3. G.A. Hurt, and A. Hanutin, Organic phosphate compounds of erythrocytes from individuals with uremia, J Lab Clin Med. 64:675 (1964).
4. A.S.M. Rejman, M.A. Mansell, A.J. Grimes, and A.M. Joekes, Rapid correction of red-cell nucleotide abnormalities following sucessful renal transplantation, Br J Haematol. 61:433 (1985).
5. R.T. Smolenski, A.C. Skladanowski, M. Perko, and M.M. Zydowo, Adenylate degradation product release from the human myocardium during open heart surgery, Clin Chim Acta. 182:64 (1989).
6. F. Bontemps, and G. Van den Berghe, Mechanism of adenosine triphosphate catabolism induced by deoxyadenosine and by nucleoside analogues in adenosine deaminase-inhibited human erythrocytes, Cancer Res. 49:4983 (1989).
7. R.T. Smolenski, C. Montero, A.V. Rodgers, and H.A. Simmonds, A high performance liquid chromatographic assay for AMP-deaminase activity in the erythrocytes of healthy subjects and patients with inherited purine disorders, Biomed Chromatogr, in press
8. C-Y. Lian, and Harkness D.H., The kinetic properties of adenylate deaminase from human erythrocytes. Biochim Biophys Acta 341:27 (1974).
9. I. Rapoport, S. Rapoport, and G. Gerber, Degradation of AMP in erythrocytes of man. Evidence for a cytosolic phosphatase activity, Biomed Biochim Acta. 5:317 (1987).
10. N. Ogasawara, H. Goto, Y. Yamada, I. Nishigaki, I. Hasegawa, and K.S. Park, Defficiency of AMP-deaminase in the erythrocytes, Hum Genet. 75:15 (1987).

NUCLEOTIDE CATABOLISM IN RED BLOOD CELLS OF RABBIT

Iris Rapoport, Werner G. Siems, Andreas Werner, and Gerhard Gerber

Institute of Biochemistry, Medical Faculty (Charité), Humboldt University, Hessische Str 3/4, O-1040 Berlin, F.R.G.

INTRODUCTION

The three metabolites ATP, ADP and AMP may be considered as a common pool connected by adenylate kinase, which in eucaryotic cells, including red blood cells, is present in sufficiently large amounts as to establish an equilibrium among the adenine nucleotides. The immediate substrate for the adenine nucleotide degradation is AMP. In red blood cells there exist two pathways for the breakdown of AMP: (i) one yielding adenosine which may be deaminated to inosine or rephosphorylated to AMP and (ii) one producing IMP by deamination and inosine by dephosphorylation of IMP. From inosine there is formed hypoxanthine by the action of nucleoside phosphorylase. For mature erythrocytes, the existence of both pathways has been established. Under physiological conditions the degradation via IMP is the predominant pathway.[1-3] In this study we continue the quantifications of adenine nucleotide breakdown in mature erythrocytes and reticulocytes. We have investigated conditions of glucose starvation which lead to an accelerated adenine nucleotide breakdown.

MATERIAL AND METHODS

Erythrocytes and reticulocytes of rabbits were obtained as described.[4] Experiments with reticulocytes were carried out between 6th to 8th day of bleeding with an average reticulocytosis of 35%. The blood was centrifuged immediately, plasma and buffy coat were carefully removed. The cells were washed thrice with isotonic triethanolamine-NaCl buffer, pH 7.4. The triethanolamine concentration in the first washing step was 100 mM, that in the next two steps 50 mM. In the case of physiological conditions the buffers contained in addition 1 mM inorganic phosphate and 15 mM glucose. Cells were resuspended in the buffer used for the last washing step. The incubation temperature was 37 °. Determinations of metabolites were carried out as described before.[3,5]

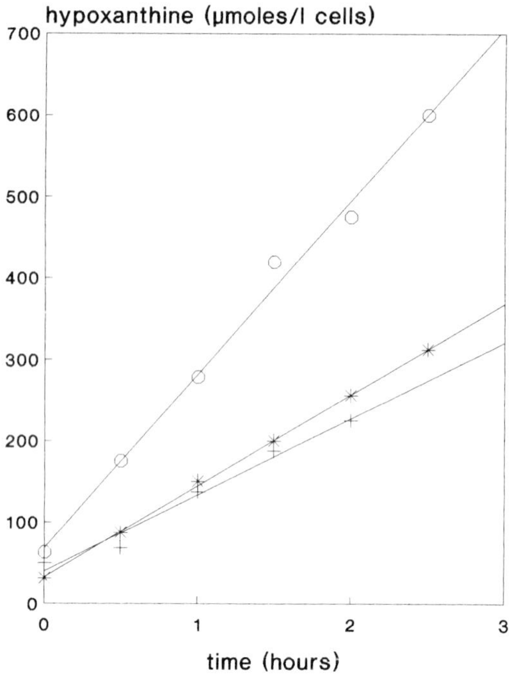

Fig. 1. Formation of hypoxanthine in reticulocyte (18.2%)-rich cell suspensions in presence of glucose (-✱- control; -+- + coformycin 1.35 µM; -o- + 5′-amino-5′-deoxyadenosine 0.6 mM).

RESULTS AND DISCUSSION

Under nearly physiological conditions the rate of breakdown of adenine nucleotides of rabbit erythrocytes amounts to 23 ± 12 µmoles (sum of adenine nucleotides)/l cells x h (n=4). This corresponds to a formation rate of hypoxanthine of 22 ± 9 µmoles/l cells x h (n=4). In reticulocytes, the rate of breakdown of adenine nucleotides is found to be markedly increased. It amounts to 331 ± 45 µmoles/l reticulocytes x h (n=3). The formation rate of hypoxanthine is 374 ± 84 µmoles/l reticulocytes x h (n=10). Coformycin at a concentration of 1 µM which inhibits the adenosine deaminase specifically has neither an influence on the formation of hypoxanthine in erythrocytes (not shown) nor in reticulocyte-rich cell suspensions under these conditions (Fig. 1). Therefore the conclusion can be drawn that under physiological conditions the main pathway of degradation of AMP follows the sequence AMP --> IMP --> inosine --> hypoxanthine in both types of cells. In addition, there seems to exist an intensiv "futile cycle" involving the sequence AMP --> adenosine --> AMP. Inhibition of the adenosine kinase by 5′-amino-5′-deoxyadenosine leads to a 3.5-fold increase of the formation

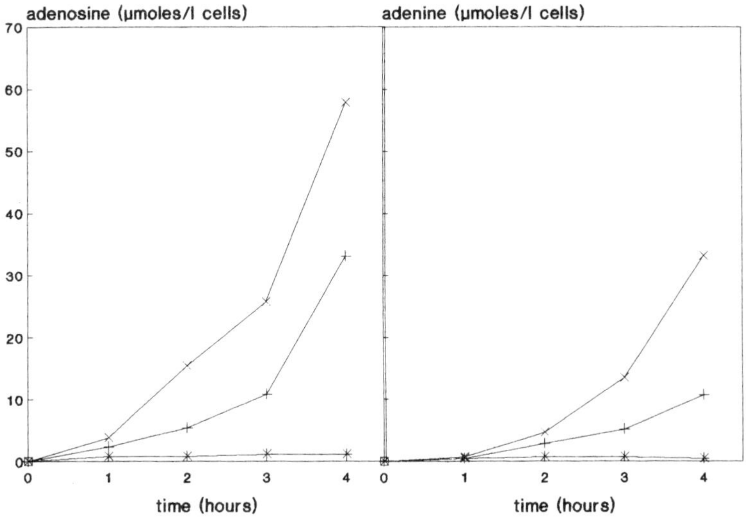

Fig. 2. Adenosine (left side) and adenine (right side) accumulation in glucose-deprived reticulocyte (35%)-rich cell suspensions (-✱- control; -+- + coformycin 1 µM; -x- + iodotubercidin + coformycin 1 µM).

of hypoxanthine in erythrocytes (not shown) and to a twofold increase in reticulocytes (Fig. 1). Glucose starvation at simultaneous inhibition of the respiratory chain by oligomycin (10 µM) causes a drastic decrease of ATP to below 25% of the initial concentration within two hours. Nevertheless, within the first two hours the concentration of ATP is high enough to permit the "futile cycling" (k_m of adenosine kinase = 800 µmoles/l).

There is no accumulation of adenosine (Fig. 2, control), and the formation of hypoxanthine within the first two hours does not differ significantly from that in the presence of glucose (Fig. 3, control). At ATP concentrations below 250 µmoles/l cells the "futile cycle" breaks down. Under these conditions the adenosine deaminase competes successfully with the adenosine kinase leading to a drastic increase (threefold) in the formation of hypoxanthine (Fig. 3, control). This increase can be prevented by coformycin (Fig. 3). Thus, in the absence of active adenosine kinase, the pathway of AMP degradation via AMP --> adenosine --> inosine --> hypoxanthine predominates. In agreement with the fact that the "futile cycle" breaks down after two hours of incubation in the absence of glucose, inhibition of adenosine kinase by use of iodotubercidin produces an increase of hypoxanthine production only during this initial phase (Fig. 3). After the second hour the hypoxanthine formation rates are nearly identical with the control. If adenosine kinase and adenosine deaminase are inhibited or kept without substrate the concentration of adenosine increases (Fig. 2). As described for human red blood cells,[6] this increase is followed by an increase of adenine (Fig. 2). Our results indicate that erythrocytes and reticulocytes differ quantitatively but not qualitatively in their pathways of breakdown of adenine nucleotides.

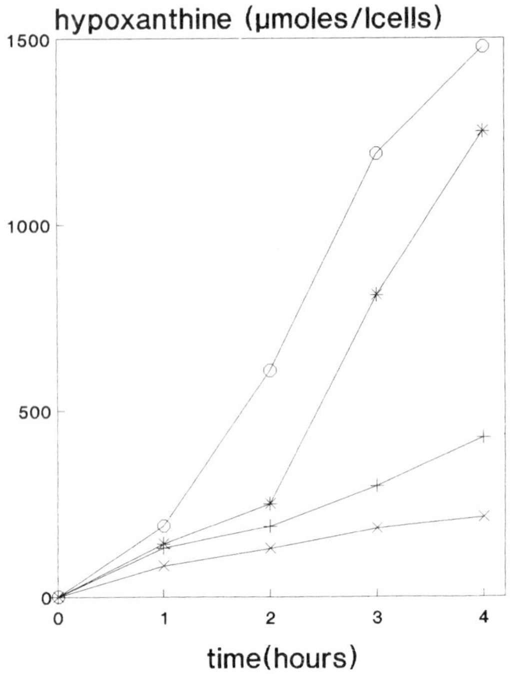

Fig. 3. Hypoxanthine accumulation in glucose-deprived reticulocyte (35%)-rich cell suspensions (-*- control; -+- + coformycin 1 µM; -x- + iodotubercidin + coformycin 1 µM; -o- + iodotubercidin).

REFERENCES

1. F. Bontemps, G. van den Berghe, and H. G. Hers, Pathways of adenine nucleotide catabolism in erythrocytes, J. Clin. Invest., 77:824 (1986).
2. D. E. Paglia, W. N. Valentine, M. Nakatani, and R. A. Brockway, Mechanisms of adenosine 5′-monophosphate catabolism in human erythrocytes, Blood, 67:988 (1986).
3. I. Rapoport, S. M. Rapoport, and G. Gerber, Degradation of AMP in erythrocytes of man. Evidence for a cytosolic phosphatase activity, Biomed. Biochim. Acta, 46:317 (1987).
4. I. Rapoport, I. Drung, and S. M. Rapoport, Catabolism of adenine nucleotides in rabbit blood cells, Biomed. Biochim. Acta, 49:11 (1990).
5. T. Grune, W. Siems, A. Werner, G. Gerber, M. Kostic, and H. Esterbauer, Aldehyde and nucleotide separations by high-performance liquid chromatography - Application to phenylhydrazine-induced damage of erythrocytes and reticulocytes, J. Chromatogr., 520:411 (1990).
6. A. Werner, I. Rapoport, and G. Gerber, HPLC analysis of purine metabolism in human erythrocytes deprived of glucose, Biomed. Biochim. Acta, 46:777 (1987).

THE EFFECT OF ASPIRIN ON BLOOD CELL NUCLEOTIDES *IN VIVO*

L.L. Herbert*, K.E. Herbert*, R. Jacobs*, D.L. Scott* and D. Perrett[$]

Departments of *Rheumatology & [$]Medicine, The Medical College of St Bartholomew's Hospital, West Smithfield, London EC1A 7BE

INTRODUCTION

Erythrocytes from rheumatoid arthritis (RA) patients have significantly lower levels of ATP compared to normals (our unpublished observations; 1093+182nmol/ml in RA compared to 1371+292nmol/ml for controls). These findings have important implications for cell function in RA and not exclusively for erythrocytes. Severe depletion of ATP has been linked to programmed cell death in resting lymphocytes (1), adenosine deaminase deficient nucleated cells (2), and may also play a vital role in the formation of mucosal injury caused by the non-steroidal anti-inflammatory drugs (NSAIDs) (3). We considered our original observations may relate to NSAIDs which the patients take regularly; salicylates have been known to inhibit cellular ATP synthesis for over thirty years (4). To further investigate this phenomenon we studied changes to erythrocyte nucleotides in normal subjects taking enteric-coated aspirin daily (900mg) for three weeks.

SUBJECTS

Four healthy volunteers (one female, 3 males, median age= 30, range 25-40) not currently receiving medication and without a history of gastrointestinal disorders participated in the study. Local ethics committee approval was given for this investigation.

METHODS

Enteric-coated aspirin (kindly supplied by Dr Peter Crowther, of Eli Lilly) was taken three times daily (300mg after meals) from day 0 for 21 days by four subjects.

Venous blood samples were collected from the subjects studied at days -7, 0, +7, +14, +21. The blood (2ml) in lithium heparin tubes was centrifuged immediately for 10 min at 2000g. The buffy layer and top 20% of red blood cells was aspirated using a water vacuum pump. Nucleotides were extracted from 0.5ml of the red cell layer using 1ml cold (4°C) trichloroacetic acid (12% w/v) containing cXMP (40uM) as internal standard. Following 10min incubation at room temperature and centrifugation at 2000g for 5min excess trichloroacetic acid was removed from the clear supernatant by back extraction with water-saturated diethyl-ether until neutralised. Nucleotides were analysed by anion-exchange gradient HPLC with uv detection at 254nm (5).

RESULTS and DISCUSSION

There were no changes in AMP, ADP and ATP concentrations in red blood cells at any time after beginning aspirin intake. Consequently there was no change in adenine nucleotide energy charge values and the ATP/ADP ratio calculated from these nucleotides (Table 1). Energy charges were consistently within the range 0.87-0.91 throughout the study, values indicative of normal cells (Table 1) (6). No changes were observed in other meaningful nucleotides eg GTP, NAD and IMP.

Subsequently two healthy subjects (one male, one female) taking aspirin (900mg/day) were studied in more detail. Blood was taken daily and nucleotides were analysed in order to discount early changes that may have been missed in the original study. However in this study also the nucleotide levels, energy charges and ATP/ADP ratios remained constant throughout the study.

TABLE 1. EFFECT OF ASPIRIN ON ADENINE NUCLEOTIDES IN ERYTHROCYTES

	MEAN (SD) VALUES FOR SUBJECTS 1-4				
DAY	[AMP]	[ADP]	[ATP]	EC	ATP/ADP RATIO
	(nmol/ml)				
-7	110.9 (36.8)	315.3 (192.5)	1681.1 (215.1)	0.87 (0.06)	6.75 (3.36)
0	99.7 (18.1)	265.5 (71.4)	1759.4 (388.8)	0.89 (0.01)	6.72 (0.57)
+7	106.9 (52.5)	190.7 (121.7)	1826.9 (337.1)	0.89 (0.03)	7.21 (2.13)
+14	104.8 (19.2)	237.2 (16.6)	1668.7 (242.8)	0.89 (0.02)	7.07 (1.19)
+21	89.5 (25.3)	207.4 (25.2)	1519.7 (166.6)	0.90 (0.01)	7.36 (0.72)

We conclude that aspirin at a therapeutic dose has no effect on nucleotides and energy charge of erythrocytes *in vivo*. Therefore, we can find no evidence that anti-inflammatory drugs would account for the low purine nucleotide energy charge values found in rheumatoid arthritis red blood cells. Arthritic patients are often taking a variety of drugs some of which may cause ATP depletion in erythrocytes not observed here with aspirin. A study in man showed that aspirin (3g/day) caused a small but significant decrease in erythrocyte ATP levels after 5 days; however HPLC detection was not used and ADP and AMP were not quantified (7). As yet we can not discount hypotheses stating that the decreased ATP levels in erythrocytes are part of, or due to, the disease process in RA. Indeed, in muscle biopsies taken from RA patients a significant decrease in ATP was observed, related to a fall in total adenine nucleotide levels (8), again suggesting that adenine nucleotide metabolism is altered in RA.

ACKNOWLEDGEMENTS

We thank the Joint Research Board of St. Bartholomew's Hospital, North East Thames Region Health Authority Locally Organised Research Scheme and the Arthritis and Rheumatism Council for financial support. DLS is Muir Hambro Fellow of the Royal College of Physicians. We thank Eli Lilly for kindly donating the enteric-coated aspirin.

REFERENCES

1 Schraufstatter IU., Hyslop PA., Jackson JH., and Cochrane CG., Oxidant-induced DNA damage of target cells, J Clin. Invest., 82:1040 (1988).
2 Carson DA., Seto S., Wasson DB., and Carrere CJ., DNA strand breaks, NAD metabolism, and programmed cell death, Exp. Cell. Res., 164:273 (1986).
3 Rainsford KD., and Whitehouse MW., Biochemical gastro-protection from acute ulceration induced by aspirin and related drugs, Biochem. Pharmacol., 29:1281 (1980)
4 Smith MJH., Salicylates and matabolism, J. Pharm. Pharmacol., 11:705 (1959).
5 Perrett D., Rapid anion-exchange chromatography of nucleotides in physiological fluids, J. Chromatographia., 16:211 (1982).
6 Atkinson D., The energy charge of the adenylate pool as a regulatory parameter. Interaction with feedback modifiers, Biochem., 7:4030 (1968).
7 Monti M., and Bucur I., Red cell organic phosphates during administration of salicylates, Scand. J. Haematol., 16:295 (1976).
8 Nordemar R., Lovgren O., Furst p., Harris RC., and Hultman E., Muscle ATP content in rheumatoid arthritis -a Biopsy study, Scand. J. Clin. Lab. Invest., 34:185 (1974).

AICARIBOSIDE INHIBITS GLUCONEOGENESIS IN ISOLATED RAT HEPATOCYTES

M. Françoise Vincent*, Paul Marangos‡, Harry E. Gruber‡, and Georges Van den Berghe*

*International Institute of Cellular and Molecular Pathology UCL 7539, B-1200 Brussels, Belgium; ‡Gensia Pharmaceuticals, Inc., San Diego, CA 92121, USA

AICAriboside (5-amino-4-imidazolecarboxamide riboside, Z-riboside) is the nucleoside corresponding to AICAribotide (AICAR or ZMP), an intermediate of the "de novo" biosynthesis of purines. AICAriboside is taken up and metabolized by various mammalian tissues, in which it is converted into ZMP by adenosine kinase (1,2). Development of AICAriboside for the treatment of human cardiovascular disease (3) led to the observation of reduction of blood glucose following large infusion (4). Investigations of the effects of AICAriboside on liver were therefore undertaken.

METHODS

Isolated hepatocytes. Isolated hepatocytes (40-70 mg of cells/ml) were prepared as described before (5) from adult male Wistar rats fasted overnight, and incubated at 37°C in Krebs-Ringer-bicarbonate containing 1 % bovine serum albumin. AICAriboside was added immediately before lactate/pyruvate. Glucose, lactate, nucleotides, AICAriboside and allantoin were measured on perchloric acid extracts (5). Triose phosphates, fructose 1,6-bisphosphate (fructose 1,6-P_2) and hexose phosphates were determined in extracts collected through silicone. The analytical methods used are given in (6). Experiments shown are representative of at least three studies that gave similar results.

Measurements of fructose-1,6-bisphosphatase. These were performed on Sephadex G-25 filtrates of 125,000 x g supernatants of rat liver, as in (7).

RESULTS AND DISCUSSION

Effect of AICAriboside on gluconeogenesis. As shown in Fig. 1, increasing concentrations of AICAriboside decreased dose-dependently the rate of formation of glucose from 2 mM lactate/0.2 mM pyruvate; the concentration required for 50% inhibition was 100 μM. AICAriboside inhibited the utilization of lactate in close parallelism with the suppression of the

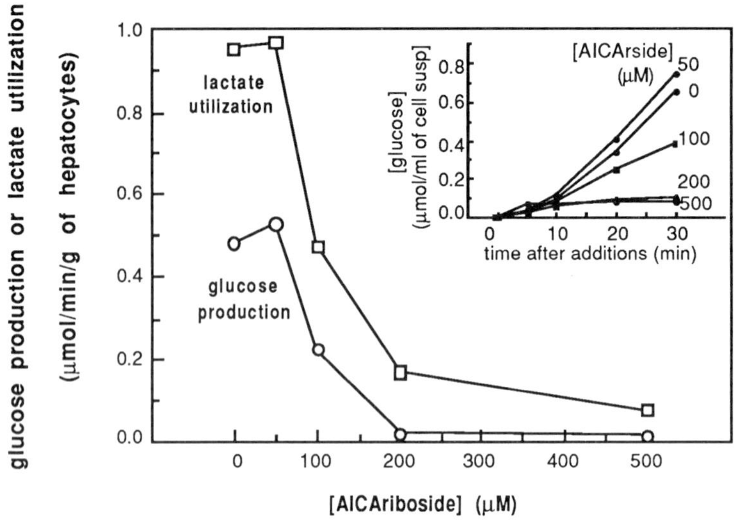

Figure 1. Influence of AICAriboside on the production of glucose and utilization of lactate. Cells were incubated with 2 mM lactate / 0.2 mM pyruvate and various concentrations of AICAriboside. Productions were calculated over the 10 to 30 min interval following the additions of AICAriboside and lactate / pyruvate at time 0 (insert).

production of glucose. The time-course of the experiments (Fig. 1, insert) revealed that these inhibitions occurred after a 10 min latency.

Effect of AICAriboside on gluconeogenic intermediates. Triose phosphates and fructose 1,6-P_2 increased about 2-fold upon addition of 500 μM AICAriboside, whereas fructose 6-P and glucose 6-P decreased by more than 80 %. This indicated that AICAriboside provoked an inhibition of gluconeogenesis at the level of fructose-1,6-bisphosphatase.

Metabolism of AICAriboside. The rate of utilization of AICAriboside increased as a function of its concentration, up to approx. 600 nmol/min per g of cells at 500 μM (not shown). Half maximal rate was obtained at approx. 150

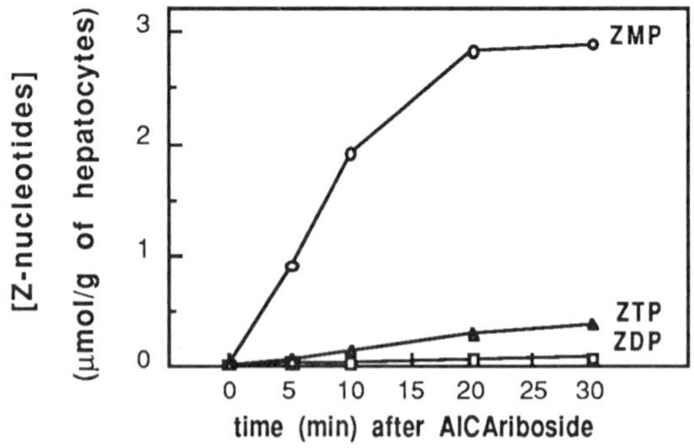

Figure 2. Formation of Z-nucleotides after the addition of 500 μM AICAriboside to isolated hepatocytes.

µM AICAriboside. Utilization of AICAriboside was accompanied by a dose-dependent buildup of its phosphorylated derivatives, ZMP, ZDP and ZTP (Fig. 2). The accumulations of all three Z-nucleotides persisted for 60 min (not shown), without very much decline. The concentrations of ATP, GTP and IMP and the production of allantoin were not significantly modified over a wide (10 - 500 µM) range of AICAriboside additions. This indicates that, in marked contrast with other cell types (2), there is no or little conversion of hepatic ZMP into its formyl derivative, and hence into purine nucleotides and their catabolites.

Effect of ZMP on the activity of liver fructose-1,6-bisphosphatase. The observation that the effect of AICAriboside required its phosphorylation (not shown), together with the cross-over study, led to an investigation of the influence of ZMP on the activity of fructose-1,6-bisphosphatase. ZMP inhibited the activity of the enzyme with an apparent Ki of 370 µM (Fig. 3). Also shown is the effect of AMP (apparent Ki ~ 12 µM), which is the enzyme's main inhibitor, together with fructose 2,6-P_2 (7). Although much less pronounced than that of AMP, the inhibition by ZMP was observed over the range of concentrations of the Z-nucleotide reached in isolated hepatocytes incubated in the presence of AICAriboside.

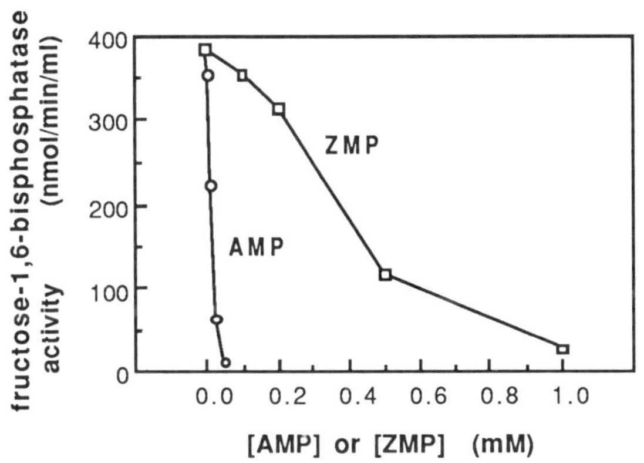

Figure 3. Influence of ZMP and AMP on the activity of liver fructose-1,6-bisphosphatase.

The inhibitory effect of ZMP on fructose-1,6-bisphosphatase can be ascribed to its structural similarity with AMP. Indeed, as pointed out by Sabina et al. (8), the rotatable nature of the bond between C-4 and C-6 in ZMP, allows this nucleotide to assume, besides its classically depicted configuration which is similar to that of GMP, a second one which resembles AMP.

PATHOPHYSIOLOGICAL IMPLICATIONS

Two conditions have previously been reported to provoke accumulation of ZMP in eukaryotic cells: (i) treatment of cancer cells with methotrexate (9) which causes inhibition of the conversion of AICAR into FAICAR, catalyzed by AICAR transformylase; (ii) deficiency of hypoxanthine-guanine phosphoribosyltransferase wich results in a marked acceleration of "de novo" purine synthesis, and renders the transformylase limiting in erythrocytes (10).

Whether both conditions result in elevation of hepatic ZMP, and influence gluconeogenesis, remains to be determined.

Uncontrolled gluconeogenesis is a major cause of hyperglycemia in diabetes (11) and a major therapeutic effect of insulin is to decrease hepatic production of glucose (12). Inhibition of gluconeogenesis by non-hormonal compounds has therefore been extensively investigated (reviewed in ref. 13). Our observations warrant a search for potent inhibitors of fructose-1,6-bisphosphatase, which may have therapeutic potential in this disorder.

ACKNOWLEDGEMENTS

We thank Ms T. Timmerman for expert technical assistance. This work was supported by grant 3.4539.87 of the Fund for Medical Scientific Research (Belgium), by the Belgian State-Prime Minister's Office for Science Policy Programming, and by a grant from Gensia Pharmaceuticals, Inc., San Diego, CA, USA. G. Van den Berghe is Director of Research of the Belgian National Fund for Scientific Research.

REFERENCES

1. Zimmerman TP, Deeprose RD: Metabolism of 5-amino-1-ß-D-ribofuranosylimidazole-4-carboxamide and related five-membered heterocycles to 5'-triphosphates in human blood and L5178Y cells. *Biochem Pharmacol* **27** : 709-16, 1978
2. Sabina RL, Patterson D, Holmes EW: 5-Amino-4-imidazolecarboxamide riboside (Z-riboside) metabolism in eukaryotic cells. *J Biol Chem* **260** : 6107-14, 1985
3. Gruber HE, Hoffer ME, McAllister DR, Laikind PK, Lane TA, Schmid-Schoenbein GW, Engler RL: Increased adenosine concentration in blood from ischemic myocardium by AICA riboside. Effects on flow, granulocytes, and injury. *Circulation* **80** : 1400-11, 1989
4. Dixon R, Gourzis J, McDermott D, Fujitaki J, Dewland P, Gruber HE: AICA-riboside : safety, tolerance and pharmacokinetics of a novel adenosine regulating agent. *J Clin Pharmacol* **31** : 342-7, 1991
5. Van den Berghe G, Bontemps F, Hers, HG: Purine catabolism in isolated rat hepatocytes. Influence of coformycin. *Biochem J* **188** : 913-20, 1980
6. Vincent MF, Marangos P, Gruber HE, Van den Berghe G: Inhibition by AICAriboside of gluconeogenesis in isolated hepatocytes. *Diabetes*, in press
7. Van Schaftingen E, Hers, HG: Inhibition of fructose-1,6-bisphosphatase by fructose 2,6-bisphosphate. *Proc Natl Acad Sci USA* **78** : 2861-3, 1981
8. Sabina RL, Holmes EW, Becker MA: The enzymatic synthesis of 5-amino-4-imidazolecarboxamide riboside triphosphate (ZTP). *Science* **223** : 1193-5, 1984
9. Allegra CJ, Hoang K, Yeh GC, Drake JC, Baram J: Evidence for direct inhibition of de novo purine synthesis in human MCF-7 breast cells as a principal mode of metabolic inhibition by methotrexate. *J Biol Chem* **262** : 13520-6, 1987
10. Sidi Y, Mitchell, BS: Z-Nucleotide accumulation in erythrocytes from Lesch-Nyhan patients. *J Clin Invest* **76** : 2416-9, 1985
11. Shafrir E, Berman M, Felig P: The endocrine pancreas: diabetes mellitus. In *Endocrinology and Metabolism, 2nd ed.* Felig P, Baxter JD, Broadus AE, Frohman LA, Eds. New York, McGraw-Hill, 1986, p. 1043-178
12. Brown PM, Tompkins CV, Juul S, Sönksen PH: Mechanism of action of insulin in diabetic patients : a dose-related effect on glucose production and utilisation. *Br Med J* **I** : 1239-42, 1978
13. Sherratt HS: Inhibition of gluconeogenesis by non-hormonal hypoglycaemic compounds. In *Short-term regulation of liver metabolism.* Hue L, van de Werve G, Eds. Amsterdam, Elsevier/North-Holland, 1981, p. 199-227

Z-NUCLEOTIDES FORMATION IN HUMAN AND RAT CELLS

H.E. Gruber, R. Jimenez and J. Barankiewicz

Gensia Pharmaceuticals, Inc., 11025 Roselle Street
San Diego, CA 92121, USA.

Little is known about the formation and role of Z-nucleotides (ZMP,ZDP,ZTP) in rat and human cells. In cells having an active purine de novo biosynthesis pathway, ZMP (AICA-riboside monophosphoribonucleoside) is formed naturally as one of the final intermediates in IMP biosynthesis. ZMP, when synthesized by this pathway, is then efficiently formylated by 5-amino-4-imidazole-carboxamide 5'-ribonucleotide transformylase (EC 2.1.2.3) and therefore the intracellular level of ZMP is usually very low and is not detectable by HPLC (1-3). ZMP can also be formed when cells with an active or absent de novo purine biosynthesis pathway are exposed to acadesine (AICA-riboside) or AICA-base. ZMP is formed from acadesine after phosphorylation, whereas formation from AICA-base requires the transfer of the phosphoribosyl group from PRPP. It has already been established that adenosine kinase (AK) is responsible for the phosphorylation of acadesine to ZMP, and AK inhibited or AK deficient cells are unable to metabolize acadesine (4-7). ZMP, when formed from acadesine, can be further metabolized by alternative pathways depending on its intracellular concentrations (2). At low ZMP concentrations, metabolism of ZMP follows predominantly the steps of de novo synthesis of IMP and other purines. At intermediate concentrations of ZMP, net retrograde flux through adenylosuccinate lyase results in sZMP (5-amino-4-imidazole-N-succinocarboxamide ribonucleotide) formation, while at high ZMP concentrations, accumulation of ZMP, ZDP and ZTP becomes significant.

There are only a few reports concerning metabolism of AICA-base, showing that AICA-base can be converted to purine nucleotides in human erythrocytes (8), and converted to uric acid and hypoxanthine after intravenous administration in man (9). The incorporation of AICA-base into purine metabolites in human lymphoblasts and fibroblasts has also been reported (6), and adenine phosphoribosyltransferase (EC 2.4.2.7) has been found to be responsible for the conversion of AICA-base to ZMP (6).

In this paper, we compared the formation and accumulation of Z-nucleotides from acadesine in different cells of human and the rat origin.

Acadesine is not metabolized in cells lacking adenosine kinase activity

The importance of adenosine kinase activity in acadesine metabolism has been studied in a variety of cells. We have reported (5) that acadesine was not metabolized in human B-lymphoblasts (W1-L2 cells) when adenosine kinase activity was inhibited by 5-iodotubercidin, or in AK deficient human B-lymphoblasts. These data indicate the importance of adenosine kinase in the initial step of acadesine metabolism.

Our observations support previous reports which suggest the involvement of adenosine kinase in acadesine metabolism in human tumor cells of type H. Ep. No. 2 (7) and the lack of acadesine conversion into ZMP or purine compounds in adenosine kinase deficient Chinese hamster ovary cells (4). More recent studies have shown an almost complete inhibition of acadesine incorporation into purine nucleotides in human B-lymphoblasts which are deficient both in adenosine kinase and hypoxanthine-guanine phosphoribosyltransferase (6).

Purine and Pyrimidine Metabolism in Man VII, Part B
Edited by R.A. Harkness *et al.*, Plenum Press, New York, 1991

Acadesine metabolism in human and rat lymphoblasts and lymphocytes

In both human B and T lymphoblasts, acadesine, when used at low concentrations (200 μM), was primarily metabolized to purine compounds, mainly to hypoxanthine, inosine and ATP (5). Incorporation of 200 μM acadesine into purine nucleotides exceeded many times its incorporation into Z-nucleotides (Table 1). However, increased accumulation of ZMP was observed in human lymphoblasts when they were incubated with acadesine at mmolar concentrations (3). It seems that at a low concentration of acadesine, Z-nucleotide accumulation was low because ZMP was efficiently metabolized into purine nucleotides, nucleosides and bases. When concentrations of acadesine were higher, synthesis of ZMP exceeded ZMP conversion to purine compounds, resulting in the intracellular accumulation of ZMP to mmolar levels (5).

Table 1 Acadesine incorporation into purine nucleotides and Z-nucleotides in human B and T-lymphoblasts and in human and rat peripheral blood lymphocytes (PBL).

Nucleotides	Human lymphoblasts B	Human lymphoblasts T	Human PBL	Rat PBL
	Distribution of radioactivity between nucleotides (%)			
AMP+ADP+ATP	81.9	70.4	29.9	31.1
GMP+GDP+GTP	0.8	12.4	18.2	14.6
ZMP+ZDP+ZTP	6.9	11.8	43.3	50.6
IMP	0.3	5.2	6.6	3.4

Human or rat cells (1 x 10^6/0.1 ml) were incubated in RPMI medium with 200μM radiolabelled [2-^{14}C] acadesine (1μCi) for 4 hrs. Incorporation of radioactivity into purine nucleotides and Z-nucleotides was measured after their separation by TLC.

In both human and rat peripheral blood lymphocytes incubated with 200μM acadesine, incorporation of acadesine into Z-nucleotides exceeded acadesine incorporation into adenine nucleotides (Table 1). Incorporation of acadesine into human peripheral blood lymphocytes has been reported to be about 8 times slower as compared to lymphoblasts (5). This may indicate that in nondividing lymphocytes, with a less active de novo purine biosynthesis, the metabolism of acadesine is not only slower, but also ZMP accumulation is much higher than that observed in lymphoblasts.

Acadesine metabolism in rat isolated cardiomyocytes and rat heart slices

In rat isolated cardiomyocytes in culture and rat heart slices, acadesine (200 μM) was metabolized primarily to purine compounds, and only low accumulation of radiolabelled ZMP was observed (Table 2).

Table 2. Incorporation of acadesine into purine nucleotides and Z-nucleotides in rat heart cultured cardiomyocytes and in neonatal rat heart slices.

Nucleotides	Cultured Cardiomyocytes	Neonatal heart slices
	Distribution of radioactivity between nucleotides (%)	
AMP+ADP+ATP	71.4	68.7
GMP+GDP+GTP	20.1	6.2
ZMP+ZDP+ZTP	5.5	18.4
IMP	2.0	6.1

Rat cultured cardiomyocytes (1x10^6/0.5 ml) or neonatal rat hearts slices (40mg tissue, wet weight) were incubated in RPMI medium with 200 μM radiolabelled [2-^{14} C] acadesine (1μCi) for 4 hrs. Incorporation of radioactivity into nucleotides was measured after their separation by TLC.

However, at higher (mmolar) concentrations of acadesine accumulation of radiolabelled Z-nucleotides was observed (10) to be similar to that seen in lymphoblasts. This data indicates that in rat cardiomyocytes in culture the last steps of purine biosynthesis de novo pathway are active and convert the ZMP formed from acadesine into purine nucleotides.

It has been reported (11, 12) that acadesine, when infused into dogs, is converted into ZMP and via IMP to other nucleotides in both skeletal and cardiac tissues. It has been also suggested that accumulation of ZMP can inhibit adenylosuccinate lyase (EC 4.3.2.2.) (11), resulting predominantly in IMP catabolism and decreased IMP interconversion to adenylate nucleotides.

Acadesine metabolism in human platelets

In human platelets, 200 μM acadesine was metabolized to both Z-nucleotides and purine nucleotides. The conversion of acadesine into purine nucleotides exceeded the conversion into Z-nucleotides, however, the accumulation of Z-nucleotides was significant and represented almost one fourth of the total purine nucleotides (Table 3). The conversion of acadesine into purine nucleotides may indicate that at least the last steps of purine biosynthesis de novo are active in human blood platelets. The significant accumulation of ZMP may indicate highly active adenosine kinase in human platelets.

Table 3. Incorporation of acadesine into purine nucleotides and Z-nucleotides in human and rat red blood cells (RBC) and human platelets.

Nucleotides	Human RBC	Rat RBC	Human platelet
	Distribution of radioactivity between nucleotide		
AMP+ADP+ATP	9.2	11.8	24.1
GMP+GDP+GTP	8.9	9.7	28.1
ZMP+ZDP+ZTP	71.2	75.2	37.9
IMP	9.5	3.2	9.5

Human or rat RBC ($5x10^6$/0.1 ml) or human platelets (10 x 10^6/0.1 ml) were incubated in RPMI medium with 200μM [2-^{14}C] acadesine (1μCi) for 4 hrs. Incorporation of radioactivity into purine nucleotides and Z-nucleotides was measured after their separation by TLC.

Acadesine metabolism in human and rat red blood cells

Metabolism of acadesine was very similar in rat and human red blood cells (Table 3). Acadesine at low (200 μM) (Table 3) or high (1 mM) concentrations (results not shown) was metabolized almost totally to Z-nucleotides, mainly to ZMP, which then accumulated intracellularly. The accumulation of ZMP was associated with ZTP accumulation, but the ZTP level was always several times lower than the ZMP (5,10). The absence of acadesine conversion into purine nucleotides in red blood cells results from the lack of de novo purine biosynthesis pathway in these cells. Therefore, ZMP accumulates in large amounts in red blood cells after intravenous infusion of acadesine to dogs (13).

Intracellular role(s) of Z-nucleotides

Our studies show that acadesine, especially when used in high (mmolar) concentrations, can lead to accumulation of Z-nucleotides not only in red blood cells but also other cells. Intracellular accumulation of ZMP can reach mmolar concentrations and exceed the intracellular levels of ATP (5).

Our knowledge about the intracellular role of Z-nucleotides is still very limited. There are indications that ZMP, when formed from acadesine may serve as a precursor for purine nucleotides, especially ATP. However, acadesine might only substitute for de novo purine synthesis, rather than accelerate ATP synthesis. In fact, ZMP can inhibit adenylosuccinate lyase and therefore inhibit the accelerated flow of metabolites from IMP to AMP. ZMP accumulation in folate-deficient bacteria was associated with ZTP accumulation and that led to the hypothesis that ZTP plays a role as an "alarmone", providing signals that govern the metabolic economy of the cells during folate deficiency (14). Significant ZTP formation has also been observed in many mammalian cells including cardiac and skeletal muscle (11), fibroblasts (4), lymphoblasts (5) and erythrocytes (5,15). It has been proposed that synthesis of ZTP from acadesine utilizes PRPP in a one step reaction (16), and that could explain the low level of ZDP usually observed (4,5,11). ZMP accumulation has been reported to correlate with the depletion of the pyrimidine nucleotide pool (4).

Accumulation of Z-nucleotides in erythrocytes from Lesh-Nyhan patients has also been found and discussed as being the result of elevated purine biosynthesis de novo in the patients (1).

Finally, an important observation has been reported (5) that red blood cells, after accumulation of ZMP following incubation with acadesine, can release back acadesine into the extracellular environment, indicating the dephosphorylation of ZMP (3). Acadesine might be a regulator of extracellular adenosine concentrations (17), while ZMP and other Z-nucleotides in red blood cells may represent a storage form of acadesine.

REFERENCES

1. Sidi, Y., Mitchell, B.S. J. Clin. Invest. 76, 2416,
2. Vincent, M.F., Marangos, P., Gruber, H.E., Van den Berge, G., Diabetes, 1991 (in press).
3. Allegra, C.J., Hoang, K., Chao Yeh, G., Drake, J.C., Baram, J., J. Biol. Chem. 262, 13520, 1987.
4. Sabina, R.L., Patterson, D., and Holmes, E.W., J.Biol.Chem. 260, 6107, 1985.
5. Jimenez, R., Gruber, H.E., Barankiewicz, J., Int. J. Pur. Pyr. Res. 1,51, 1990.
6. Page, T., Int. J. Biochem. 21, 1377, 1989.
7. Schnebli, H.P., Hill, D.L., Bennett, L.L., J. Biol. Chem. 2422, 1997, 1967.
8. Lowy, B.A., Williams M.K., Pediat. Res. 11, 691, 1972.
9. Wyngaarden, J.B., Seegmiller, J.E., Laster, L., Blair, A.E., 8, 455, 1959.
10. Barankiewicz, J. and Jimenez, R. (1991) (in preparation).
11. Sabina, R.L., Kernstine, K.H., Boyd, R.L., Holmes, E.W., Swain J.L., J. Biol. Chem. 257, 10178, 1982.
12. Mauser, M., Hoffmeister, H.M., Nienaber, C., Shaper, W. Circ.Res. 56, 220, 1985.
13. Dixon, R., Gourzis, J., McDermott, D., Fujitaki, J., Dewland, P., Gruber, H., J. Clin. Pharmacol. 31, 342, 1991.
14. Bochner, B.R., Ames, B.N., Cell 29, 929, 1982.
15. Zimmerman, T.P., Deeprose, R.D., Biochem. Pharmacol. 27, 709, 1982.
16. Sabina, R.L., Holmes, E.W., Becker, M.A. Science 223, 1193, 1984.
17. Gruber, H.E., Hoffer, M.E., McAllister, D.R., Laikind, P.K., Lane, T.A., Schmid-Schoenbein, G.W. and Engler, R.L., Circulation 80, 1400, 1983.

PURINE METABOLISM IN REGENERATING LIVER-BEARING RATS

Brunetta Porcelli, Antonella Tabucchi, Patrizia Valerio**, Daniela Vannoni, Massimo Molinelli* and Roberto Pagani

Inst. of Biochemistry and Enzymology, *Inst. of Pediatrics, **Inst. of Dentistry - University of Siena, Italy

INTRODUCTION

Many biochemical events have been described in the regenerating liver , and several studies have focused on the control of RNA, DNA and protein metabolism (1,2). The gluconeogenetic function of the regenerating liver has also been studied revealing a fall in glycaemia, a rapid recovery (3)and gluconeogenesis from lactate or aminoacids (4,5). Little is known about nitrogen metabolism under such conditions, especially purine and pyrimidine metabolism. A number of enzyme studies have been performed. Adenine phosphoribosyltransferase has been reported to increase in regenerating rat liver extracts for 4 days after partial hepatectomy, and to still be above control values after 14 days (6). Hypoxanthine-phosphoribosyltransferase activity has been found not to increase in extracts of regenerating rat liver until the second day after the operation, and to begin to decrease on the fourth day (6). The activity of xanthine oxidase does not undergo any particular variation up to 96 hour after operation (7). However, the behavior of other enzymes involved in the de novo synthesis or catabolism of purine nucleotides, purine nucleotide content, and the pattern of plasma and urinary purine bases -all important parameters of purine nucleotide metabolism- have not been extensively investigated.

In order to improve our knowledge of the behavior of purine nucleotide metabolism in regenerating liver-bearing rats, we studied the behavior of three catabolic products of purine metabolism, free oxypurines, uric acid and allantoin, in the urine of regenerating liver-bearing rats.

MATERIALS AND METHODS
Animals and treatment

Male Albino Wistar rats (Nossan Company, Corezzano, Milano) were kept under controlled temperature (21-23 °C) and light periods (07.00 - 19.00 h). They received water ad libitum and a standard pellet diet (Nossan Company) with 22% protein.

The rats were partially hepatectomized according to the method of Higgins and Anderson (8). The operations were carried out between 9.00 and 11.00 a.m. with the animals under diethyl-ether anaesthesia, in aseptic conditions. 65-75% of the liver was removed, leaving the right lateral lobe and the small caudate lobe. The animals were immediately allowed a normal diet.

The postoperative effects of the anaesthesia wore off very rapidly (1 h), and this was taken as zero time for subsequent observations. Sham operated rats having normal food intake were used as controls.

Food intake was standardized in both groups of animals. They were given the same diet in aliquots which according to our experience, were completely eaten:

10 gr of the standard pellets during the first 24 h after partial hepatectomy; 20 gr of pellets during the subsequent 24 h; after 48 h, pellets were allowed ad libitum. Temperature and light period remained above. Urine was collected every 24 hours and analyzed for oxypurines, uric acid and allantoin.

Uric acid was determined according to Praetorius & Poulsen (9). Allantoin was estimated by Rimini 's reaction (10). Oxypurines were evaluated by high pressure liquid chromatography. We used a Beckman Instrument mod 332, equipped with an UV detector at 254 nm and a Supelcosil LC-18 column (250x4.6mm). Isocratic elution, was performed with 10mM KH_2PO_4 (pH 5.5) and the oxypurines were identified on the basis of retention time, when coeluted with internal standards and treated with xanthine oxidase.

RESULTS AND DISCUSSION

The results are shown in the following figures and tables. Figure 1 showes a typical chromatogram of urinary oxypurines, and the same chromatogram with internal standards.

A strong decrease in the urinary excretion of hypoxanthine and xanthine was observed. This could be explained by the fact that the regenerating liver shows a very high activity of APRT (6), the enzyme of the purine salvage pathway. The reutilization of adenine would decrease the formation of hypoxanthine and xanthine through adenase and xanthine oxidase.

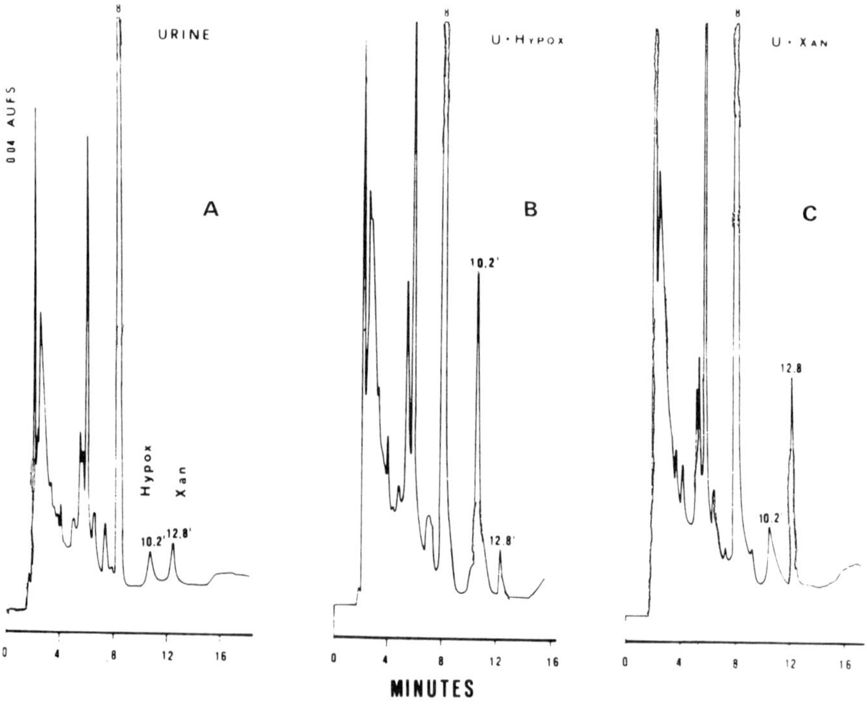

Fig. 1. A: Typical chromatogram of urinary oxypurines. B and C: the same chromatogram with internal standards.

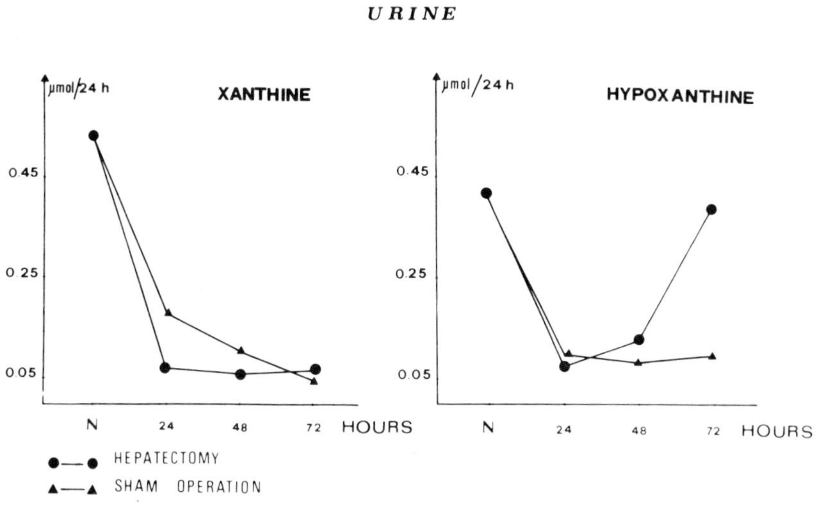

Fig. 2. Behavior of urinary oxypurines during liver regeneration. N= normal rats.

Figures 2 and 3 show the pattern of uric acid and allantoin in the urine of normal sham-operated and regenerating liver-bearing rats, at different times. These data also indicate a lower rate of purine nucleotide catabolism in regenerating rat liver, and this could favour the synthesis of nucleotides, RNA and DNA. However, since the same result was obtained in the sham-operated rats, it means that the decrease in urinary oxypurines is related more to stress than to the operation. The other important finding was that uric acid and allantoin excretion were normal both in sham-operated and hepatectomized rats. It is difficult to interpret this result, since it is generally assumed that most allantoin and uric acid are produced in the liver.

These findings suggest that:

1) the distribution of xanthine oxidase and uricase in rat tissues should

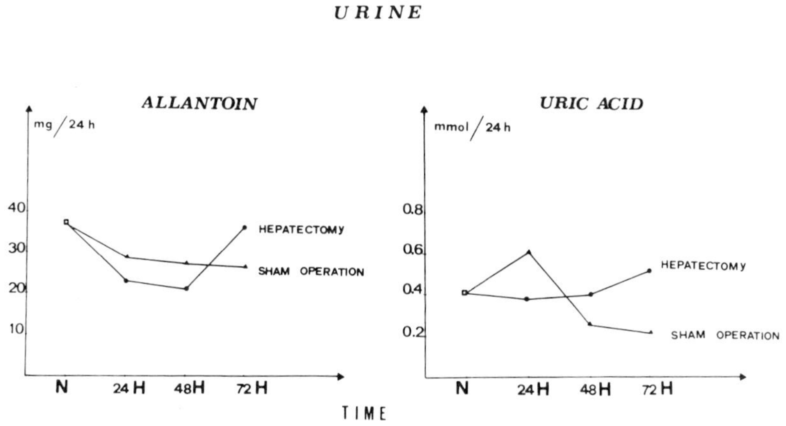

Fig. 3. Behavior of allantoin and uric acid during liver regeneration. N= normal rats

be investigated with more sensitive and specific techiques;
2) the remaining liver increases the formation of the two catabolic compounds, one of the many compensatory phenomena occuring in the regenerating liver.

REFERENCES

1. F.J. Bollum and V.R. Potter, Nucleic acid metabolism in regenerating rat liver. Soluble enzyme concert thymidine to thymidine phosphates and DNA, Cancer Res. 19: 561 (1959)
2. O.A. Scornik and V. Botbol, Role of changes in protein degradation in the growth of regenerating livers, J. Biol. Chem. 251: 2891 (1976)
3. C.B. Wood, S.J. Karran and L.H. Blumgart, Metabolic changes following varying degrees of partial hepatectomy in the rat, Brit. J. Surg. 60: 613 (1973)
4. N. Katz, Correlation between rates and enzyme levels of increased gluconeogenesis in rat liver and kidney after partial hepatectomy, Eur. J. Biochem, 98: 535 (1979)
5. N.J. Curtin and K. Snell, Enzymic retrodifferentiation during hepatocarcinogenesis and liver regeneration in rats in vivo, Br. J. Cancer, 48: 495 (1983)
6. A.W. Murray, Purine phosphoribosyltransferase activities in rat and mouse tissues and in Ehrlich ascites-tumour cells, Biochem. J., 100: 664 (1966)
7. N. Prajda, H.P. Morris and G. Weber, Imbalance of purine metabolism in hepatomas of different growth rates as expressed in behavior of xanthine oxidase, Cancer Res., 36: 4639 (1976)
8. G.H. Higgins and R.H. Anderson, Experimental pathology of the liver. Restoration of the liver of the white rat following partial surgical removal, Arch. Pathol., 12: 186 (1931)
9. E. Praetorius and H. Poulsen, Enzymatic determination of uric acid, Scand.J.Clin.Lab.Inv.,5: 273 (1953)
10. E.G. Young and C.F. Conway, On the estimation of allantoin by the Rimini-Schryver reaction, Biol. Chem. 142: 839 (1942)
11. G.P. Wheeler, J.A. Alexander, A.S. Dodson, S.D. Briggs and H.P.Morris, Searches for exploitable biochemical differences between normal and cancer cells. IX. Anabolism and catabolism of purine by hepatomas 5123 and H-35', Cancer Res., 22: 769 (1962)
12. G.P. Wheeler, J.A. Alexander, A.S. Dodson and S.D. Briggs, Searches for exploitable biochemical differences between normal and cancer cells. X. Catabolism of purines by regressing tumors. Cancer Res., 22: 1309 (1962)
13. G.P. Wheeler and J.A. Alexander, Searches for exploitable biochemical differences between normal and cancer cells. IV. Metabolism of purine in vivo. Cancer Res., 21: 390 (1961)

PURINE NUCLEOTIDE SYNTHESIS IN RAT LIVER AFTER CASTRATION

Enrico Marinello, Maria Pizzichini, Anna Di Stefano,
Lucia Terzuoli, Roberto Pagani

Institute of Biochemistry and Enzymology
University of Siena, Italy

INTRODUCTION

The synthesis of purine nucleotides in the rat liver occours via two main pathways: "de novo synthesis", a series of reactions leading from PRPP to IMP, and the "salvage pathway" by which the free bases, adenine and guanine, are condensed with PRPP to give AMP and GMP. IMP, formed by the de novo pathway, is channelled at the "inosinic branch point" into the formation of AMP and GMP. The normal regulation of de novo synthesis is poorly understood and there is little in the literature on the subject.

In the present study, we investigated the effects of the absence of testosterone on free purine nucleotide de novo synthesis in the rat liver. The esistence of receptors for testosterone in the rat liver (1), indicates that the liver is also a target organ for this androgen. However, it is not clear whether these receptors only serve to "block" the hormone, or whether they serve some specific function.

We specifically examined the synthesis and hormonal regulation of AMP, the most important purine nucleotide. We followed the process and its overall rate, by studying the incorporation of ^{14}C-formate into acid-soluble adenine (2,3,4). Adenine content and its specific radioactivity were determined at different times after administration of tracer.

Analysis of data included linear regression of specific activity vs time, and comparison with our previous results (5) in which only one time of administration was considered.

MATERIAL AND METHODS

Nine-week old male albino Wistar rats were castrated under ether and then kept at 25°C and allowed water and a normal diet ab libitum. Control rats were sham-operated. The rats received a single intraperitoneal dose (10 μCi/100 g b.w.) of ^{14}C-formate (54.8 mCi/mmol) and were killed by decapitation 17′-30′-60′-154′ after the injection. The livers were rapidly excised, omogenized in 5% TCA and centrifuged. The acid soluble supernatant was utilized for analysis of total acid-soluble nucleotides, which were hydrolyzed with 1 N H_2SO_4 (1 h at 100°C). The free bases were precipitated with mercuric acetate (6), redissolved in water, adsorbed onto cationic resin (AG 50W-X8, 200-400 mesh, hydrogen form) and further separated as previously reported (7).

The adenine-containing peak was studied. Identification of the base was carried out by 1) spectral analysis, 2) comparison with suitable standards, 3) coelution with standards. The concentration of base was determined by UV adsorption. The radioactivity was determined on a Nuclear Chicago Scintillation Counter, using Instagel as scintillation cocktail.

All reagents were of the highest commercially available purity, and were obtained either from Merck or Fluka. Standard compounds were obtained from Sigma. Resin was from Bio Rad. ^{14}C-formate (54.8 mCi/mmol) was an Amersham product.

The concentration and the specific radioactivity of adenine were evaluated. Linear regression of specific activity vs time and one-way analysis of variance (significance level: $p < 0.01$) were performed.

RESULTS AND DISCUSSION

The data reported in Table 1 showed an apparently unchanged adenine content after castration. The radioactivity (dpm/g) of adenine was slightly enhanced at 30´, decreasing at 60´ and 154´.

TABLE 1. The content of adenine (mg/g) and its radioactivity (dpm/g) in liver of a normal and castrated rats, at different time after ^{14}C-formate administration.

	normal rats		castrated rats	
	mg/g	dpm/g	mg/g	dpm/g
17´	0.32 ± 0.02	5507 ± 1111	0.35 ± 0.07	5912 ± 1000
30´	0.38 ± 0.02	8987 ± 1552	0.41 ± 0.03	10596 ± 1942
60´	0.36 ± 0.02	12764 ± 2703	0.36 ± 0.01	7764 ± 1671*
154´	0.31 ± 0.03	16927 ± 1930	0.33 ± 0.02	11606 ± 2980*

The values represent the mean ± S.E. of four different rats.
* $p < 0.01$ referred to normal rats.

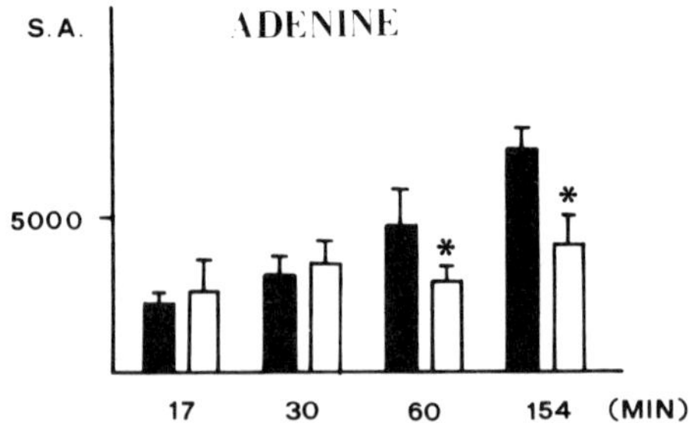

Figure 1. Pattern of specific radioactivity of adenine in normal (■) and castrated (□) rats, at different times after ^{14}C-formate injection

The values were perfectly compatible when expressed as specific radioactivity or dpm/g (Figure 1).

Linear regression analysis revealed a decreasing trend after castration (figure 2). This means that the formation of AMP was lower in the absence of testosterone, decreasing after castration.

This finding is apparently in contrast with our previous observation (5) of a reduced turnover of all nucleotides (IMP, GMP and AMP) in castrated prepubertal rats, and accelerated adenine turnover in castrated adult rats. It is difficult to explain this discrepancy, especially in view of the fact that all experimental conditions and parameters (strain, age, sex of the rats, etc.) were identical in the present study. In the past experiment, it was therefore not possible to follow the incorporation of the tracer into adenine at different times and no plots or linear regression could be performed.

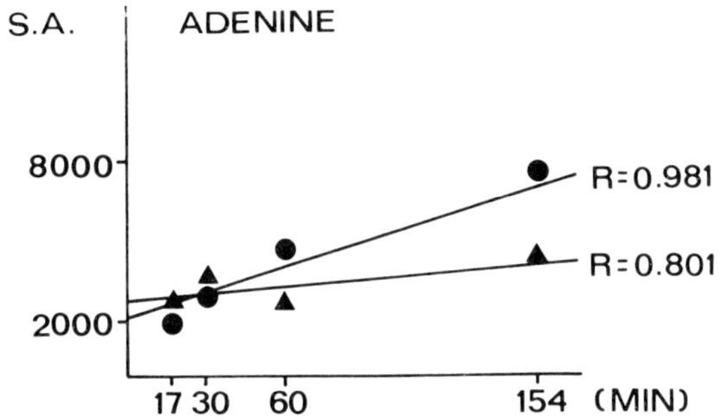

Figure 2. Linear regression of specific radioactivity of adenine vs time after ^{14}C-formate injection in normal (●) and castrated (▲) rats.

Clearly a point time experiment gives less information than a time plot and therefore the present experiments give a better indication of the pattern of adenine synthesis in the rat liver after castration, and on its regulation by testosterone.

REFERENCES

1) Lefebvre Y.A., Morante S.J.: Binding of dihydrotestosterone to a nuclear envelope fraction from the male rat liver. Biochem. J., 202: 225, (1982).

2) Sheehan T.G., Tully E.R.: Purine biosynthesis de novo in rat skeletal muscle. Biochem. J., 216: 605, (1983).

3) Zoref-Shani E., Kesslker-Icekson G., Wasserman L., Sperling O.: Characterization of purine nucleotide metabolism in primary rat cardiomyocyte cultures. Biochim. Biophys. Acta, 804: 161, (1984)

4) Welch M.M., Rudolph F.B.: Regulation of purine biosynthesis and interconversion in the chick. J. Biol. Chem., 257(22): 13253, (1982).

5) Di Stefano A., Pizzichini M., Resconi G., Tabucchi A., Leoncini R., Marinello E.: The hormonal regulation of purine nucleotide turnover. Influence of testosterone in rat liver. It. J. Biochem., 39(4): 216, (1990).

6) Di Stefano A., Pizzichini M., Resconi G., Marinello E.: Determination of allantoin in mammalian tissues. Adv. Exp. Med. Biol., 253B: 511, (1989).
7) Pizzichini M., Di Stefano A., Marinello E.: The regulation of purine ribonucleotide biosynthesis by glucocorticoid hormones. It. J. Biochem., 34(5): 305, (1985).

INCORPORATION OF PURINE RIBONUCLEOTIDES INTO NUCLEIC ACIDS OF RAT LIVER AFTER CASTRATION

Maria Pizzichini, Anna Di Stefano, Giuliano Cinci,
Daniela Vannoni, Roberto Pagani, Enrico Marinello

Institute of Biochemistry and Enzymology
University of Siena, Italy

INTRODUCTION

It is well known that testosterone has a positive effect on RNA synthesis in sex organs (1,2) and a similar effect on kidney RNA polymerase has been described (3,4). However, the effects of the hormone in these organs and in the liver has not yet been clarified.

Little is known about the relationships and transformation of purine ribonucleotides to RNA in the liver of animals after castration, and after androgen treatment. Nor is the pattern of RNA content in the liver under such conditions well known.

In the present study we compared the metabolism of free purine nucleotides and nucleic acids in the liver of adult male Wistar rats before and after orchiectomy. We have followed the specific radioactivity of adenine and guanine from RNA and DNA after injection of the tracer ^{14}C-formate.

MATERIALS AND METHODS

Nine-week old male albino Wistar rats were castrated under ether and then kept at 25°C and allowed water and a normal diet ad libitum. Control rats were kept in the some conditions but were sham-operated. All animals received a single intraperitoneal dose (10 μ Ci/100 g b.w.) of ^{14}C-formate (54.8 mCi/mmol) and were killed by decapitation 17´, 30´, 60´ and 154´ after injection. The livers were rapidly excised, homogenized in 5% TCA and centrifuged.

Nucleic acids were separated according to Munro and Fleck (5). Essentially TCA extracts were centrifuged, the precipitate washed twice with distilled water and suspended in 0.3 N KOH (1 h at 37°C). DNA and proteins were separated from RNA by precipitation with perchloric acid and centrifuged.

DNA was obtained by treating the precipitate with 0.7 N $HClO_4$ (15´ at 90°C). Ribonucleotides and deoxyribonucleotides underwent acid-hydrolysis (1 h at 100°C in H_2SO_4 1 N final) to obtain the free bases. These were precipitated with 1 N $AgNO_3$ according to a previously reported procedure and redissolved in 1 N HCl, adsorbed onto cationic resin (Dowex 50 W X8, 200-400 mesh) and further separated (6). The concentration of the purine bases was determined by UV adsorption. Radioactivity was determined on a Nuclear Chicago Scintillation Counter, using Instagel as scintillation cocktail.

Purine and Pyrimidine Metabolism in Man VII, Part B
Edited by R.A. Harkness *et al.*, Plenum Press, New York, 1991

All reagents were of the highest commercially available purity, and were obtained from Merck or Fluka. Standard compounds, nucleotides (IMP, AMP and GMP) and bases (adenine, guanine) were obtained from SIGMA. Resin was from Bio-Rad. ^{14}C-formate (54.8 mCi/mmol) was an Amersham product. The concentration and specific radioactivity of adenine were evaluated.

Linear Regression of specific activity vs time, and one-way analysis of variance (significance level : $P<0.01$) were performed.

RESULTS AND DISCUSSION

The content of adenine and guanine was similar and did not show any variations in RNA or DNA. It is not reported here for the sake of brevity.

When the values were expressed in dmp/g or as specific radioactivity (dpm/µmol), adenine showed a decreasing trend whereas guanine and G/A ratio slightly increased after castration (Table 1). The pattern was less clear in the case of DNA (Table 2).

When the linear regression of specific radioactivity vs time was plotted after castration it revealed an evident decrease in RNA and DNA synthesis (Fig.1). An interesting point emerging from these results is that the sythesis of RNA, as indicated by the incorporation of the tracer ^{14}C-formate into the bases adenine and guanine, was lower after castration. This is in agreement with the observation that overall purine nucleotide metabolism seem depressed in the absence of testosterone.

Since receptors for testosterone have been demonstrated in the rat liver (1), this organ must also be a target organ for the androgen. However, it is not clear whether these receptors only serve to "block" the hormone or whether they serve some specific functions.

Table 1. Radioactivity content (dpm/g) and specific radioactivity (dpm/µmol) of adenine and guanine from RNA in the liver of normal (N) an castrated (C) rats, at different times after ^{14}C-formate administration.

Time		ADENINE		GUANINE		G / A
		dpm/g	S.A.	dpm/g	S.A.	dpm/g
17´	N	481 ± 97	209 ± 33	195 ± 13	33 ± 1	0.43 ± 0.06
	C	440 ± 105	208 ± 47	295 ± 78	53 ± 12	0.66 ± 0.06*
30´	N	860 ± 173	451 ± 65	255 ± 33	49 ± 5	0.32 ± 0.07
	C	782 ± 228	392 ± 112	371 ± 118	69 ± 23	0.38 ± 0.06
60´	N	1834 ± 368	550 ± 184	356 ± 65	57 ± 12	0.32 ± 0.07
	C	884 ± 122*	378 ± 67	511 ± 88	94 ± 24*	0.58 ± 0.07*
154´	N	3236 ± 423	1466 ± 139	686 ± 163	138 ± 35	0.20 ± 0.02
	C	1520 ± 316*	658 ± 182*	463 ± 72	82 ± 16*	0.31 ± 0.03

The values represent the mean ± S.E. of four different rats.
* $p < 0.01$ with respect to controls

Table 2. Radioactivity content (dpm/g) and specific radioactivity dpm/μmol) of adenine and guanine from DNA in the liver of normal (N) and castrated (C) rats, at different times after ^{14}C-formate administration.

time		ADENINE dpm/g	ADENINE S.A.	GUANINE dpm/g	GUANINE S.A.	G / A dpm/g
17´	N	71 ± 9	60 ± 8	53 ± 4	27 ± 2	0.75 ± 0.04
	C	89 ± 22	65 ± 11	69 ± 19	37 ± 7	0.76 ± 0.07
30´	N	122 ± 28	111 ± 13	62 ± 9	33 ± 7	0.53 ± 0.07
	C	185 ± 65	112 ± 16	114 ± 43	41 ± 9	0.58 ± 0.04
60´	N	227 ± 112	172 ± 49	179 ± 77	78 ± 19	0.86 ± 0.14
	C	127 ± 12	131 ± 12	93 ± 25	60 ± 15	0.70 ± 0.14
154´	N	385 ± 66	344 ± 34	137 ± 28	72 ± 7	0.35 ± 0.02
	C	198 ± 31*	194 ± 33*	106 ± 10	64 ± 3	0.55 ± 0.05*

The values represent the mean ± S.E. of four different rats
* $p < 0.01$ with respect to control.

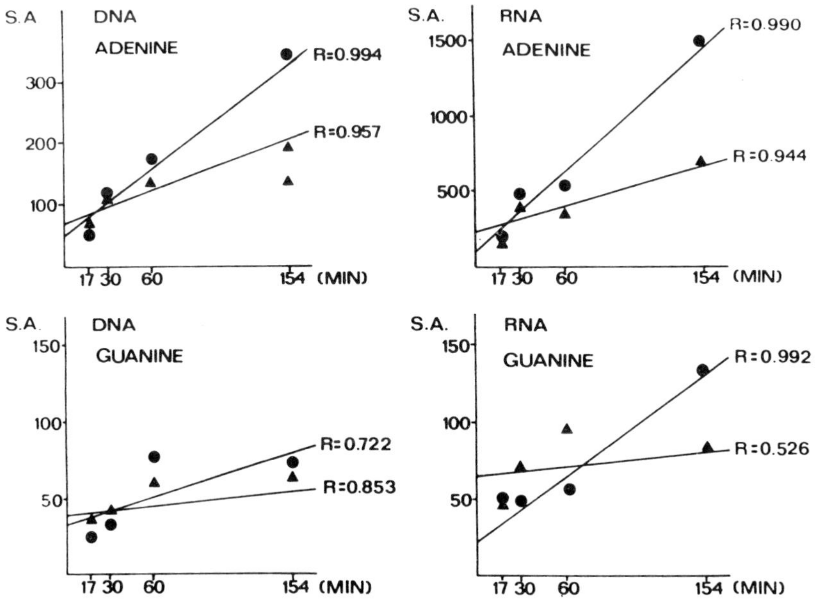

Figure 1. The linear regression of the specific radioactivity of adenine and guanine vs time, after ^{14}C-formate administration in normal (●) and castrated (▲) rats.

In the case of purine metabolism, such a function should be studied at the enzymatic level to determine which enzymes of purine metabolism are affected by testosterone administration or absence of testosterone.

Whether the action of testosterone on RNA is direct (RNA polymerase), or only a conseguence of its effects on purine nucleotide metabolism, should be also investigated. It is interesting to note that contradictory data exist in the literature. Testosterone is reported to depress nucleic acid synthesis in adipose tissues (8), but to enhance DNA metabolism in sex organs.

We conclude by remarking that the G/A ratio of RNA is enhanced after castration, indicating concomitant stimulation of GMP synthesis; the change in the ratio more likely is due to the decreased formation of AMP than to an increased formation of GMP, which only occurs when cell duplication is enhanced. However, this has jet to be demonstrated in the liver of castrated rats.

REFERENCES

1) Higgins S.J., Burchell J.M.: Effects of testosterone on Messenger Ribonucleic Acid and Protein Synthesis in rat seminal vesicle. Biochem.J., 174:543, (1978)
2) Avdalovic N., Bates M.: The influence of testosterone on the synthesis and degradation rate of various RNA species in the mouse kidney. Biochim. Biophys. Acta, 407: 299, (1975).
3) Mainwaring W.I.P., Mangan F.R., Peterken B.N.: Studies on the solubilized ribonucleic acid polymerase from rat ventral prostate gland. Biochem. J., 123: 619, (1971).
4) Davie P., Griffiths K.: Stimulation of ribonucleic acid polymerase activity in vitro by prostatic steroid-protein receptor complexes. Biochem. J., 136: 611, (1973).
5) Munro H. N., Fleck A.: Enzymatic determination of uric acid. Analyst. 91: 78, (1965).
6) Pizzichini M., Di Stefano A., Marinello E.: The regulation of purine ribonucleotide biosynthesis of glucocorticoid hormones. It. J. Biochem., 34(5): 305, (1985).
7) Lefebvre J. A., Morante S. J.: Binding of dihydrotestosterone to a nuclear-envelope fraction from the male rat liver. Biochem. J., 202: 225 (1982).
8) Haug A., Spydevold Φ., Hostmark A. T.: Effect of orchiectomy and testosterone substitution on the enzyme activities and DNA content in rat liver and epididymal fat. Int. J. Biochem., 17(1): 31, (1985).

THE EFFECT OF PYRROLINE-5-CARBOXYLATE ON R5P AND PRPP GENERATION IN MOUSE LIVER IN VIVO

Pnina Boer and Oded Sperling

Department of Clinical Biochemistry, Beilinson Medical Center, Petah-Tikva, and Department of Chemical Pathology, Sackler Faculty of Medicine Tel Aviv University, Tel-Aviv, Israel

INTRODUCTION

The role of the availability of ribose-5-phosphate (R5P) in the regulation of PRPP synthesis and therefore also of nucleotide synthesis, has been the subject of much debate [1-10]. R5P is produced mainly by the hexose-monophosphate-pentose pathway (HMPP). Pyrroline- 5-carboxylate (P5C), an intermediate in the interconversions of proline, ornithine and glutamate, was found to markedly stimulate the oxidative branch of the HMPP in erythrocytes and cultured fibroblasts [11-14], resulting in increased PRPP content and nucleotide formation, attributed to increased generation of R5P [15-17].

In the present study we investigated the consequences of the administration of P5C into mice, in vivo [1].

MATERIALS AND METHODS

ICR male mice (20-25 g) fed ad libitum on a commercial diet were used. Mice were injected into the tail vein with P5C (250 nmoles in 0.9 % NaCl/g body wt.). After 2.5-20 min, the mice were killed by cervical dislocation and the livers were excised and freeze clamped immediately in liquid nitrogen. PRPP and R5P were extracted from the liver tissue by heat [18] and assayed as described before [1].

RESULTS AND DISCUSSION

In agreement with previous findings in human erythrocytes and cultured fibroblasts in vitro [15-17], we could demonstrate also in mice in vivo, that i.v. administration of P5C resulted in increased PRPP content in the liver tissue (Fig 1). An increase in liver R5P generation following the administration of P5C is taken for granted, in view of the known effect of P5C on the oxidative branch of the HMPP, as demonstrated in erythrocytes and cultured fibroblasts [13,15]. The finding in these tissues of the P5C-induced

Purine and Pyrimidine Metabolism in Man VII, Part B
Edited by R.A. Harkness *et al.*, Plenum Press, New York, 1991

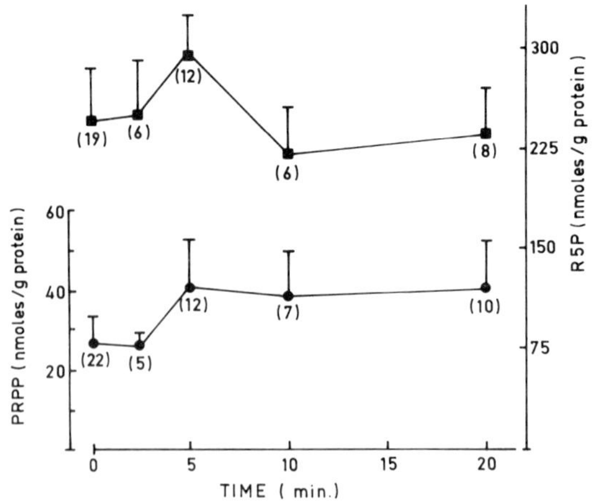

Fig 1. The effect of P5C administration (i.v.) on liver content of R5P and PRPP in mice. Numbers in parentheses repsresent number of animals.

increase in PRPP and purine nucleotides, is indeed compatible with increased R5P generation. In agreement with the above, we could actually demonstrate an increase in liver R5P content following P5C administration (Fig 1). The elevation in R5P content (a 19.3% increase; $p<0.0025$) was smaller than that in PRPP content, and transient, probably reflecting the high turnover rate of this intermediate in the liver tissue. The difference in the time-related increases in R5P and PRPP may reflect the fact that the increase in PRPP is subsequent and depending on the increase in R5P content of the tissue. Furthermore it may also reflect the different turnover rates of these metabolites.

The results of the present study support the suggestion that P5C enhances PRPP generation through the acceleration of R5P production. The finding that P5C, a naturally occuring activator of the oxidative HMPP, is a potent stimulator of PRPP synthesis in a central metabolic organ like the liver, furnishes strong evidence for a physiological role of R5P availability in the regulation of PRPP and nucleotide synthesis. This finding is compatible with several observations. In rat liver and in human lymphoblasts, fibroblasts and erythrocytes, the intracellular R5P concentration is about 10 fold lower than the Km of PRPP synthetase for this substrate [2-7]. Moreover, at physiological Pi concentration, this Km is about 5 fold higher than that at high nonphysiological Pi concentration, so that the tissue R5P concentration is 1:50 in relation to the Km [4]. An additional evidence for the connection between activation of the oxidative branch of the HMPP, increased generation of R5P and enhanced rate of PRPP and nucleotide synthesis is furnished by the effects of insulin. This hormone has been shown to enhance the carbon flux through the oxidative HMPP in rat hepatocytes and adipocytes [14]; to stimulate the rate of PRPP production in mouse liver in vivo [8]; and to restore the level of purine and pyrimidine nucleotides in rat liver [9]. Positive correlations between

R5P, PRPP and purine nucleotide formation were also found in human lymphoblasts starved for an essential amino acid [9], in phytohaemagglutinin activated lymphocytes [10], and in rat liver under different dietary and hormonal conditons [3]. In agreement with the above is the finding of high levels of R5P in cultured fibroblasts derived from two purine overproducing gouty subjects [5]. The results of the present study are of importance in understanding the connection between carbohydrate and purine metabolic pathways and in the understanding of the pathophysiology of purine overproduction in some inborn errors of carbohydrate metabolism, such as glycogen storage disease type I [19,20].

REFERENCES

[1] Boer, P. and Sperling, O., Biochem. Med. Metabolic Biol., in press (1991)
[2] Casazza, J.P. and Veech, R.L., Biochem. J. 234, 635 (1986).
[3] Kunjara, S., Sochor, M., Siti Aisah Ali, A., Greenbaum, A.L. and Mclean, P., Biochem. J. 244, 101 (1987).
[4] Boss, G.R., J. Biol. Chem. 259, 2936 (1984).
[5] Becker, M.A., J. Clin. Invest. 57, 308 (1976).
[6] Fox, I.H. and Kelley, W.N. J. Biol. Chem. 247, 2126 (1972).
[7] Roth, D.G., Shelton, E. and Deuel, T.F., J. Biol. Chem. 249, 291 (1974).
[8] Lalanne, M. and Henderson, J.F., Canad. J. Biochem. 53, 394 (1975).
[9] Weber, G., Lui, M.S., Jayaram, H.N., Pillwein, K., Natsumeda, Y., Faderan, M.A. and Reardon, M.A., Adv. Enz. Regul. 23, 81 (1985).
[10] Pilz, R.B., Willis, R.C. and Boss, G.R., J. Biol. Chem. 259, 2927 (1984).
[11] Yeh, G.C. and Phang, J.M., Biochem. Biophys. Res. Comm. 94,450 (1980).
[12] Phang, J.M., Dawning, S.J., Yeh, G.C., Smith, R.J. and Williams, J.A., Biochem. Biophys. Res. Comm. 87, 363 (1979).
[13] Yeh, C.G., Roth, E.F. Jr., Phang, J.M., Harris, S.c., Nagel, R.L. and Rinaldi, A., J. Biol. Chem. 259, 5454 (1984).
[14] Febregat, I., Revilla, E. and Machado, A., Biochem. Biophys. Res. Comm. 146, 920 (1987).
[15] Yeh, G.C. and Phang, J.M., Biochem. Biophys. Res. Comm. 103, 118 (1981).
[16] Yeh, G.C. and Phang, J.M., J. Biol. Chem. 263, 1383 (1988).
[17] Phang, J.M. and Downing S.J., Clin. Res. 33, 573A (1985).
[18] Lalanne, M. and Henderson, J.F., Analytical Biochem. 62, 121 (1974).
[19] Howell, R.R., Arthritis Rheumatism 8, 780 (1965).
[20] Kelley, W.N., Rosenbloom, F.M., Seegmiller, J.E. and Howell, R.R., J. Pediatr. 72, 488 (1968).

EFFECTS OF ORAL RIBOSE ON MUSCLE METABOLISM DURING BICYCLE ERGOMETER IN PATIENTS WITH AMP-DEAMINASE-DEFICIENCY

Daniel R. Wagner, Ursula Gresser, Irmingard Kamilli, Manfred Gross and Nepomuk Zöllner

Medizinische Poliklinik der Universität
Munich, Germany

INTRODUCTION

AMP-deaminase-deficiency, a frequent enzyme defect of skeletal muscle first described by Fishbein et al. in 1978, is characterized by post- exertional muscle stiffness and cramps. The disruption of the purine nucleotide cycle induced by the deficiency of AMP-deaminase is probably the cause of the muscular disorder (Sabina et al., 1980). So far the only successful therapy has been the oral administration of ribose as first demonstrated by Patten et al. in 1982.

The aim of our study was the investigation of the changes in muscular metabolism during therapy with ribose in patients with AMP-deaminase-deficiency.

The data of the control subjects have already been published (Wagner et al., 1991) and the data of the patients with AMP-deaminase-deficiency after oral ribose therapy are in press (Wagner et al.).

MATERIALS AND METHODS

3 patients with AMP-deaminase-deficiency (diagnosed by ischemic forearm exercise test and histochemical/biochemical examination of muscle biopsies) performed exercise on a bicycle ergometer with increasing work load without ribose on the first day and with ribose on the second day (oral administration of 3 g dissolved in table water every 10 minutes, beginning one hour before exercise until the end). In two patients an additional neuromuscular disorder was found (in one case polyneuropathy and in the other case muscledystrophy type Becker-Kiener). The patients performed exercise until heart rate was 200 minus age.

Plasma ammonia and plasma lactate concentrations were measured by quantitative enzymatic determination; inosine and hypoxanthine were measured by high performance liquid chromatography (HPLC).

Six healthy volunteers served as control subjects. Because of ethical reasons ribose was not administred to the control subjects.

RESULTS

Without ribose the patients complained of post-exertional muscle stiffness and cramps. The muscular performance was normal in comparison with control subjects, except for the patient with muscledystrophy type Becker-Kiener. After ribose the maximum capacity was not increased, but two of three patients had no more post-exertional muscular symptoms, whereas the patient with additional polyneuropathy still complained of muscle stiffness and cramps.

Plasma ammonia concentrations increased during exercise without ribose from 13 µmol (10-16) to 24 µmol (21-26) and during exercise with ribose from 16 µmol (12-23) to 22 µmol (15-28).

Plasma lactate concentrations increased during exercise without ribose from 0.7 mmol/l (0.7-0.8) to 4.1 mmol/l (1.5-5.5) and during exercise with ribose from 0.8 mmol/l (0.6-1.0) to 5,2 mmol/l (3.5-6.9).

Plasma inosine concentrations remained unchanged during exercise without ribose (0.2 µmol/l) and increased during exercise with ribose from 0.3 µmol/l (0.2-0.5) to 0,6 µmol/l (0.5-0.7).

Plasma hypoxanthine concentrations increased during exercise without ribose form 1.4 µmol/l (0.8-1.8) to 3,8 µmol/l (1.8-5.5) and during exercise with ribose from 5,3 µmol/l (1.3-12.6) to 7,5 µmol/l (3.8-13.7).

DISCUSSION

Since the studies of Segal and Foley in 1958 it is known that after absorption ribose is converted both to glucose via pentose phosphate pathway or to nucleotides via ribose-5-P and phosphoribosyl pyrophosphate (PRPP).

The increased lactate levels after oral administration of ribose suggest that ribose may serve as an energy source. However the rapid utilization of ribose seems to be insufficient to account for the beneficial effect of ribose in AMP-deaminase-deficiency for two reasons: 1. the amount of energy contained in 3 g of ribose is poor, 2. there is no explanation why other sugars are not successful in that case.

The increased inosine and hypoxanthine levels after oral administration of ribose suggest that ribose may enhance the de novo synthesis of purine nucleotides. In former experiments we have shown that in AMP-deaminase-deficiency purine nucleotides may be lost from muscle cells in form of adenosine (Wagner at al., 1991).

In addition to the results of Zöllner at al.(1986) and Gross et al. (1989, 1991) our results suggest that in AMP-deaminase-deficiency ribose may not only serve as an energy source, but may also enhance the de novo synthesis of purine nucleotides.

REFERENCES

1.Fishbein W.N., Griffin J.L., Armbrustmacher V.W., 1978, Myoadenylate deaminase deficiency: a new disease of muscle. Science 200:545-8

2.Gross M., Reiter S., Zöllner N., 1989, Metabolism of D-ribose administred continuously to healthy persons and to patients with myoadenylate deaminase deficiency. Klin Wochenschr 67:1205-13

3.Gross M., Kormann B., Kamilli I., Gresser U., Reiter S., 1989, Influence of D-ribose administration on the exercise-induced increase in hypoxanthine excretion. Ann Nutr Metab 33:215

4.Gross M., Kormann B., Zöllner N., 1991, Ribose administration during exercise: effects on substracts and products of energy metabolism in healthy subjects and a patient with myoadenylate deaminase deficiency Klin Wochenschr 69:151-5

5.Patten B.M., 1982, Beneficial effect of D-ribose in patient with myoadenylate deaminase deficiency. Lancet i:1071

6.Sabina R.L., Swain J.L., Patten B.M., Ashizawa T., O'Brien W.E., Holmes E.W., 1980, Disruption of the purine nucleotide cycle. A potential explanation for muscle dysfunction in myoadenylate deaminase deficiency. J. Clin. Invest 66:1419-23

7.Segal S., Foley J., 1958, The metabolism of D-ribose in man. J. Clin. Invest. 37:719-735

8.Wagner D.R., Felbel J., Gresser U., Zöllner N., 1991, Muscle metabolism and red cell ATP/ADP concentration during bicycle ergometer in patients with AMPD-deficiency. Klin Wochenschr 69:251-255

9.Wagner D.R., Gresser U., Zöllner, Effects of oral ribose on muscle metabolism during bicycle ergometer in AMPD-deficient patients. Ann Nutr Metab (in press)

10.Zöllner N., Reiter S., Gross M., Pongratz D., Reimers C.D., Gerbitz K., Paetzke I., Deufel T., Hübner G., 1986, Myoadenylate deaminase deficiency: successful symptomatik therapy by high dose oral administration of ribose. Klin Wochenschr 64:1281

INDEX